Simulation of
Manufacturing Systems

Simulation of Manufacturing Systems

Allan Carrie

Department of Design, Manufacture and Engineering Management, University of Strathclyde

JOHN WILEY & SONS

Chichester · New York · Brisbane · Toronto · Singapore

Library of Congress Cataloging-in-Publication Data:

Carrie, Allan.
   Simulation of manufacturing systems / Allan Carrie.
     p. cm.
   Includes bibliographies.
   ISBN 0 471 91574 2 : (U.S.)
   1. Manufacturing processes—Computer simulation.   I. Title.
TS183.C37 1988

British Library Cataloguing in Publication Data:

Carrie, Allan
   Simulation of manufacturing systems
   1. Systems. Simulations. Applications of
   computer systems.
   I. Title
   003'.0724

   ISBN 0 471 91574 2

Printed and bound in Great Britain by Anchor Brendon Ltd, Tiptree, Essex

Contents

Preface

Computer simulation has been applied to manufacturing for about 30 years, and often found to be of immense value. But for most of that time it has been within the province of a few specialists, remote from, and not well understood by, the manufacturing engineers. To many, simulation has been one of those techniques described as "a solution looking for a problem".

With the advent of computer-integrated flexible manufacturing systems, which involve very large capital costs and complex interactions, simulation has found a very real and challenging problem.

Although much has been said about simulation in recent years, there are still many misconceptions. For example, the publicity for spreadsheet packages, which emphasises the ease with which "what-if" modelling may be done in a problem area of comparatively trivial complexity, has raised expectations to a level which is only realisable with a considerable investment in time and cash. This book sets out to help by looking inside the black box, which to many simulation still is.

This book will be of direct relevance to the many University, Polytechnic and College courses in Manufacturing Engineering and related disciplines, and to the academics who teach them. It will also be of interest to industrial engineers who wish to acquire more than a black box level of understanding of the subject. It should also be useful to management scientists who, already knowing something of simulation, wish to apply simulation in manufacturing companies.

Chapter 1 is introductory, giving a brief statement about simulation and its relevance to manufacturing. It includes a paper-and-pencil simulation of a simple machine shop.

Chapters 2, 3 and 4 present the main theoretical background to simulation. They give a sufficient understanding to enable readers to proceed to applications of simulation, but are not exhaustive in the way which would be expected of a specialist text for management scientists. Chapter 2 introduces the main forms of simulation, and the way the flow of time is managed in discrete, continuous and combined simulation. Chapter 3 describes stochastic simulation, and the Monte Carlo method of simulated sampling. An inventory control example is presented. Chapter 4 describes discrete simulation and introduces the key terms and concepts. A hand simulation of a simple job shop is presented.

Chapter 5 introduces package software for simulation. The job shop model is defined on a computer with the aid of a model generator package, which produces a computer program in a simulation language. This model is run and the results discussed.

Chapter 6 presents a review of simulation software. It is not a point by point comparison of commercial packages. Instead it describes the main types of package, and places them in the context of historical development and sophistication and user-friendliness in the help they give the user in model development and obtaining results.

Chapter 7 describes flexible manufacturing systems. It describes the various elements which may be in them, gives examples of systems in industry and describes some of the planning problems which they present.

Chapters 8 to 12 present a series of models of flexible manufacturing systems of increasing complexity. Each chapter introduces additional features and discusses how these features can be modelled. In many cases, experiments are run to assess the sensitivity of the system and its performance to the features being discussed. In most cases the ECSL language is used, since it is a high level language and can be read almost like English language text. Some models using other software packages are presented.

Chapter 13 comments on the conduct of simulation projects, while chapter 14 reviews some areas of development in the field, such as the application of expert systems to simulation.

Inevitably, there are limitations to the depth and breadth with which such a wide field can be covered in a short text. This book is not intended to give an exhaustive treatment of either the theory of simulation or of Flexible Manufacturing Systems. Sufficient coverage of both is given to enable us to treat, in reasonable depth, the application of the former to the latter. For publishing reasons it has not been possible to include full colour illustrations, without which it is difficult to present the full impact of visual interactive simulation packages. The range of software now available for simulation is so wide that it is not possible to give a full review. Also, the scene is changing very rapidly. In fact, chapter 6 was re-written twice during the preparation of the book, to reflect enhancements in certain packages. Other changes will have been made since the text was completed.

I would like to acknowledge the assistance received from many friends both in manufacturing industry and among software vendors. They are too numerous to mention them all by name, however the names of most of their companies are quoted in the illustration acknowledgements, and many of the individuals are named in the references. I am grateful for the information received, and hope that my remarks are correct. I apologise for any errors which have crept in.

Special thanks must go to George Timson and his colleagues and former colleagues at Anderson Strathclyde. My association with that company, mainly through the Teaching Company Scheme, gave me the opportunity to become involved in a major FMS simulation project, at a level of detail and over a period of time that few academics are privileged to obtain.

I am grateful to my colleagues Ronnie McMillan, for reading the draft and for making many helpful suggestions, Felix Chan, for his help with the MAST and SIMAN models in chapter 11, and to Ron Barron, for assistance with the material on the continuous modelling system mentioned in chapter 2.

Some sections of the text in chapters 7 and 13 are based on papers which I have already published. I am grateful for permission to re-use the material in papers published by Mechanical Engineering Publications Ltd, Bury St Edmunds, Elsevier Science Publishers – North-Holland, Amsterdam, or Taylor and Francis Ltd, London. The references are quoted in the text.

Finally, I will feel that I have succeeded if this book helps even just a few to embark on simulation with more confidence and less trepidation, or to understand more fully the complexities of advanced manufacturing systems. To them all: good luck!

CHAPTER 1

Introduction to simulation

1.1 THE MEANING OF SIMULATION

Almost everyone has heard of simulation in some context or other. Perhaps you have a coat made of simulated fur or leather. In this sense the term conforms to the definition given by the *Oxford English Dictionary*:

Simulate:
To assume falsely the appearance or signs of (anything); to feign, pretend, counterfeit, imitate; to profess or suggest (anything) falsely.

Imitation fur is usually cheaper than the real thing, and less destructive to the species concerned.

More recently we have become familiar with graphical animated pictures on television of space craft flying past planets, often with the word "simulation" displayed at the bottom of the screen. While this use of the term also has a sense of pretence about it, it comes much closer to the meaning of simulation which we shall adopt in the context of manufacturing systems. Indeed, modern editions of dictionaries, such as the *Oxford English Dictionary*, include more appropriate definitions:

Simulation:
The technique of imitating the behaviour of some situation or system (economic, military, mechanical, etc.) by means of an analogous situation, model or apparatus, either to gain information more conveniently or to train personnel.[1]

Thus from the space craft pictures we "gain information" about what it might be like to fly past the rings of Saturn. Strictly we gain information only about certain aspects of the flight, we learn nothing about life-support systems, weightlessness, and so on. Despite this the simulation achieves its objective: showing what the planets might look like as we fly past; it does not intend to inform us about life support systems. This is a fundamental point about simulators: each is designed to serve specific, usually limited, purposes.

Simulator:
An apparatus for reproducing the behaviour of some situation or system; esp. one that is fitted with the controls of an aircraft, motor vehicle, etc., and gives the illusion to an operator of behaving like the real thing.[1]

1

We have also seen on television film of pilot training simulators in which images of the airport at which the pilot is trying to land are displayed on the windows of the pilot's cockpit. These simulators are very sophisticated examples of apparatus for reproducing the behaviour of the aircraft concerned in a physical way.

1.1.1 Models

In planning the layout of a factory it is common to use scale models of the buildings, machines and so on, so that we can see the layout of plant and be sure that enough space has been provided for movement of materials and workers. An example is shown in Figure 1.1, which is a model of the Flexible Manufacturing System (FMS) installed at Hattersley Newman Hender Ltd. While these iconic models can give an image of what the factory will look like they can tell us very little, if anything, about how it will operate. For example, have we provided sufficient space for work in progress beside the machines? How long will it take to process the customers' orders? How long might a job wait for processing on a certain machine? These questions concern the operation of the factory not its appearance. Manufacturing system simulators could be constructed to look like an image of the factory, but, since the objective is to study the operation of the system not its appearance, the extra information gained would probably not justify the expense involved. Pilots require immediate information in visual, tactile and audible forms, but with manufacturing systems there is seldom the same urgency. Consequently manufacturing sytem simulators usually present their information in the form of diagrams, charts, printouts and reports similar to those which the personnel concerned would encounter in the course of their duties. For the same reason the "apparatus" concerned is most conveniently a computer system, the "model" is in the logic of the program, and the information is supplied by printout or visual display. Thus we are concerned not with physical models but abstract ones which describe in logic the behaviour of the system.

Many students of management science will have encountered logical models which express the relationships between variables in the form of equations. Often these equations can be solved to give the optimum solution to the problem concerned. The Economic Order Quantity model is a well known example of such a model. These models are optimising in the sense that they yield the one best value of the function concerned. Many aspects of manufacturing systems (in the broadest sense) can be studied by that kind of model, but they depend on explicit analytical formulae describing known relationships. For example, if certain conditions exist we could use equations from queueing theory to predict how long jobs will wait for service at machines and how large the queues are likely to be. Generally, however, in manufacturing systems as a whole the necessary conditions do not apply, there are no such analytical relationships, and the behaviour of the system cannot be described in this way.

On the other hand we can state with certainty various rules about the internal behaviour of the things making up the system being modelled. For example, we know that if a job arrives at a machine for which some other jobs are waiting then the job will have to wait and the length of the queue will increase by one. We can also be sure that whenever a job leaves the queue and is processed by the machine the number of jobs in the queue will decrease by one. These relationships hold whatever the length of the queue at any point in time. If we know the queue length at one time and record

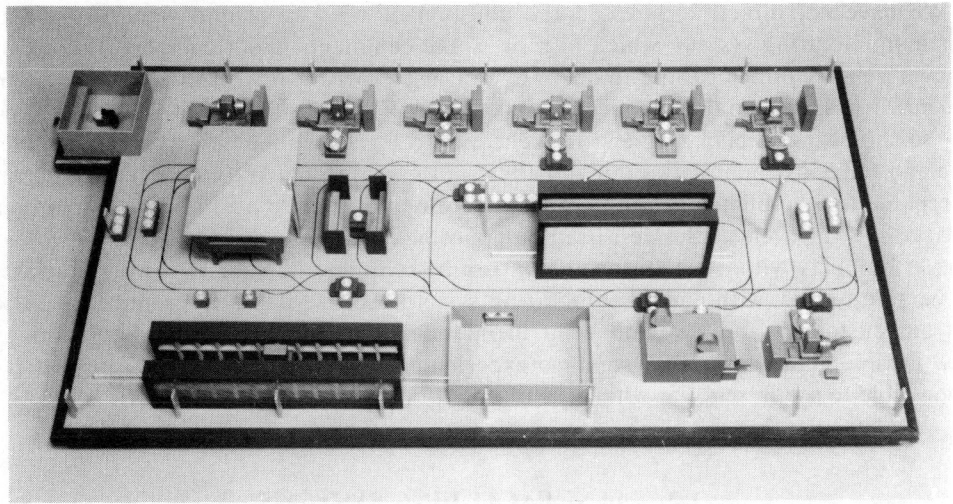

Figure 1.1   Model of flexible manufacturing system at Hattersley Newman Hender (reproduced by permission of HNH and KTM Ltd, Brighton, England).

all the arrivals and departures thereafter we can build up an exact record of the queue length at all subsequent times. A logical model of a complex system may be built by incorporating many such simple relationships and hence we may be able to predict the behaviour of the whole system. If we examine the behaviour of the system over a sufficiently long time we may be able to draw general conclusions about the system which we can use to predict its performance. Simulation modelling therefore is a means of studying the behaviour of a system as a whole by defining in detail how its various components interact with each other. The more complex the system the more inappropriate theoretical equations become and the more appropriate simulation becomes. Since most manufacturing systems are indeed complex, simulation is a most suitable tool.

It will be evident that the "model" which we refer to so frequently is an abstract logical model. An appropriate definition is:

Model:
A simplified or idealized description of a system, situation, or process, often in mathematical terms, devised to facilitate calculations and predictions.[1]

One of the main factors contributing to the complexities of the interactions between the various components of a model is the way some parameters of the system vary with time. For example, any system involving human beings is subject to random variation, because we all vary our rate of work, suffer illness, and so on. The demand for a product might be forecast to be 500 per week. If it is not exactly 500 every week how will this variation affect the performance of the system? If we know how sales vary around the estimate we may be able to model the variability. How simulation handles this stochastic variation will be described in a later chapter.

We have not directly addressed the question of why we should want to simulate our manufacturing system, which in view of the comments about complex interactions sounds a pretty difficult thing to do. Why not just experiment with the real system and see how it responds to the changes we make? For the same reason that we buy artificial fur coats and build pilot trainers: it is cheaper to buy a simulated fur coat than a real one and to train a pilot on a simulator than on the real thing. We may not have a spare aeroplane for the pilot to train on, and to take one of them out of service would incur a large loss in revenue. Besides, until our pilot has learned certain basic skills we dare not let him play with a real aeroplane in case he scratches it. Similarly, we are unlikely to be able to play with our manufacturing system because it has to produce the goods for our customers and we don't want to interrupt that work. We may be planning a new flexible manufacturing system, so experimenting with it is impossible, just as it is impossible to see in real life what it is like to fly past Saturn's rings.

1.2   MANUFACTURING SYSTEMS

Simulation can be applied to many different activities. When we narrow down consideration to a particular field we can define the characteristics of the system with more precision. Every manufacturing system exhibits much the same characteristics, although differing in detail. Because of this general similarity we can delve deeper into the subject and develop models which can be applied to different manufacturing systems with little modification. Every manufacturing system comprises products and the facilities which are used to produce them, such as machines, operators, handling devices, storage locations, pallets, fixtures, tools and so on. Consequently, simulation models of them will have common features. The differences between them will reflect the different ways in which the various facilities are combined to form a particular system. In fact a system may be defined as:

System:
A set or assemblage of things connected, associated, or interdependent, so as to form a complex unity; a whole composed of parts in orderly arrangement according to some scheme or plan.[1]

Manufacturing systems epitomise this definition since they comprise so many different elements, mechanical, electronic, human, information, procedures and so on. Indeed it is their complexity that makes simulation such a valuable tool in planning them and in analysing their behaviour, and in ensuring that we maintain the "orderly arrangement" and get the best use of them.

Simulation can be applied to many aspects of manufacturing systems, however a two areas stand out in particular:

1. In job shops, the simulation of dispatching rules and the assessment of the effect of different rules on the shop's ability to meet delivery dates and utilise the machines.
2. In flow lines, to investigate the extent to which inter-stage buffer storage can minimise the loss of output of the line due to breakdowns at workstations.

There is a substantial literature on both these areas, but a detailed commentary is outside the scope of this text. In this text we shall concentrate on those types of manufacturing system known as Flexible Manufacturing Systems, usually abbreviated to FMSs. As will be seen they represent very sophisticated combinations of elements. There is, as yet, no universally accepted definition of a Flexible Manufacturing System but the following would be widely accepted:

Flexible Manufacturing System:
Two or more computer numerically controlled units interconnected with automated workhandling equipment and supervised by an executive computer having random scheduling capabilities.[2]

These are standard production engineering terms, and it is not proposed to go into them in detail. Numerically controlled machines are ones which can read and follow a pre-prepared digital program of movements. Computer numerically controlled machines can receive the program from a remote computer over a communications link and store the program in their memory, rather than merely read each block of instructions from a punched tape. The executive controls all the actions of the machines and handling devices and thus its random scheduling capability enables it to respond to changing circumstances such as the breakdown of one of the units or sudden changes in demand for its products.

An excellent example of a flexible manufacturing system is that at Hattersley Newman Hender (HNH) for the manufacture of valve bodies. Figure 1.1 showed a model of the system and Figure 1.2 gives a layout plan. The system has five KTM FM100 machining centres, and two more machines based on the FM100, but adapted to perform "outfacing" operations. There is a Durr washing machine and a station where rings are fitted manually into the valves. There is a load–unload area to which material is supplied from an automated store. There is also an automated store for fixtures where they are held when the component which they hold is not in production. There are several buffer storage points in the system where components can be placed temporarily if the machine for their next operation is not available. Worpieces are carried between the load–unload area and the workstations by Wagner AGVs, of which there are seven. The machines have a twin pallet shuttle mechanism. The FM100s have twin tool drums for 40 tools each and the outfacing machines have one 40-tool drum. There is a tool setting room, from which tools are supplied to the machines manually. The machines have a Siemens controller, and there are Siemens programmable controllers at the washing machine and at the stores. All of these are linked to a Siemens R30 minicomputer which controls the whole system. There are VDUs in the control room, in the tool setting room, and at the ring fitting station and load–unload area. Some 2400 different types of component will be produced by the system. The system will enable the company to produce the components in a few hours instead of about two weeks. Work in progress inventories will be considerably reduced. Batch quantities will be much smaller than before, and the company will be able to respond much more quickly to changes in the marketplace.

FMSs provide manufacturing management with new ways of tackling the organisational problems of batch manufacture, promising substantial improvements in production rates, work in progress levels, lead times, and direct and indirect costs. However they represent a large capital investment with high start-up costs and demand careful planning by high calibre staff. Their complexity makes it difficult to predict how productive they will be. Simulation is a very important tool in planning FMSs because, when an investment of millions of pounds or dollars is being considered, the expenditure

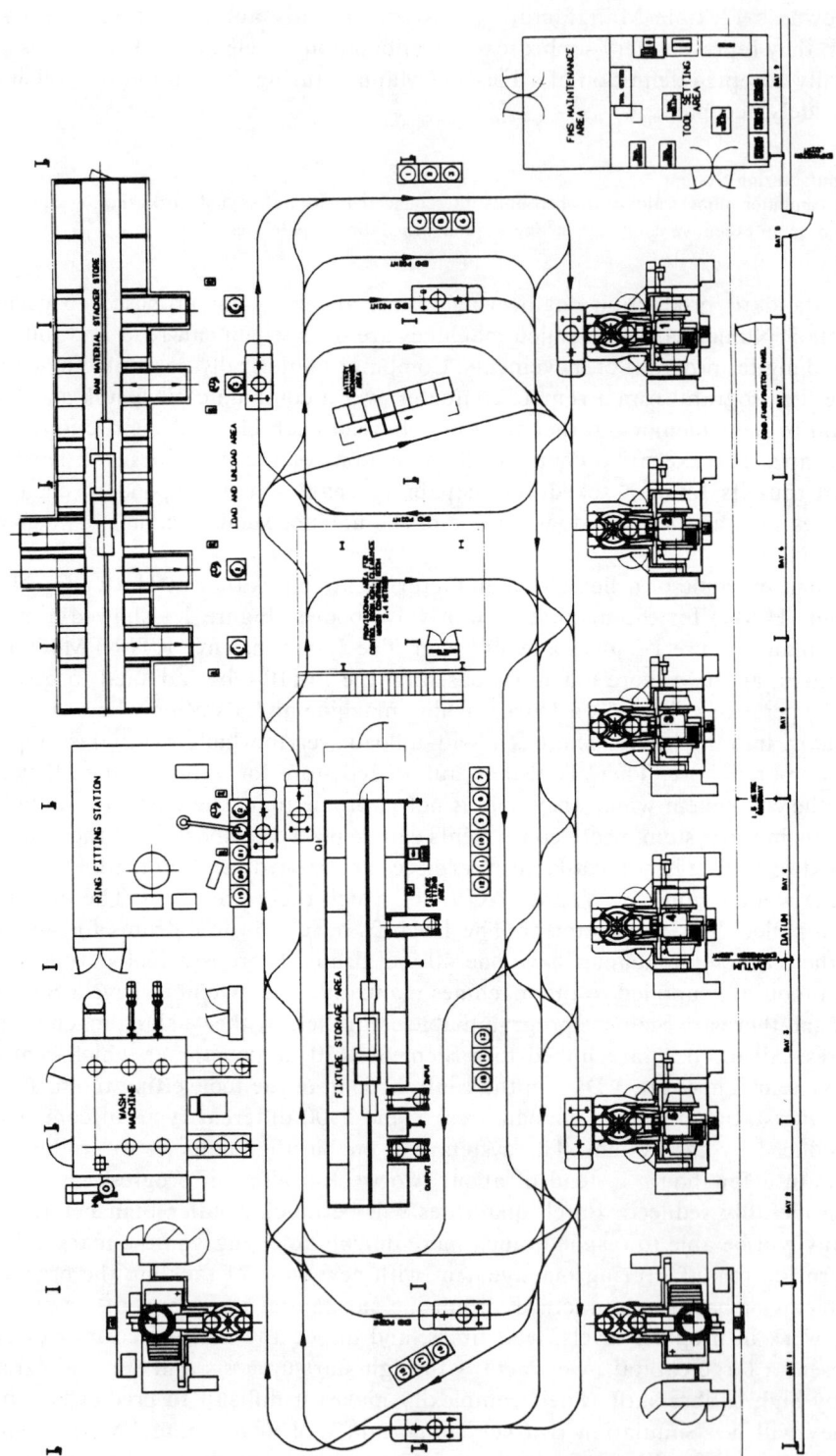

Figure 1.2 Layout plan of flexible manufacturing system at Hattersley Newman Hender (reproduced by permission of HNH and KTM Ltd, Brighton, England).

of a few tens of thousands is a small insurance premium to avoid major problems which would otherwise be unforeseen.

The main purposes of simulation of FMSs are:

— to assess the capacity and equipment utilisations in the system,
— to identify the bottlenecks in the system,
— to compare the performance of alternative designs,
— to ensure that there are no fundamental weaknesses in the FMS design,
— to develop operating strategies for work scheduling and job sequencing.

Unfortunately, the use of the technique is not necessarily straightforward. It requires specialist knowledge, and the majority of companies considering installing an FMS have only very limited expertise, if any, to carry out simulation studies. Relatively few firms who might install an FMS have an Operational Research Department possessing the necessary expertise. Nevertheless, most firms recognise the importance of simulation and seek to acquire expertise to allow them to evaluate suppliers' proposals, and, increasingly, to consider possible FMS designs prior to contacting potential suppliers. The services of consultants is therefore frequently relied upon.

The purposes of this book are to show manufacturing engineers the role of simulation in planning FMSs, to help them to talk to simulation specialists in their own language, to satisfy themselves that the assumptions and methods employed by an outside expert are valid (or not), and to help them to develop their own expertise.

1.3   INTRODUCTION TO MANUFACTURING SYSTEM SIMULATION

At its most basic, simulation involves recording everything that goes on in the system, or rather our model of the system, and making decisions whenever they are necessary. This process is rather similar to the way in which various planning charts, such as the Gantt chart, are used. In order to lay an adequate foundation for what comes later let us model a small machine shop on a Gantt chart.

The machine shop consists of five machines, labelled A through E, of different types. Figure 1.3 shows our model of the shop. A number of jobs are to be processed by the shop. Each job carries an identifying number and requires a certain number of operations. Each operation requires a particular machine and has a specified duration. Jobs arrive for processing in the machine shop at irregular intervals. We will model the passage of only five jobs through the machine shop and assign an arbitrary arrival time for each job. Table 1.1 lists the jobs, their arrival times and the sequence and duration of their operations. Note that the arrival time is the number of hours which elapse between the start of running the model and the arrival of the job. Thus job 1 arrives after 0 hours, ie at the start of hour 1, and job 2 arrives after 2 hours. We will assume that at the start of the model the shop is empty, and none of the machines are engaged in any operation.

We will assume that the jobs can be moved instantaneously from one place to another. This is a rather sweeping assumption, so perhaps we should make a mental note to have a second look at the situation without this assumption. We will also assume that the machine operators, setup men, progress staff and all the other aspects of a real-life machine shop will behave in a way which will not invalidate our model and the results which we will deduce. Again, this is perhaps difficult to accept, but let's proceed and see how we get on!

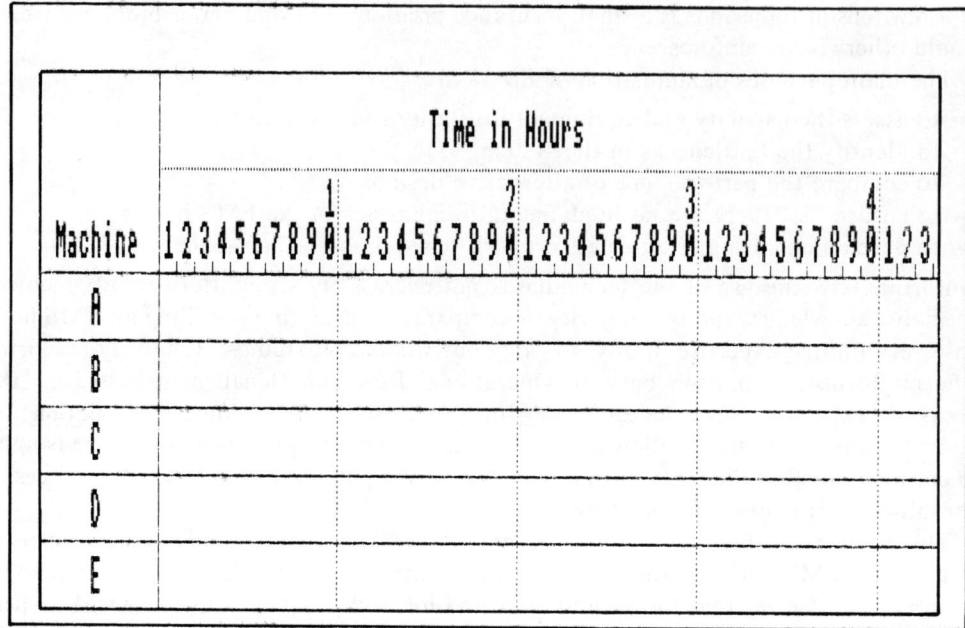

Figure 1.3   *Chart for modelling simple machine shop.*

Table 1.1   List of jobs for model of small machine shop

Job number	Arrival time	Number of operations	Operation number	Machine required	Operation time
1	0	4	1	B	18
			2	C	8
			3	A	5
			4	D	6
2	2	3	1	B	14
			2	D	2
			3	E	16
3	4	3	1	A	4
			2	E	3
			3	A	2
4	6	4	1	D	3
			2	C	4
			3	E	9
			4	A	3
5	7	4	1	C	8
			2	A	5
			3	B	8
			4	E	6

The first job to arrive is job 1, at time 0 (ie after 0 hours). Its first operation is on Machine B, so we can enter a bar on the chart to represent the operation, see Figure 1.4. The entry "j1/o1" indicates that the bar represents the time for operation 1 on job 1.

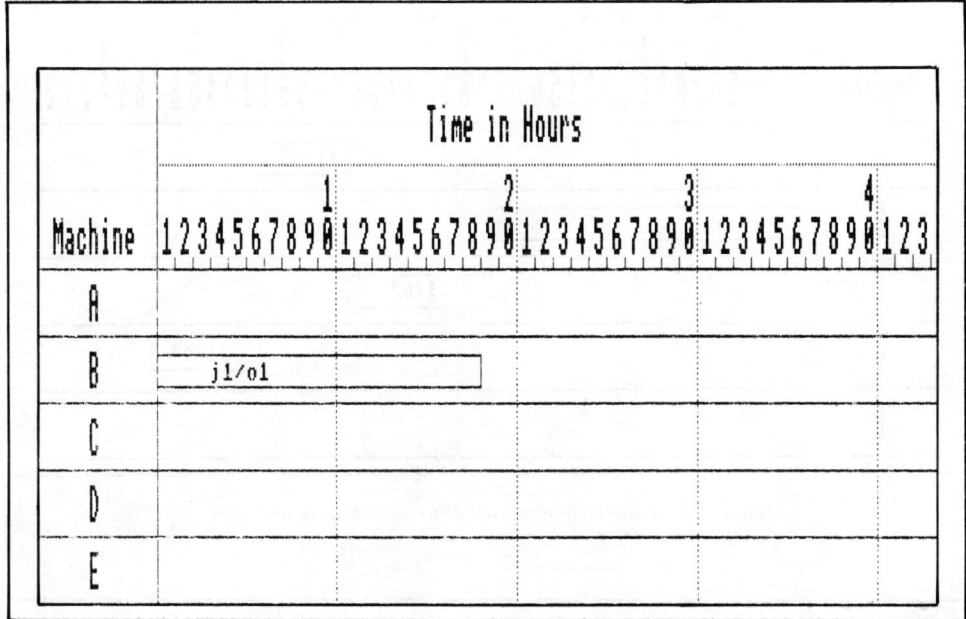

Figure 1.4   *Chart showing state of machine shop at time 0.*

Now if we were planning ahead to see whether the job could be completed by a specified due date we MIGHT proceed to enter all its operations on the chart as in Figure 1.5.

In simulation we DO NOT do this, because it assumes that machine D would be available for job 1 when it comes off machine B, and so on. In fact, if machine D were busy on another job, job 1 would have to wait in a queue. In simulation we proceed hour by hour making sure that we record everything that happens, making decisions whenever necessary but not trying to anticipate the future.

So let's start again. At time 0 job 1 arrives, its first operation starts on machine B and we enter the bar on the chart as in figure 1.4. We note that there are no other jobs waiting, since jobs 2 to 5 have not arrived yet, and so we move the clock on to the next event.

After the end of the second hour job 2 arrives. We advance the simulation "clock" to the end of the second hour, Figure 1.6. Job 1's first operation is on machine B, but at time 2 machine is busy on job 1, so job 2 has to wait.

The next event to occur will be after four hours, when job 3 arrives. Its first operation is on machine A, which is idle so the operation can begin straight away, and we enter it on the chart, as in Figure 1.7.

Two hours later job 4 arrives, requiring machine D which is available, so its first operation can proceed, Figure 1.8.

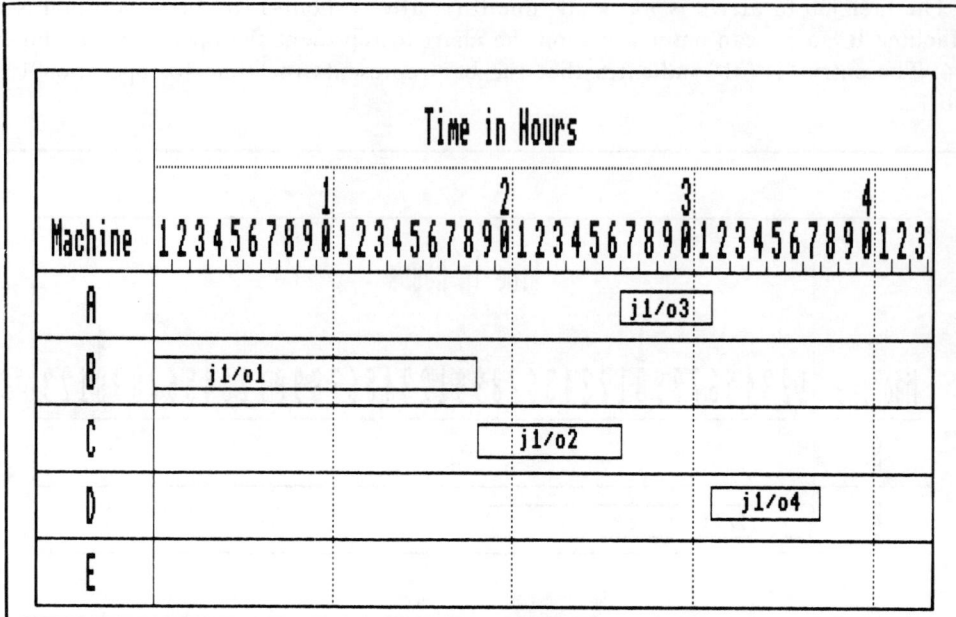

Figure 1.5   Chart showing all four operations on job 1.

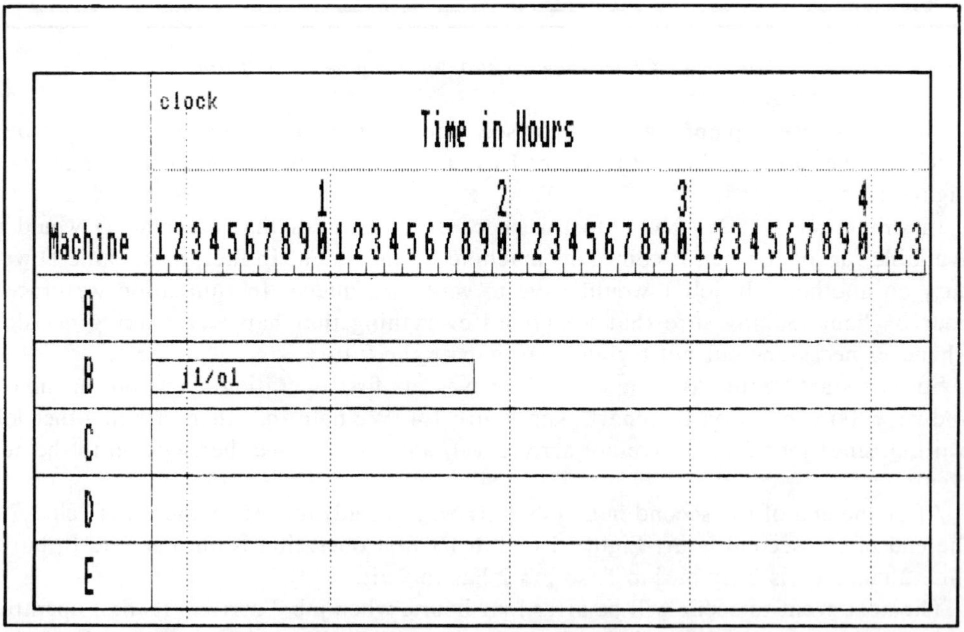

Figure 1.6   Chart showing state of machine shop at time 2.

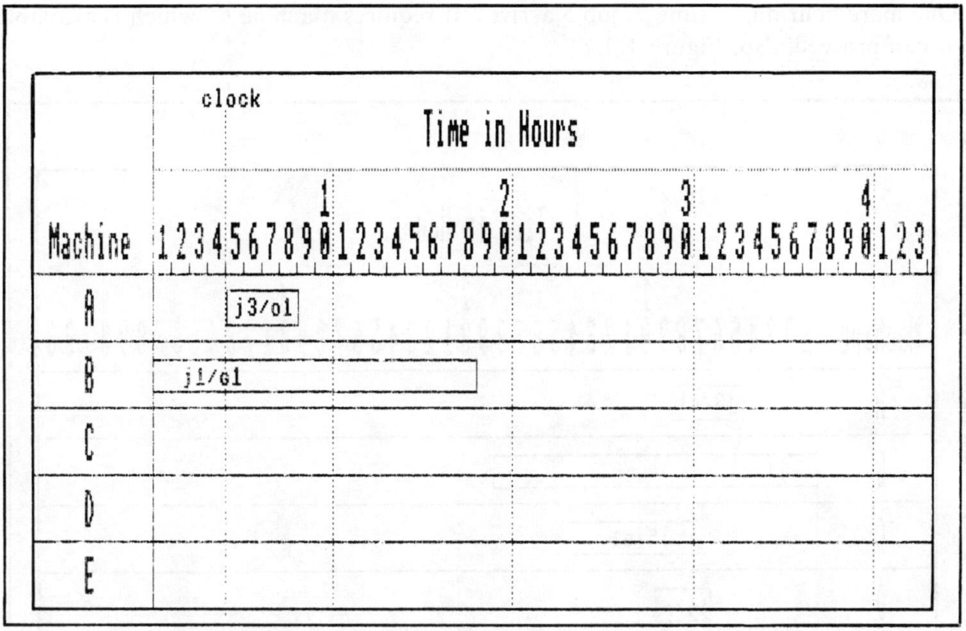

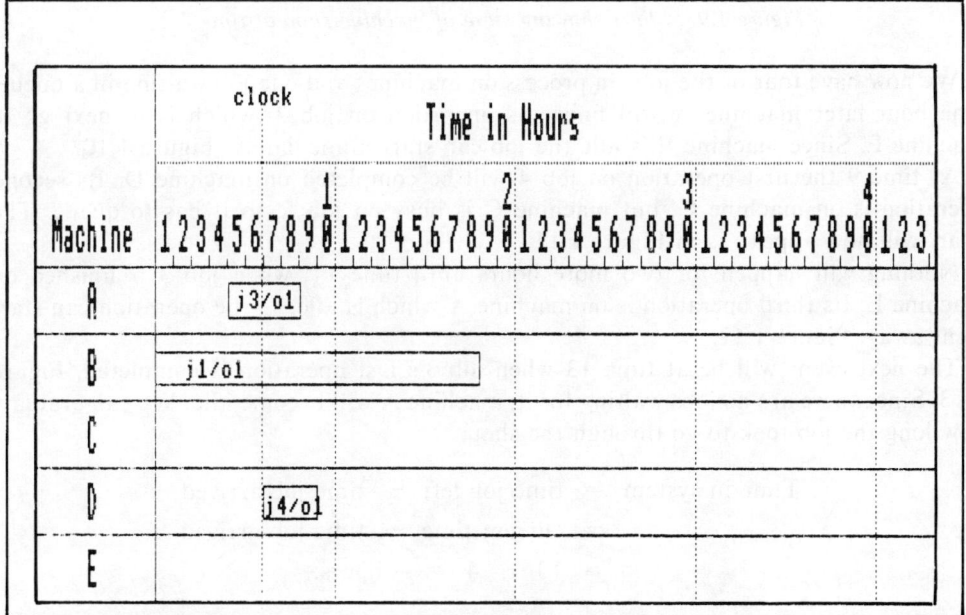

Figure 1.7   Chart showing state of machine shop at time 4.

Figure 1.8   Chart showing state of machine shop at time 6.

One more hour on, at time 7, job 5 arrives. It requires machine C, which is available, so it can proceed also, Figure 1.9.

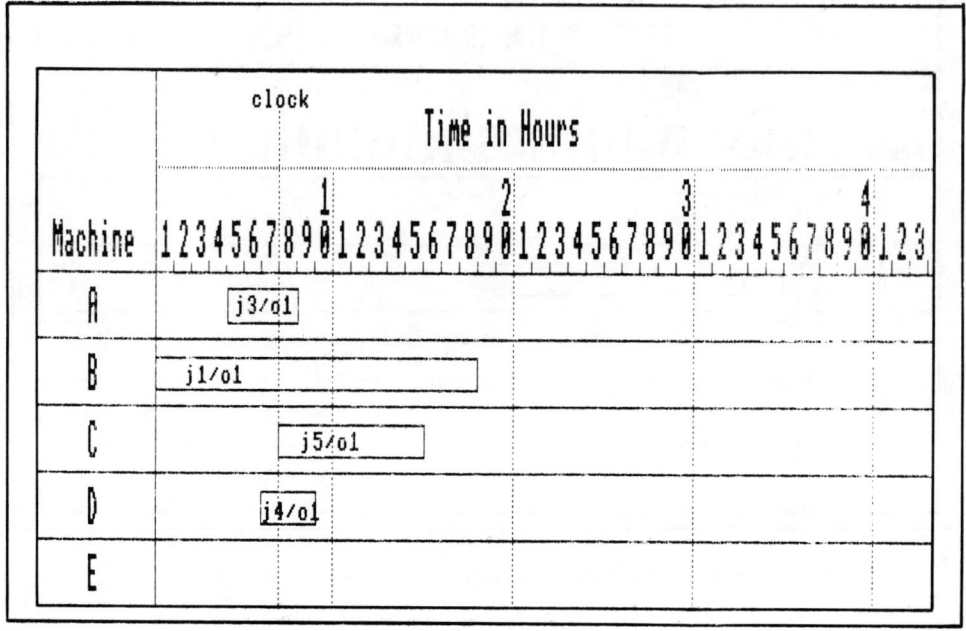

Figure 1.9   Chart showing state of machine shop at time 7.

We now have four of the jobs in process on machines and one job waiting in a queue. One hour later machine A will finish its operation on job 3, which is to next go to machine E. Since machine E is idle the job can start immediately, Figure 1.10.

At time 9 the first operation on job 4 will be completed on machine D. Its second operation is on machine C, but machine C is busy on job 5, so it has to queue. The chart will now appear as in Figure 1.11.

Nothing will happen for two more hours until time 11, when job 3 is finished on machine E. Its third operation is on machine A which is idle, so the operation can start right away, Figure 1.12.

The next event will be at time 13 when job 3's last operation is completed, Figure 1.13. Since there are no jobs waiting for it, machine A will become idle. We can evaluate how long the job took to go through the shop:

$$\text{Time in system} = \text{time job left} - \text{time job arrived}$$
$$= \text{current time} - \text{time job arrived}$$
$$= 13 - 4 = 9 \text{ hours.}$$

When time moves on another two hours job 5 will be completed on machine C. It will next go to machine A for its second operation, which can begin immediately because machine A is idle. Now that machine C is free job 4 can come out of the queue and its second operation commence. The chart will now look as in Figure 1.14. Notice that

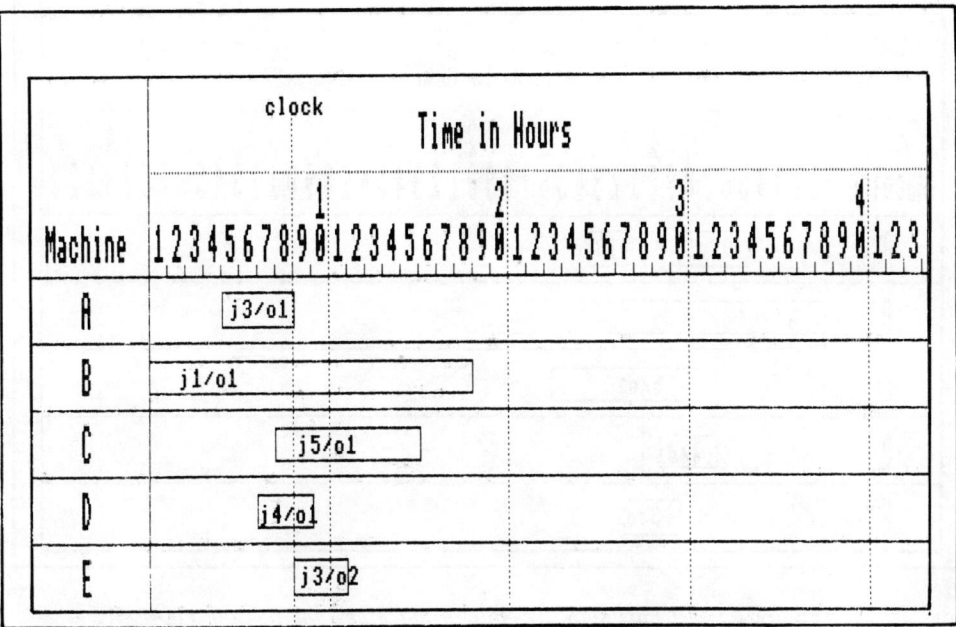

Figure 1.10   Chart showing state of machine shop at time 8.

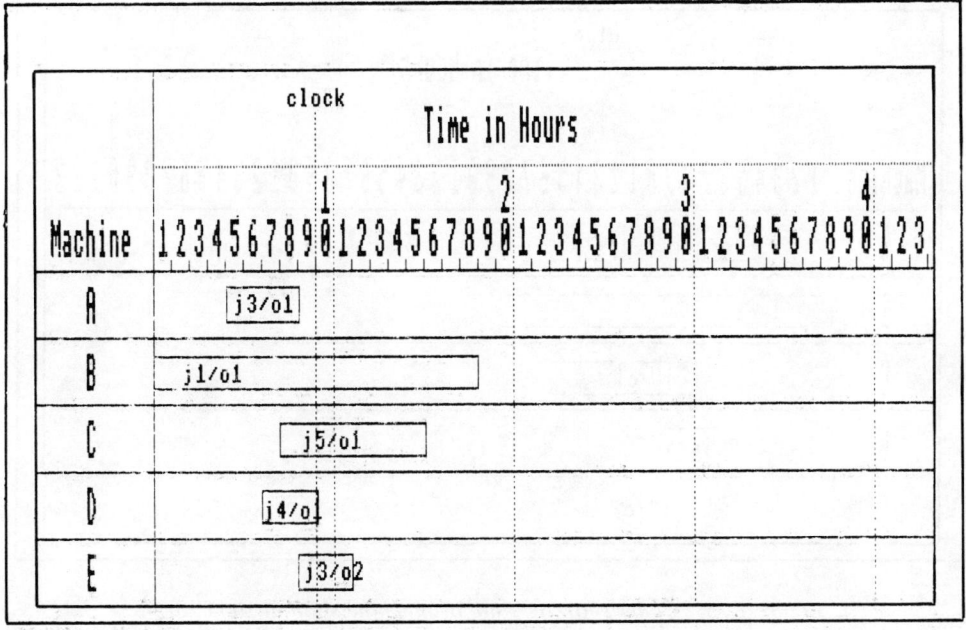

Figure 1.11   Chart showing state of machine shop at time 9.

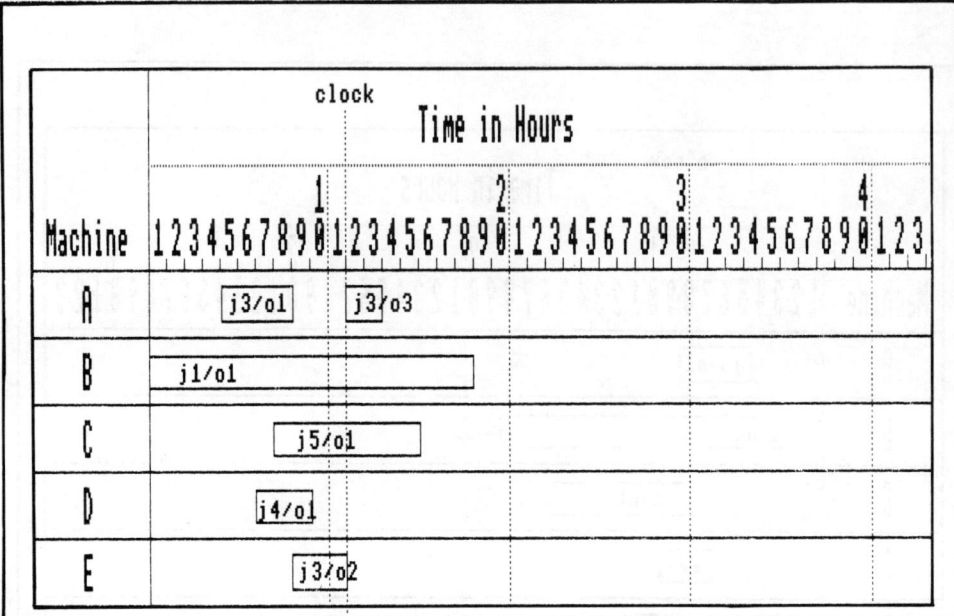

Figure 1.12   Chart showing state of machine shop at time 11.

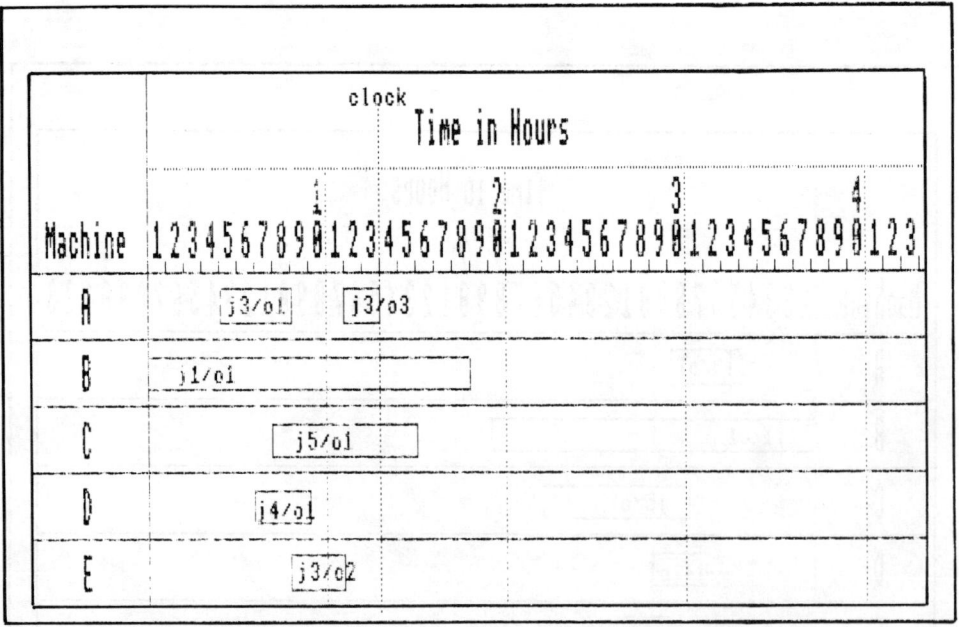

Figure 1.13   Chart showing state of machine shop at time 13.

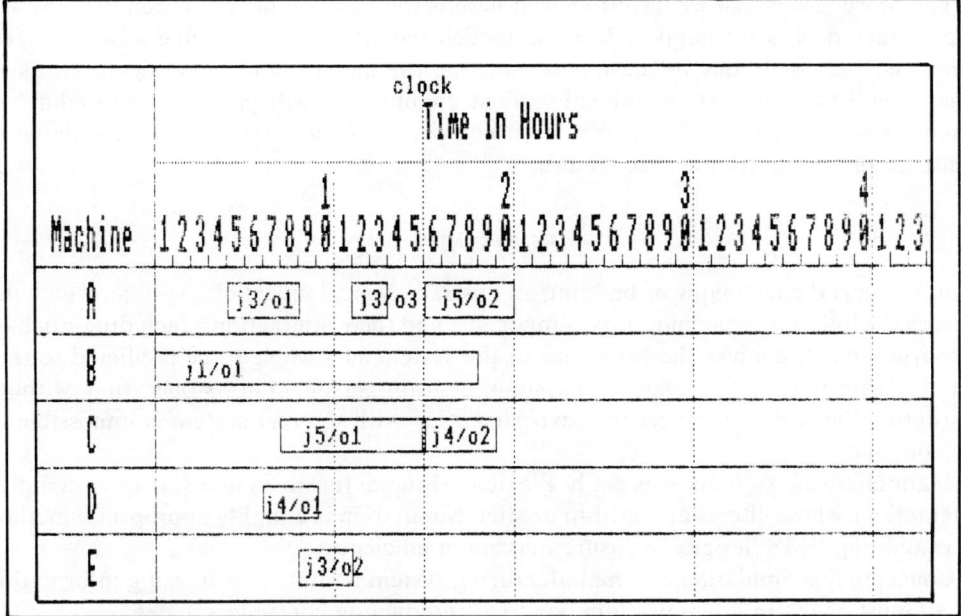

Figure 1.14   Chart showing state of machine shop at time 15.

on this occasion two operations began, because not only did the job whose operation was completed move on to another machine immediately, but a job which was waiting for the machine which the job left could get on the machine and be processed.

At any instant we could note which machines are working and which are idle. This would enable us to compute the utilisation of the machines, a statistic normally obtained from any manufacturing system simulation. This may be calculated from the formula:

$$\text{Machine utilisation} = \frac{\text{cumulative working time}}{\text{total observed time}}$$

For example, we can look at machine A and see that it has been idle for 4 hours, busy for 4, idle for 3, busy for 2 and idle for 2. Its utilisation is therefore

$$(4 + 2)/(4 + 4 + 3 + 2 + 2) = 6/15 = 0.40$$

Rather than do this calculation in detail, we do not have to count both the busy time as well as the idle time because they must add up to the current clock value. All we have to do is, every time an event occurs, to add the time since the last event to a counter for each busy machine. We shall encounter this method later when we come to consider computer techniques.

We can continue in this way as long as we seem to be gaining information about the system. In this case we would discover that, sure enough, when job 1 wants to go to machine C for its second operation the machine is busy. Therefore events would not unfold as anticipated in figure 1.5.

The procedure which we have followed here is the basis of all simulation models. We kept a record of what happened in the model and whenever a change took place we examined the conditions in the model, such as the queues, and made decisions as to what would happen next. In the subsequent chapters we will place this procedure on a more systematic basis, implement it on computers, discuss the software available to assist us and present some case studies.

SUMMARY

Simulation is the technique of building an abstract, logical model of a system, which describes the internal behaviour of its components and their interactions, including stochastic variability. It enables the behaviour of the system as a whole to be predicted so that we may gain information about the system, or train personnel in its operation, without disrupting the real system, because experimenting with the real system is impossible or uneconomic.

Manufacturing systems, especially Flexible Manufacturing Systems, involve complex interactions whose effects are hard to predict. Simulation is a highly appropriate method of evaluating FMS designs to ensure maximum efficiency.

Conceptually simulation of a manufacturing system is similar to entering information on a Gantt chart, in a manner familiar to all production controllers.

EXERCISES

1. Complete the simulation of the machine shop until all the jobs are complete and leave the shop. Observe the throughput time for each job, and compute the utilisation of each machine.

2. Repeat the simulation, this time assuming that one hour must be allowed for moving a job between one operation and the next, and also to the machine for the first operation and from the machine of the last operation. Was there any occasion when two or more jobs were moving? If so, what would have been the effect on the results have been if there had been only one fork truck?

REFERENCES

1. *Shorter Oxford English Dictionary*, Oxford University Press.

2. Findlay, R. C., FMS concepts realised, in *Proceedings of 3rd International Conference on FMS*, Boeblingen, West Germany, 1984, H.-J. Warnecke, editor, IFS (Publications) Ltd., Bedford, England, pp 397–405.

CHAPTER 2

Types of simulation and the flow of time

2.1 HOW THINGS CHANGE

The method which we used in chapter 1 to model the simple machine shop is that used in most manufacturing system simulations. However there are different types of simulation, which are designed to handle different types of system. A complex manufacturing model will probably involve a combination of methods. Consequently, before proceeding further it will be necessary to devote some consideration to these alternative forms of simulation and the circumstances when each might be appropriate. In a later chapter we will review the main computer simulation languages. On the whole these languages are of different types designed to model different types of real-life system, according to the way in which the system being modelled changes over time. Consequently we need to observe in some detail how these changes over time occur.

In the machine shop model we observed that the system changed (in the sense that the machines began or completed processing jobs) only at distinct points in time. Between these instants no change took place. This type of change is known as discrete change. This can be illustrated by considering the number of machines which are working at any point in time. Time-dependent variables can be plotted as a graph against time. A graph of the number of machines working in the model is shown in Figure 2.1. Observe that the graph is made up of horizontal and vertical lines. The horizontal sections describe the periods between points in time when changes occur and the vertical sections describe the change taking place. Notice that in this case the changes are in steps of one, because the smallest change that can occur is an increase or decrease of one machine in the number of machines working. The change could be a multiple of 1, because it is conceivable that two or more machines could begin or complete an operation at the same instant. Similarly, two opposing changes could cancel each other out. At time 11 one operation ends and another begins, so there is no change in the value of the statistic even although there has been a change in the system.

In real life, on the other hand, change is going on all the time, such as the gradual but continuous change of night and day, of the seasons, of daylight and of temperature. These forms of change are known as continuous change. A graph of a continuously varying time-dependent variable might be a smooth curve. For example, the displacement from the equilibrium position of a weight supported by a spring after it has been set in motion might be as shown in the graph in Figure 2.2. The graph of a continuous

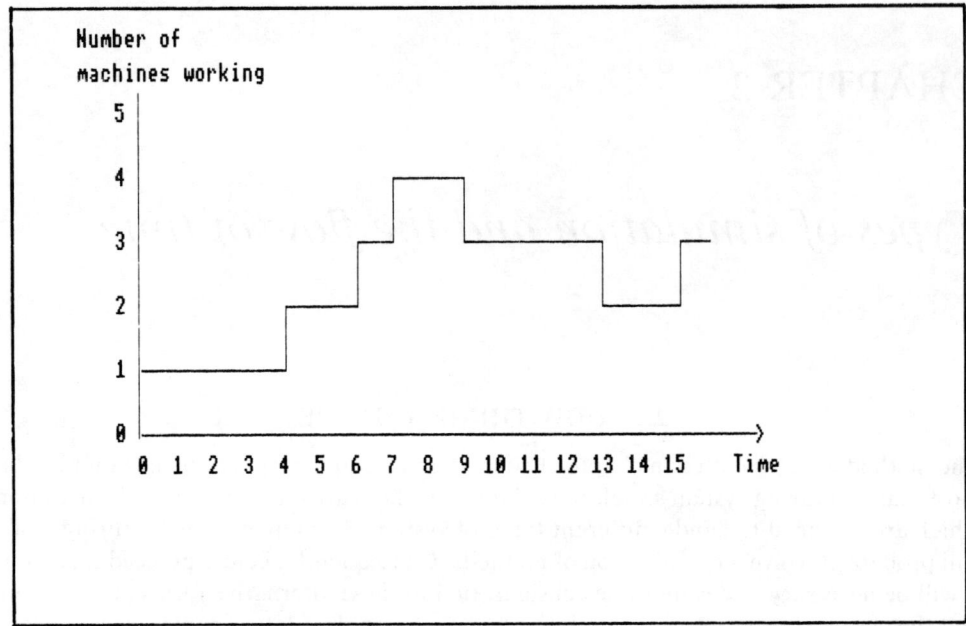

Figure 2.1   Graph of number of machines working in machine shop model.

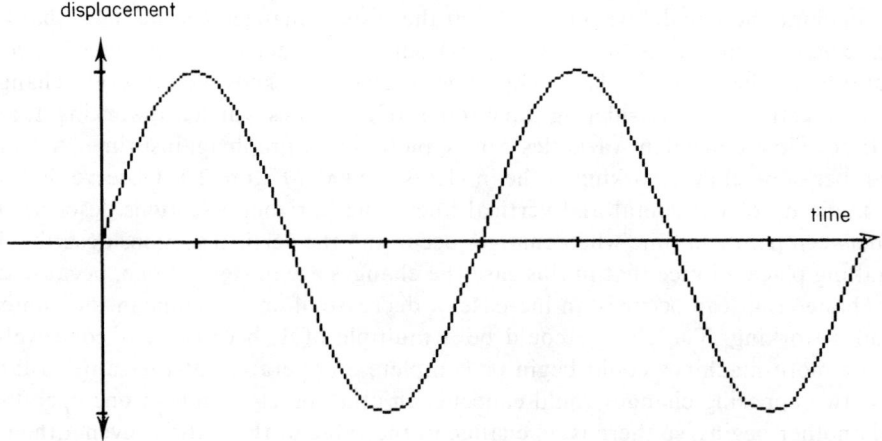

Figure 2.2   Graph of displacement of a weight supported by a spring.

variable can also be quite irregular in shape, not defined by some mathematical form. It might also be a straight line, as in the case of the volume of liquid in a tank into which the liquid flows at a constant rate.

2.2   CONTINUOUS, DISCRETE AND COMBINED TYPES OF SIMULATION

Corresponding to these two types of change there are two basic types of simulation model, and, as we shall observe in a later chapter simulation software packages are frequently written to facilitate modelling systems which predominantly exhibit one or the other type of change. However, both forms of change exist side by side in real life, and many systems will require a simulation package which can handle both forms. Manufacturing systems contain both types of change. In addition to discrete changes such as the starting or stopping of machining operations, there are gradual changes such as the extent of wear of a tool. Some variables can be handled by either method, often depending on the purpose of the simulation model. For example, the amount of stock in a stockroom might be treated as a discrete variable, since in reality stock is issued and received in discrete quantities, or it might be treated as a continuous variable if the simulation were to study the long-term effects of a stock control policy.

2.2.1   Discrete simulation

Discrete simulation is used when discrete change predominates in the system being modelled. The components of the system can be in any one of a number of discrete states at any point in time, and change from one state to another instantaneously. A machine may be either idle or working. It will change from idle to working at the instant when it starts to process a job. It will remain in the working state until the operation is completed, at which instant it becomes idle once more. Two basic elements of discrete simulation are the rules which determine when the next event will occur, and the rules for changing the state of the model when an event does occur. For example, one of the rules for the machine shop model could be stated as follows:

    If the event is "end of an operation"
      then
        put the job into the queue for its next operation
      and
        if there are no jobs waiting for the machine
          then set the machine to idle
          else select a job which is waiting for the machine
            and schedule another "end of operation" event.

In order to apply these rules the life history of each individual job and machine (and any other item) must be modelled explicitly, and its state is recorded throughout the time during which the model is being operated.

  In discrete modelling many changes are considered to occur instantaneously even though in real life they may take a short period of time. This is permissible so long as the period involved is small relative to the general time scale of the model. For example, the loading and unloading of a job into a machine may actually take a few seconds, or minutes perhaps, but the model will probably assume that loading is done instantaneously at the moment when loading actually starts, and unloading is done instantaneously at the moment when unloading in reality finishes.

2.2.2 Continuous simulation

Continuous modelling techniques are used when continuous change predominates. We can frequently describe the way the variables fluctuate over time by mathematical equations. For example, the displacement of the weight in figure 2.2 can be described by the equation of a simple harmonic motion. In the early days of computing, analogue computers, in which the values of variables were represented by voltages at points in an electrical circuit, were used to model continuous systems.

For example, in mechanical vibrations the movement of a vibrating mass can be described by a differential equation. An electrical circuit could be built up containing elements which cause the voltage at an appropriate point to vary according to the equation. These voltages could then be observed and displayed on an oscilloscope. Figure 2.3 shows a weight of mass M suspended on a spring of stiffness K whose movement is inhibited by a dashpot having a viscous damping coefficient of R and which is subjected to an applied force P at time T. The variables are the displacement of the mass, x, its velocity, dx/dt, and its acceleration d^2x/dt^2. The equation of motion is as follows:

$$M\frac{d^2x}{dt^2} + R\frac{dx}{dt} + Kx = P(t)$$

In recent years software packages have been developed on digital computers to model continuous systems. An example of a digital simulation package is TUTSIM.[2] This package contains subroutines to achieve the functions of the elements in the electrical circuit. Instead of the circuit, a block diagram is developed in which each block represents a subroutine. A block diagram for modelling the vibrating mass system of figure 2.3 is shown in Figure 2.4. Whereas in an analogue computer voltages would vary continuously, as they do in real life, in digital computers time has to be treated as a succession of tiny increments. This is analogous to plotting a curve on a dot-matrix printer in which the dots can appear only at discrete points. The higher the resolution of the printer (or the display screen) the closer to true life the result appears. Figure 2.5 shows a printout of the results. The output consists of the plot itself and tables defining how the plot is to be interpreted. Under "TIMING" two values are given. The first is the time step between successive calculations of the values of the functions, set to 0.5 units. The second value is the length of time over which the values are to be calculated, which was set to 200 units. "OUTPUT BLOCKS AND RANGES" specifies the variables being plotted, which block of the diagram they are measured at, and the lower and upper values on the scale for that variable. X1 is the horizontal axis, representing time, and the scale runs from 0 at the left-hand end to 200 at the right-hand end, corresponding to the time for which the calculations are to be done. Y1 is the output of block 3, and represents the velocity of the mass. On its scale the bottom of the graph has the value − 1, and the top is at +1. Zero for this variable is therefore at the mid-point of the scale. Y2, the output of block 4, represents the displacement of the mass from its equilibrium

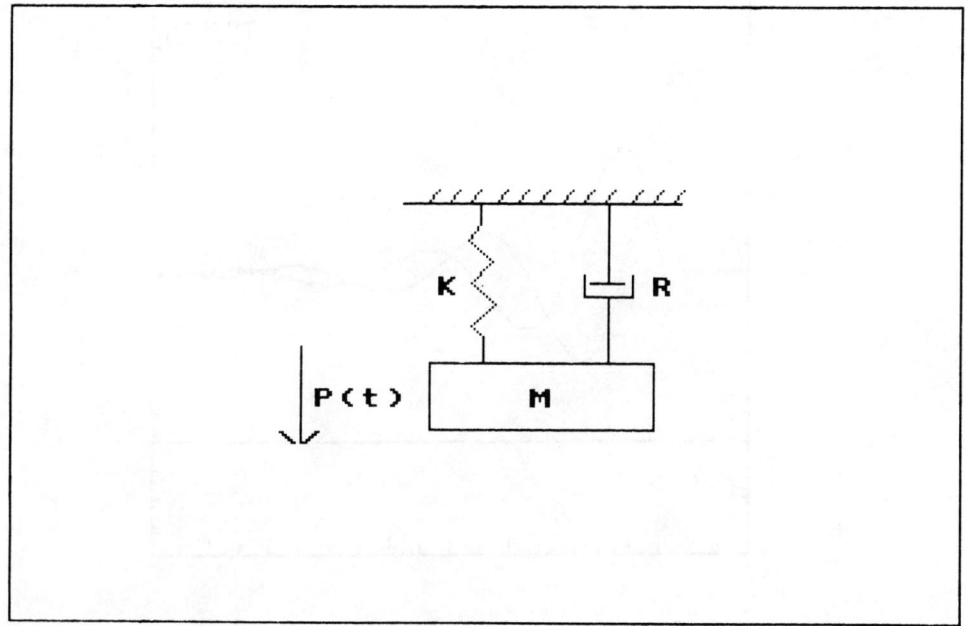

Figure 2.3 *Weight of mass M suspended by a spring of stiffness K with dashpot of viscosity R given an impulse P.*

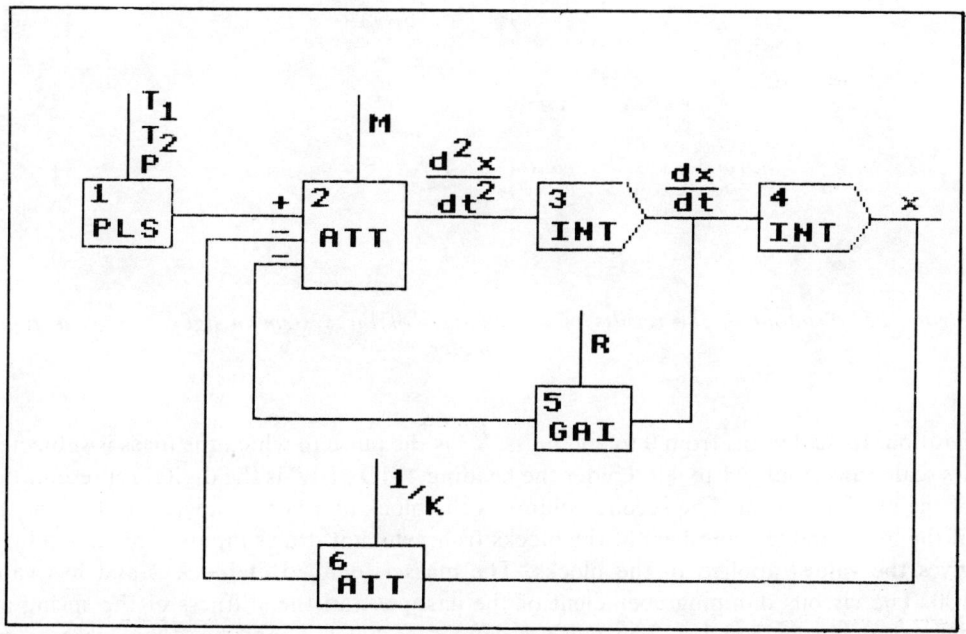

Figure 2.4 *Block diagram for modelling the continuous system in figure 2.3 on a digital computer.*

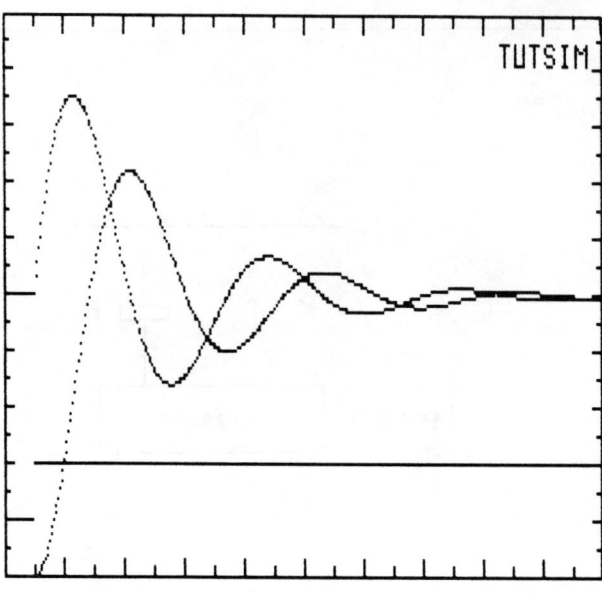

```
TIMING
  0.50000E+00    0.20000E+03

OUTPUTBLOCKS AND RANGES
  X1:      0        0.00000E+00    0.20000E+03
  Y1:      3       -0.10000E+01    0.10000E+01
  Y2:      4        0.00000E+00    0.20000E+02
  Y3:      1       -0.10000E+01    0.90000E+01

MODEL
  0.10000E+02         1 PLS
  0.20000E+03
  0.10000E+01
  0.10000E+02         2 ATT    1    -5    -6
  0.00000E+00         3 INT    2
  0.00000E+00         4 INT    3
  0.50000E+00         5 GAI    3
  0.10000E+02         6 ATT    4
```

Figure 2.5   Printout of the results of the analysis of the system in figure 2.3 on a digital computer.

position. Its scale runs from 0 to 200 units. Y3 is the pulse to which the mass is subjected. Its scale runs from −1 to +9. Under the heading "MODEL" is the digital representation of the block diagram. The second column is the block number, followed by the function of the block and the numbers of the blocks from which it draws inputs. The first column gives the values applied to the blocks. The mass is defined at block 2 and has value 100. The viscous damping coefficient of the dashpot and the stiffness of the spring are defined at blocks 5 and 6. The pulse is defined at block 1, and has three parameters. The first of these is the time at which the pulse is applied, ie after 10 time units, the second is the time at which the pulse is reversed, in this case set to 200 units which is at the extreme end of the plot and therefore not shown, and the third parameter is the magnitude of the pulse and has value 1. This information is sufficient to indicate that the pulse is plotted as two horizontal lines in the lower part of the plot. The plot of velocity starts at the mid-point of the vertical scale and is constant at zero until the pulse is applied, after which it oscillates. Similarly, the displacement is at zero, at the bottom of the plot, until the pulse is applied when it begins to oscillate. With a package of this sort it is a simple matter to assess the effect of different spring stiffnesses and damping coefficients on the time until the mass settles down and the displacement at which it does so.

Pritsker[1] gives a simulation model involving the level of oil in a storage tank. The level will vary depending on whether the pump supplying oil to the tank and the pump delivering oil from the tank are switched on or off. If oil were pumped out of the tank at a constant rate, then to find the quantity of oil which had been consumed over a period of time it would only be necessary to multiply the rate of flow out of the tank by the length of the time period involved, as illustrated in Figure 2.6(a). However, if the rate of flow varied continuously, as in Figure 2.6(b), such a simple calculation would not be possible. Then it would be necessary to integrate the function defining the rate of flow over the time interval concerned. Consequently, integration algorithms are key elements of simulation packages for continuous modelling on digital computers.

Whereas in discrete simulation each entity is modelled in detail, in continuous simulation we are only concerned with the level of variables and the rate at which they change. For example, in a production-inventory system simulation we would consider the total number of items in the stock and not the life history of each individual item.

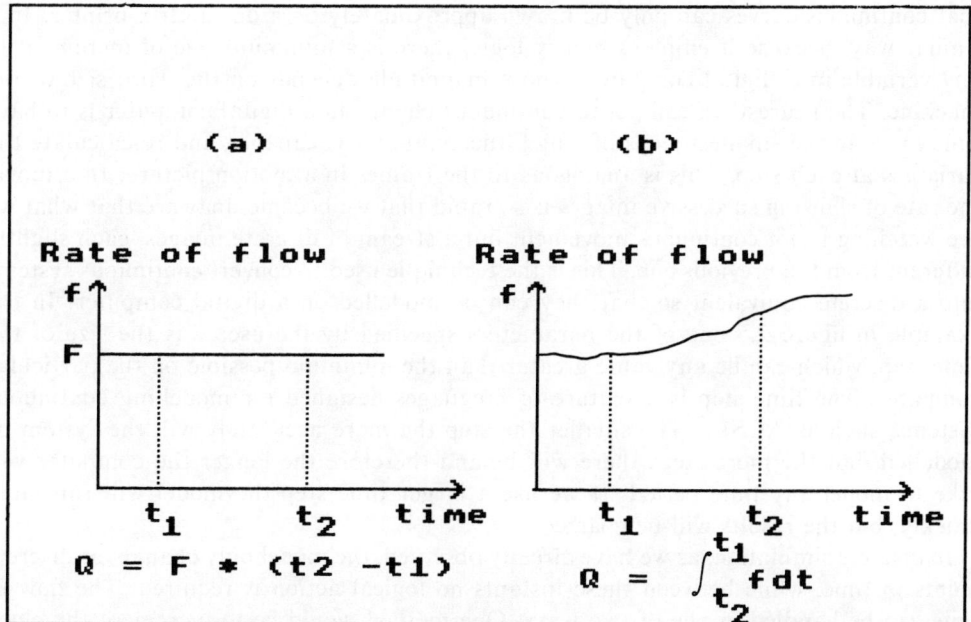

Figure 2.6   *Computing the quantity of oil flowing through a pipeline, (a) when the rate of flow is constant, (b) when the rate of flow varies continuously.*

2.2.3   Combined simulation

Combined simulation involves both continuous and discrete change. Many real systems involve both types of change. For example, in a model of oil in a storage tank one of the rules might be that when the level reaches an upper limit the flow into the tank must be stopped until the level has fallen somewhat. Similarly, when the level reaches a lower limit then the flow out of the tank must be suspended until more oil has been supplied to the tank. Thus the level of a continuous variable (the level of oil in the tank) may cross a threshold value thereby triggering a discrete change (turning the supply or discharge off or on). In a manufacturing system the amount of wear on a tool might be modelled as a continuous variable and the movement of work through the system and the machines discretely. While a machine is in operation the tool is cutting and the rate of tool wear could be set according to some equation. When the machine is idle the rate of wear would be set to zero. When the wear reaches some value it will be necessary to stop the operation to enable the operator to change the worn tool for a fresh one. This is another example of a continuous variable triggering a discrete event.

2.3   HANDLING THE FLOW OF TIME

It follows from this discussion that the flow of time in a simulation model can be handled in different ways depending on the nature of the system being modelled. We must examine this topic in more detail because it is fundamental to the simulation process. In real life, time flows continuously at a steady rate. Unfortunately, in a digital computer it is impossible to have variables change continuously. Figure 2.5 illustrated that continuous curves can only be drawn approximately on a dot matrix printer. In a similar way, because it employs binary logic, there is a minimum size of increment in any variable in a digital computer, whose magnitude depends on the word size of the machine. The nearest we can get to continuous change in a digital computer is to have time move in the smallest steps of which the computer is capable, and re-calculate the variables at each step. This is analogous to the frames in a motion picture. In a movie the rate of showing successive images is so rapid that we become unaware that what we are watching is not continuous movement but a stream of discrete images, each slightly different from the previous one. This is the technique used to convert continuous systems into a discrete equivalent so that they can be modelled on a digital computer. In the example in figure 2.5 one of the parameters specified by the user was the size of the time step, which can be any value greater than the minimum possible on the particular computer. The time step is a feature of languages designed for modelling continuous systems, such as ACSL.[3] The smaller the step the more accurately will the system be modelled, but the more steps there will be and therefore the longer the computer will take to model any time period. If we use a larger time step the model will run more quickly, but the results will be coarser.

   In discrete simulation, as we have already observed, the model only changes at discrete points in time, while between these instants no logical action is required. The flow of time can be handled in one of two ways. One method would be to increment the clock in small steps, or time slices, and check each step whether any changes had to be made in the model. However, since changes are only necessary at intervals this would be a rather inefficient process. The alternative involves maintaining a list of the events which are going to occur. We can then identify the event which will occur first, wind the clock on to the time of that event in a single jump, then make whatever changes are to take place. These two approaches are known as the "time slice" and "next event" methods respectively.

We could have used either method in the machine shop model. The commentary was based on the next event technique because it made remarks such as "nothing will happen until time 11 ..." or "when time moves on another two hours ...". In this method the maintenance of the list of events is of crucial importance. Changes in the model usually involve adding a new event to the list. At the start of the run (at time 0) the list contained five events corresponding to the time of arrival of the five jobs. As time progressed these events were encountered, removed from the list and new ones, such as the completion of an operation, were added to the list. Alternatively, we could have used the time slice method using steps of one hour, since all the arrival times and operation times were multiples of one hour. The step could just as well have been one minute or one second, but the smaller the step the more checks we would have had to do. If we had used a very small time unit, activities of very short duration could have been modelled, and the smaller the unit the closer to continuous simulation we would come. There are many packages designed for discrete modelling, most of which use the next event technique, since it is generally more efficient. However, most modern discrete simulation packages, eg GASP IV[1] and ECSL,[4] also have facilities for modelling continuous variables using the time step approach.

Between the two extremes, time flowing in very small increments as in continuous simulation or in substantial jumps in discrete simulation, there is a large class of systems, such as production planning and inventory control systems, in which it is sensible to treat time as a series of quite large time steps. Depending on the purpose of the simulation, steps of a day, week or even month might be used. Extending the analogy with motion pictures this method is analogous to time-lapse photography. Provided the interval between "exposures" is not too long we get a good representation of the subject. The period selected has to be a compromise between efficiency in time and cost versus accuracy of results. If we increase the interval we save time and cost, but risk missing some detail. If the intervals are smaller than necessary we may gain accuracy, but waste time and money. The term "digital" simulation is sometimes used to describe this method of simulation, because it exploits the method by which continuous models may be implemented on a digital computer. In this method of handling time, the differential equations of continuous modelling are replaced by what might be termed difference equations, specifying the changes in the value of the variables between one time period and the next. In an inventory control system for example, although material can be issued at frequent and irregular intervals over the day, we may be able to model the system accurately enough by using daily time steps. Thus we would record the total issues during the day and set the stock level to the end-of-the-day value. Using this approach the level of stock might be defined by the following equation:

Stock level at end of period T
= Stock level at end of period $(T - 1)$
− Total issues during period T
+ Total receipts during period T

We will use this method in the next chapter when further fundamental principles of simulation are introduced.

SUMMARY

Different types of system exhibit different types of change. There are different types of simulation language for modelling different types of system. The flow of time will be handled in different ways according to the type of system. The principal features of the three systems of modelling are summarised in Table 2.1.

Table 2.1   Characteristics of different types of modelling

	Type of model		
	Continuous	Digital	Discrete
Time step:	infinitessimal	small time slices	jumps from one event to the next
Method:	differential equations	difference equations	logical relationships
Components:	aggregate	aggregate	individual entities
Variables:	levels	levels	queues, states, attributes
Changes:	rates	rates	events

REFERENCES

1. Pritsker, A. A. B., *The GASP IV Simulation Language*, John Wiley, New York, 1974.

2. *Instructional Manual for Dynamic Simulation Software, TUTSIM*, Process Automation and Computer Systems, Lymington, England.

3. *ACSL User Guide*, Mitchell and Gauthier Associates Inc., Concord, Mass., 3rd edition, 1981.

4. Clementson, A. T., *Extended Control and Simulation Language*, Cle-Com Ltd, Birmingham, 1985.

CHAPTER 3

Stochastic simulation

3.1 RANDOM VARIATION

The previous chapter dealt with the nature of change in a system, and in this chapter we wish to examine a particular type of change, namely that due to random variation. There are many forms of these variations. There are those aspects of a system about which we know little, such as external influences in the business environment, or human factors, such as illness, absenteeism, temperament, pace of work and so on. Then there are aspects which will vary between known limits, but we do not know the values which they will have on particular days or occasions. Some of the aspects just mentioned really come into this category—absenteeism, for example; if we were to record the amount of absenteeism for a year we could establish its average level and we might allow for this level in our plans. But what will happen on the days when absenteeism is higher than the expected level? From our records we could also compute the variance of the distribution, and perhaps establish that it conforms to one of the known statistical distributions. This would enable us to identify a level for which there was only a small probability that absenteeism would exceed on any day. Thus we could plan, confident that it was unlikely that absenteeism would disrupt our plans. However, because it is unlikely that absenteeism would reach this level on a regular basis, our plans would normally err to one side, and the error would gradually accumulate. What we need is a method of modelling absenteeism which takes account of its day-to-day variability as well as its overall level.

An alternative would perhaps be to make our plans assuming that absenteeism would be exactly the same every day as it was on the corresponding day last year. In other words we model our system using the same series of values as we had observed. This approach would suffer from two major defects. Firstly, there were probably some special factors applying last year which would not normally exist. Perhaps the company was the object of a takeover bid, and absenteeism was higher than normal due to the general air of uncertainty that accompanied the bid. The second defect is that we would evaluate our model against only one particular series of absenteeism values, and, although the general level of absenteeism might be the same as last year, it is inconceivable that every day it would be exactly the same as last year. Our method must allow us to evaluate our plans against any pattern of absenteeism which is within reasonable expectation.

The solution to this problem is to select values at random from the range of the possible values in such a way that the series of values obtained exhibits the same characteristics as our observed statistics on the absenteeism phenomena, using a technique known as Monte Carlo simulation.

3.2   MONTE CARLO SIMULATION

Monte Carlo simulation is the name given to the technique of taking random samples from the distribution of a variable in order to supply a series of values for use in the model. The name arises from the assertion that the roulette wheels in the casino at Monte Carlo are unbiased, and that every number is as likely to come up as any other. Since few readers may be familiar on a personal basis with the roulette wheels at Monte Carlo, let us take an everyday example: tossing a coin.

A coin has two sides, the head and the tail. When we toss the coin we know with certainty that it will come down showing one or other side and that there is no other possibility (unless it lands on its edge or disaster strikes and it rolls away down a drain, but we will assume that such events do not occur in the world we seek to model). We do not know which face will show on any specific throw. Since each throw is independent of every other throw we would expect to see as many heads as tails if we throw the coin a large number of times.

For similar reasons, if we throw a die a large number of times we would expect to get roughly an equal number of each of the six possible values, as illustrated by Figure 3.1. A probability distribution in which the probabilities of obtaining each of the possible values are equal is known as a uniform distribution. We nevertheless accept that it would be very unlikely to have exactly one of each in every six throws. The extent of the variation from uniformity is a function of the randomness of the die-throwing process. If the deviation is large we may suspect that the die is not true.

Value shown	1	2	3	4	5	6
Expected number of occurrences	6	6	6	6	6	6
Probability of occurrence	0.17	0.17	0.17	0.17	0.17	0.17

Figure 3.1   Expected outcomes when throwing a die 36 times.

In several board games the players throw two dice at a time, moving according to the sum of the values on the two dice. In this case we would not expect to get an equal number of each of the possible values, of the different ways in which the total value may be made up. There are 36 possible combinations of the dice (6 × 6) but only 11 possible values for their sum (2 through 12), and some values are more likely than others. Only one combination can give a value of 2, likewise 12, but two combinations give each of 3 and 11, and so on, until the six combinations giving a value of 7. These frequencies are illustrated in Figure 3.2. Whereas throwing one die gave a uniform distribution, throwing two gave a triangular distribution. If we were to throw several dice together, say 12, we would obtain a distribution which very closely approximates a normal distribution.

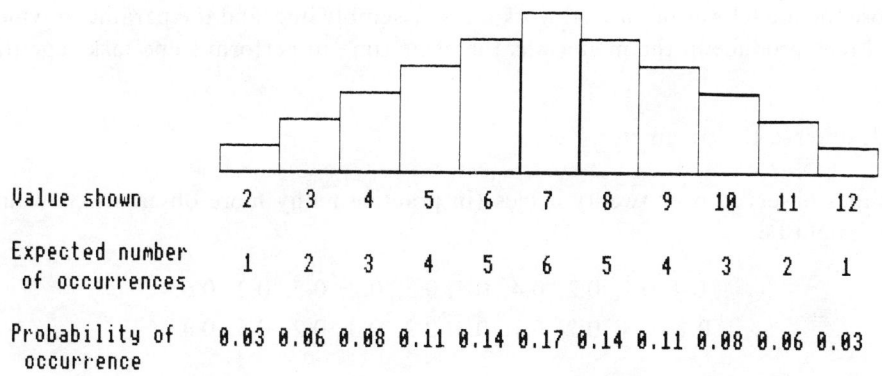

Value shown	2	3	4	5	6	7	8	9	10	11	12
Expected number of occurrences	1	2	3	4	5	6	5	4	3	2	1
Probability of occurrence	0.03	0.06	0.08	0.11	0.14	0.17	0.14	0.11	0.08	0.06	0.03

Figure 3.2   Expected outcomes when throwing two dice 36 times.

Now suppose that in our factory the number of persons absent each day varies between 2 and 12 with a mean value of 7 with the same probability as in figure 3.2. Then we could generate a series of numbers which would appear very similar to values of the number of absentees by throwing two dice each time we wanted a value for use in our model. Alternatively we could throw one 36-sided die which had one side labelled 2, two sides labelled 3, 3 sides labelled 4, and so on. In this way the values which we use, the digits 2 through 12, would occur with the correct frequency, even though we were using a uniform distribution, the 36 sides of the die, to produce the values.

When modelling, we need a method which will enable us to produce streams of numbers to suit any required probability distribution. It would be very inconvenient to have to depend on throwing dice specially made to have the desired number of faces and values on them. But it is usually easy to obtain uniformly distributed random numbers, for example from random number tables, and then to apply the method we have just devised, i.e. of generating random numbers from a uniform distribution, and assigning to each possible outcome the correct number of different random numbers.

Monte Carlo simulation, also called simulated sampling, is the name given to this technique of generating a stream of random numbers conforming to any desired probability distribution by drawing uniformly distributed random numbers and converting them to the desired probability distribution. Before we can apply the method we need to know the distribution of the values of the parameter which we wish to reproduce in our model. This is normally done by observing the parameter in practice, for example by using work measurement techniques. When applied manually the method has six steps:

1. Observe the parameter to be modelled.
2. Compute the frequency of the values.

3. Compute the cumulative frequency distribution and convert to cumulative probability distribution.
4. Associate a value of the parameter with uniform random numbers in the range 00 to 99.
5. Generate a stream of uniform random numbers using random number tables.
6. Obtain the value of the parameter corresponding to each of the uniform random numbers.

Suppose the model was of manual work on an assembly line, and the parameter which we wished to reproduce in the model was the cycle time to perform some task repetitively.

Step 1, observe the parameter:

The times observed over twenty cycles (in practice many more observations would be taken) might be:

$$0.4. \ 0.1, \ 0.2, \ 0.4, \ 0.7, 0.3, \ 0.2, \ 0.5, \ 0.2, \ 0.6,$$
$$0.1, \ 0.3, \ 0.2, \ 0.1, \ 0.3, 0.2, \ 0.3, \ 0.2, \ 0.5, \ 0.4$$

Steps 2 and 3, compute the frequency and cumulative probability distributions:

These give frequency and probability distributions as follows:

Value	Frequency	Cumulative frequency	Cumulative probability
0.1	3	3	0.15
0.2	6	9	0.45
0.3	4	13	0.65
0.4	3	16	0.80
0.5	2	18	0.90
0.6	1	19	0.95
0.7	1	20	1.00

Step 4, associate values of the parameter with each random number:

Table 3.1 gives uniformly distributed random numbers in the range 00 to 99. They can be related to the values of the parameter as shown:

Value	Cumulative probability	Range of random numbers
0.1	0.15	00 to 14
0.2	0.45	15 to 44
0.3	0.65	45 to 64
0.4	0.80	65 to 79
0.5	0.90	80 to 89
0.6	0.95	90 to 94
0.7	1.00	95 to 99

Table 3.1   Random numbers

81	78	89	59	40	33	60	67	19	21
94	51	37	19	31	89	34	62	23	42
21	75	28	18	20	10	76	43	91	69
35	88	90	88	53	00	80	77	42	96
60	94	49	75	69	26	96	24	05	91
97	84	54	83	27	28	99	94	46	19
92	10	80	60	32	63	08	71	06	14
23	70	49	30	71	21	23	26	20	76
82	07	15	38	54	15	75	99	27	84
80	75	31	64	67	97	64	06	56	81
42	10	00	37	24	33	56	28	43	89
54	94	54	43	71	87	78	60	72	06
57	36	84	56	98	10	17	89	53	25
61	37	35	11	63	87	59	64	92	62
85	17	23	11	05	56	35	36	34	52
02	84	29	56	99	02	03	35	96	70
24	64	48	50	42	79	28	99	53	12
63	20	82	33	82	22	07	33	39	93
38	94	98	52	70	50	78	00	55	08
98	76	37	41	46	27	58	22	67	65

Now, whenever a random number is obtained we can compare its value with the figures in the right-hand column until we find the range containing it and then use the corresponding parameter value in the left-hand column in our model. For example 27 is in the range 15 to 44 and would be associated with a value of 0.2. Figures 3.3, 3.4 and 3.5 show the process graphically, in the form of the data histogram, the cumulative histogram and the random numbers associated with each value of the parameter.

Steps 5 and 6, generate a stream of random numbers and obtain the coresponding parameter values:

Now obtain a series of random numbers and obtain the corresponding parameter value. We can get random numbers by reading the table down columns, across rows or in any other unbiased way. If we read down the left-hand column we would obtain the following values:

Random number	81	94	21	35	60	97	92	23	82	80	etc.
Parameter value	0.5	0.6	0.2	0.2	0.4	0.7	0.6	0.2	0.5	0.5	etc.

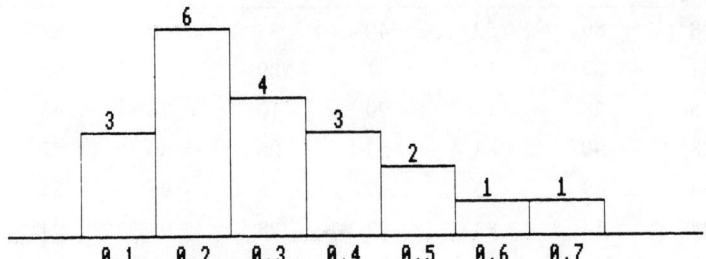

Figure 3.3   Histogram of observed cycle times.

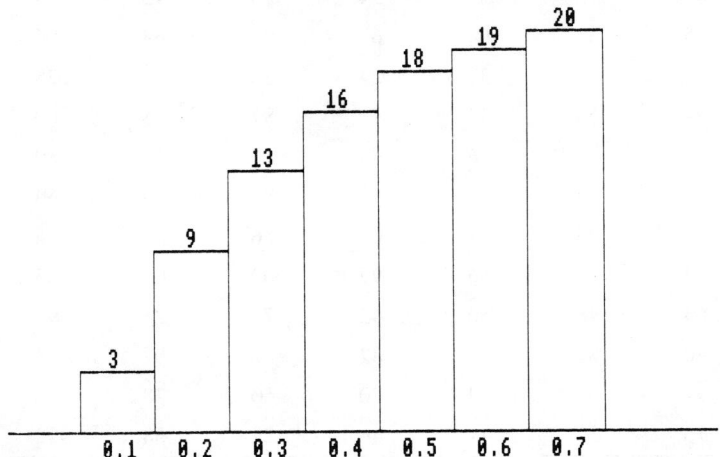

Figure 3.4   Cumulative histogram of observed cycle times.

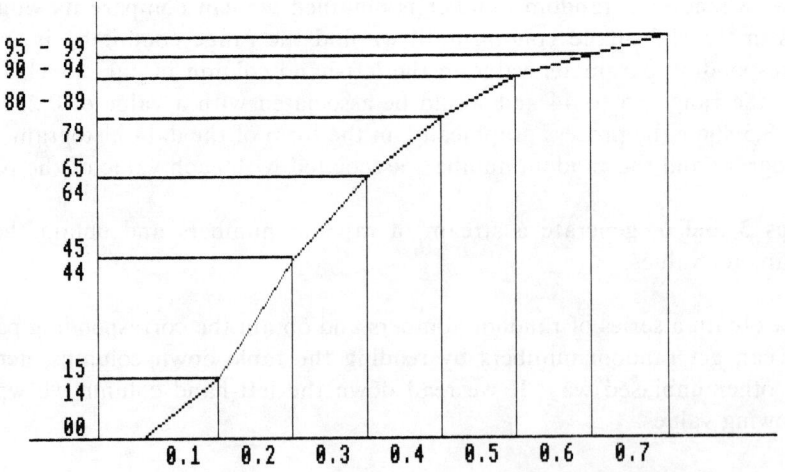

Figure 3.5   Cumulative probability curve relating random numbers to values of cycle time.

3.3   ACCURACY AND VALIDITY OF SIMULATED SAMPLING

If greater resolution is needed three-digit random numbers may be used, ranging from 000 to 999. This might be wise if there are many possible different values of the parameter, or their frequencies do not fit into percentage points without awkward rounding. In the example above, all the frequencies happened to be exact percentages, so this problem did not arise.

Various aspects of the accuracy of the process should be considered. For example, have a sufficient number of observed values been taken to adequately record the parameter? In the example above the same probability was obtained for values of 0.6 and 0.7. One would expect that 0.7 would be less likely than 0.6 because it is the more extreme value. Perhaps our observations slightly underestimated the frequency of 0.6 and overestimated the frequency of 0.7. There may be some reason why the observed figures would apply, but in general, if the frequencies do not conform to a smooth curve, insufficient observations may have been taken.

To remedy the situation, rather than go back and take more observations, we might draw the cumulative frequency and draw a smooth curve through the plots and then use the smoothed curve to get the values for simulating the system.

The example above used a discrete distribution of observed values. In reality the time for the operation will be continuously distributed over the interval 0.1 to 0.7 (or even wider) and values do not have to be multiples of 0.1. A value of 0.25 might be just as valid. Perhaps what we observed in step 1 was the number of occurrences of values between 0.05 and 0.15, 0.15 and 0.25, etc., rather than exactly 0.1, 0.2 and so on. In that case we would draw a smooth curve and use values which vary continuously.

If the parameter varies continuously then there exists the possibility that it conforms to one of the standard statistical probability distributions, such as normal, negative exponential, Poisson and so on. This would enable us to generate random values using the mathematical equation of the distribution, rather than employing a look-up table or by reading values off a graph, as above. To do this it would be necessary to compute the mean, variance and any other parameters required by the distribution concerned. Tables of random normal numbers are readily available. Most computer packages for simulation include functions for generating random variables conforming to any of the main statistical distributions.

Having ensured that the frequencies used to generate values are valid, we have to consider the validity of the samples obtained. If we had used any other column of the random number table we would have got a different set of values to feed into our model. Would we have obtained the same end results? This is mainly a matter of taking enough samples from the distribution. The values we obtained in steps 5 and 6 above seem to be biased towards larger values. We can easily check the average of the values obtained and compare it with the average of the original distribution. If the two averages are very close we may be confident that a valid set of samples has been obtained. If they are not very close then it is unlikely that enough samples have been drawn. How close they should be is an important point in determining the validity and the length of run of a simulation experiment. We will postpone further consideration of this and various other aspects of the validity of the experiment until later.

3.4 RANDOM NUMBER GENERATORS

How a computer, which is a deterministic machine, can generate random numbers is an interesting question. The simple answer is that it can't. However, it can generate pseudo-random numbers which are statistically indistinguishable from truly random ones.

3.4.1 Uniform random numbers

We can illustrate the principle by considering numbers on any page of a telephone directory. This will have many numbers, which might consist of a three-digit exchange code followed by a four-digit line number. You should observe that some exchange numbers are more popular than others. In other words the exchange numbers are not uniformly distributed. However if you look at the subscribers' line numbers a more even pattern will be found, and if the last digit only is examined the ten possible digits may appear with roughly equal frequencies. In other words the least significant digit approximates a uniform distributed random number. Similarly, if you examine mathematical tables, such as logarithms, the frequency of occurrence of the digits 0 to 9 in the least significant digit is roughly uniform, even though the most significant digit progresses steadily from 0 to 9. In the same way, the least significant digits of the result of multiplying two large numbers together give roughly a uniformly distributed series of numbers. There is a substantial literature on the methods of generating numbers which are as close to truly random as possible, and on the statistical tests which can be applied to the numbers generated, eg Tocher.[1] For our present purposes it will be sufficient to illustrate one of the most common methods used in computer simulation. We start with any value, known as the random number seed, and multiply it by a prime number. We then take the value of the required number of least significant digits as the first random number in the stream. This value is then multiplied by the same prime number, and so on. Poole and Szymankiewicz,[2] p231, give the following illustration:

For example if the starting number is 0.645329 and the prime is 317, then:

        0.645329   × 317 = 204.569, random number = 0.569
        0.569   × 317 = 180.373, random number = 0.373
        0.373   × 317 = 118.241, random number = 0.241
        0.241   × 317 = 76.397, random number = 0.397
        0.397   × 317 = 125.849, random number = 0.849

 If we require integer random numbers between 0 and 100, we multiply each random number by 100 and round up or down.

Most packages have a more elaborate method of generating uniform random numbers. For example, the function used in the ECSL package[3] is:

$$X_n \; = \; 3125 \; \times \; X_{n-1} \text{ modulo } 2^m$$

where m is one less than the number of bits in a computer word. One of the characteristics of such random number generators is that after some time the sequence of numbers repeats itself. If this cycle is short the generator is of little value. This formula has a cycle length of $2^{(m-2)}$, which is sufficient for most simulations, if not all.

3.4.2 Other distributions

Once we have a method of generating uniform random numbers it is reasonably easy to generate numbers according to other distributions. This has already been illustrated in the discussion of dice throwing. We showed that the sum of two dice thrown together had a triangular distribution, and we asserted that if twelve dice were used the sum would be normally distributed, or sufficiently close to it. An exponential distribution can be obtained by taking the logarithms of uniform random numbers. By such mathematical devices, and there is a substantial literature on this topic, see for example Naylor et al.,[4] random numbers can be generated for many statistical distributions.

3.5   AN INVENTORY CONTROL SIMULATION

To illustrate the method in a more substantial fashion let us perform a stochastic simulation. Suppose you are responsible for maintaining the stock of an item, perhaps some consumable part, such as a cutting tool used in a manufacturing system. From your knowledge of inventory control theory, you know that there are two basic procedures for ordering materials to maintain the stock, known as the Fixed Quantity System and the Fixed Period System. They are also known as the Continuous Review System and the Periodic Review System, respectively. You want to perform a simulation experiment to compare the costs of operating the two policies.

In the Fixed Quantity System (FQS) the same quantity is ordered from the supplier every time an order is placed. Orders will be placed whenever the remaining stock falls below the so-called re-order level (ROL). The interval between orders will alter depending on how quickly the stock has been used up since the last order was placed. The decision variables in the FQS are therefore the order quantity and the re-order level. Figure 3.6 illustrates the variation in level of stock over a period of time. Initially the stock is at a high level and it falls as the stock is used up. When it falls below the re-order level we place an order. The re-order level must be high enough so that the remaining stock will not have been exhausted before the order arrives. When the order is received the stock rises to a high level once more. The problem is that both the rate of consumption of the stock and the time taken by the supplier to deliver the order vary from time to time. If the demand for the item is less than usual or the supplier delivers more quickly than usual there should be plenty of stock on hand when the new material arrives. However, if the supplier takes longer than usual or the demand is higher, then there may be a serious risk that the stock will run out before the replenishment order arrives.

In the Fixed Period System (FPS) an order will be placed at regular intervals, such as weekly or monthly. The quantity ordered will be sufficient to bring the stock up to a level which will maintain the stock until the next order will be due. The order quantity will vary depending on how quickly the stock has been used up over the period since the last order was placed. In this system the decision variables are the interval between placing orders, known as the review period, and the level up to which the stock should be brought by the replenishment order, known as the maximum stock level (MSL). Figure 3.7 illustrates the situation. When an order is placed the stock on hand is brought up to the maximum stock level. Strictly this maximum stock level refers to the sum of the stock on hand plus the stock on order. As the stock is used up, both the stock actually on

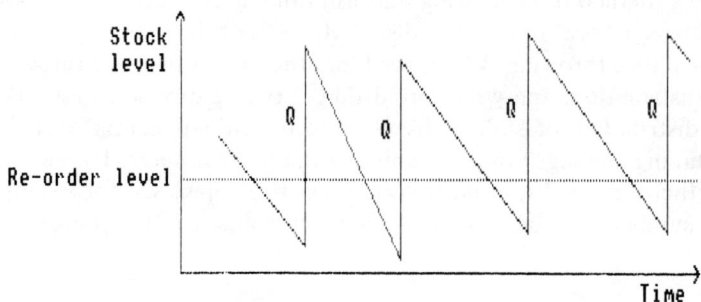

Figure 3.6   Fixed Quantity System of inventory control.

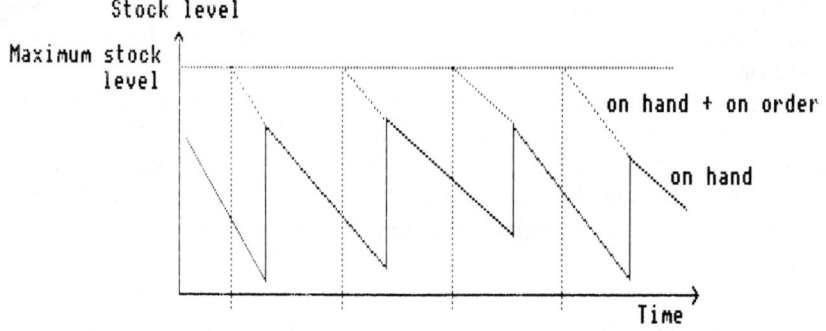

Figure 3.7   Fixed Period System of inventory control.

hand and the stock on hand plus on order gradually decline. When the order is received the stock on hand and the stock on hand plus on order become equal, and continue to decline until a further order is placed one period later. Consequently the maximum stock level has to be sufficient to cover the expected demand for the item over the review period plus the lead time.

For both cases it is customary to measure the effectiveness of the decision rule by aggregating the costs of placing the replenishment orders, the costs of holding the stock and the costs of any shortages.

In order to simulate the system we need to know the costs involved and the likely demand for the item and the lead time on filling orders. Suppose we analyse the past history of the item and discover that the demand for the last 100 days has been as in Table 3.2, and that the lead time has varied between 3 and 5 working days as shown in Table 3.3. We also obtain from our plant accountant the cost information shown in Table 3.4.

Table 3.2   Demand on past 100 working days

22	44	33	57	41	45	49	47	44	44
48	45	33	55	41	42	50	72	46	43
66	40	70	43	39	49	44	66	39	31
52	42	44	31	42	34	38	38	48	31
52	43	50	57	42	44	43	40	53	48
52	49	50	37	40	28	46	47	31	38
26	30	62	35	41	42	48	39	38	47
32	41	43	43	64	51	55	31	55	46
52	45	44	56	34	52	50	56	59	58
58	44	57	53	48	54	33	27	57	39

Table 3.3   Lead time history

Lead time	Frequency of ocurrence
3 days	28% of occasions
4 days	44% of occasions
5 days	28% of occasions

Table 3.4   Cost data

C_o = Cost of placing an order	= £100
C_i = Cost of holding stock	= £0.1 per unit per day
C_s = Cost of shortage	= £5 per unit

In order to compute the decision variables we need to analyse the past data and compute the mean and standard deviation of the distribution of the demand per day.

Mean demand = Total demand/Number of observations

$$\mu_D = \Sigma D/N$$

$$= 4523/100 = 45.23$$

The standard deviation is given by the formula:

$$\sigma_D = \sqrt{((\Sigma D^2 - (\Sigma D)^2/N)/(N - 1))}$$

$$= \sqrt{(214077 - 4523 \times 4523/100)/99}$$

$$= \sqrt{95.98}$$

$$= 9.80$$

We shall also need the mean and standard deviation of the lead time:

$$\mu_L \; = \; 28 \times 3 \; + \; 44 \times 4 \; + \; 28 \times 5/100$$

$$= \; 4$$

$$\sigma_L \; = \; \sqrt{((1656 \; - \; 400 \times 400/100)/99)}$$

$$= \; \sqrt{(56/99)}$$

$$= \; \sqrt{0.5657}$$

$$= \; 0.75$$

To apply the Monte Carlo method we need to construct a histogram of the demand per day, as shown in Figure 3.8. We have taken cells of width 5, but we could have taken cells of width 1, or 10, or any convenient width. Notice that the shape of this histogram is almost a normal distribution, but not quite. The demand has been between 40 and 44 on more days, and between 35 and 39 on fewer days, than we would have expected in a normal distribution. The same effect appears at the other end of the histogram around demands of about 60 per day. Had we taken cells of width 1 the shape would have been even more irregular because some values of demand did not occur in the last 100 days. On the other hand, if we had used cells of width 10 the shape might have been smoother.

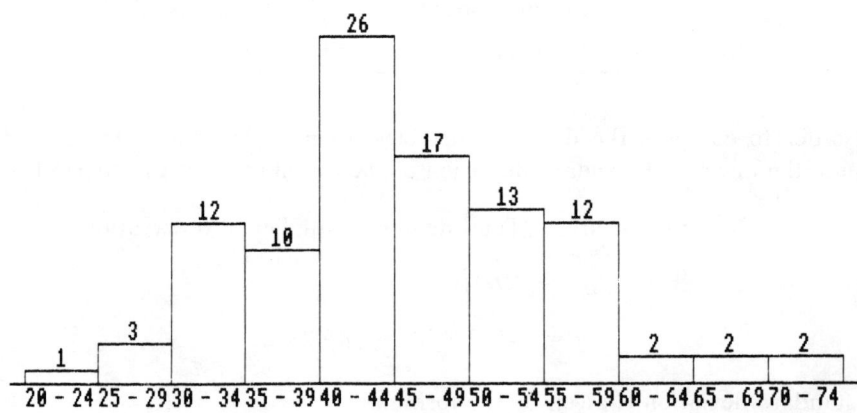

Figure 3.8   Histogram of demand per day.

Next we construct the cumulative histogram, representing the cumulative frequency distribution for daily demand, as shown in Figure 3.9. Since we have 100 days' data this is also the cumulative percentage probability distribution, and following the description of the method given above we can associate uniform random numbers in the range 00 to 99 with values of sales as shown on the vertical axis of Figure 3.10.

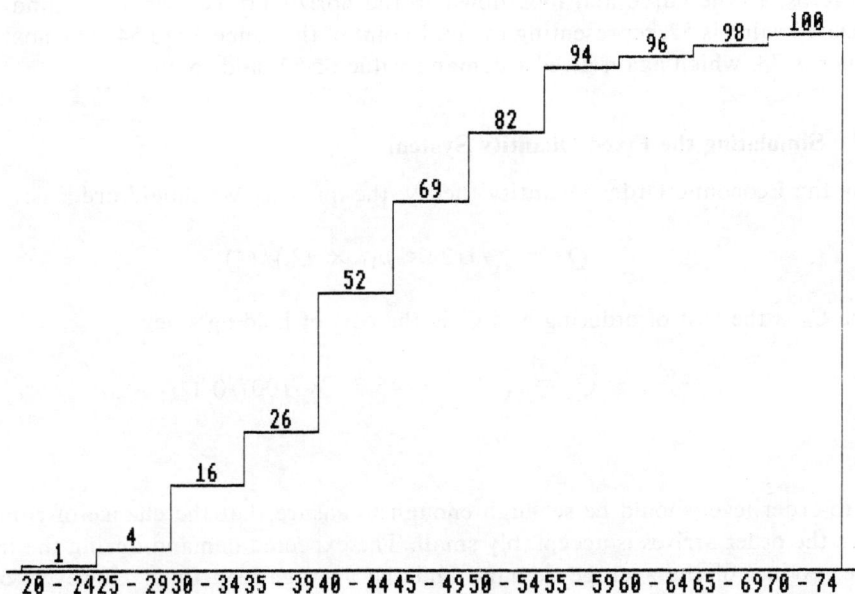

Figure 3.9   Cumulative histogram of demand per day.

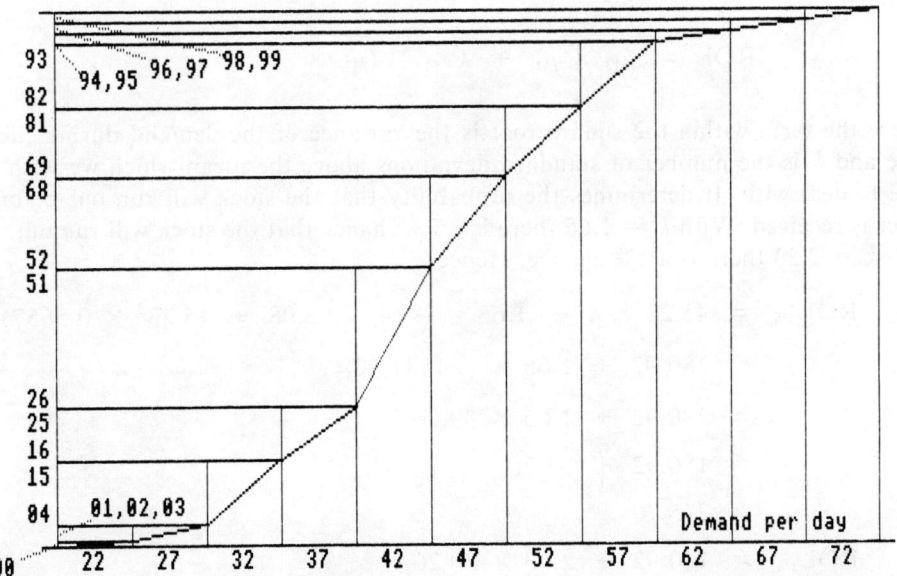

Figure 3.10   Cumulative probability curve relating random numbers to values of demand per day.

Now we refer to our random number tables, table 3.1, and working across the top row we obtain the number 81. We find this number on the vertical axis of figure 3.10, scan across to the curve and look down to the horizontal axis where we find that the associated value is 52, representing the mid-point of the range 50 to 54. The next random number is 78, which again gives a demand value of 52, and so on.

3.5.1 Simulating the Fixed Quantity System

Using the Economic Order Quantity theory, the quantity we should order is:

$$Q = \sqrt{((2 \times \mu_D \times C_o)/C_i)}$$

where C_o is the cost of ordering and C_i is the cost of holding stock.

$$Q = \sqrt{((2 \times 45.23 \times 100)/0.1}$$

$$= 300.8$$

The re-order level should be set high enough to ensure that the chance of running out before the order arrives is acceptably small. The expected demand during the lead time is the product of the average demand and the average lead time. However, on about half the occasions, demand will exceed the average and some allowance must be made for the variation of both demand and lead time, in the form of a safety stock. If the safety stock is too small we will run out of stock too often, while if it is too large the cost of holding the stock will be unnecessarily high. It is common to set the safety stock at a level which would give a certain probability of running out. Lewis[5] provides the formula:

$$ROL = \mu_D \times \mu_L + k \times \sqrt{(\mu_L \times \sigma_D^2 + \mu_D^2 \times \sigma_L^2)}$$

where the term within the square root is the variance of the demand during the lead time and k is the number of standard deviations above the mean which we wish to be able to deal with. It determines the probability that the stock will run out before the order is received. With $k = 1.65$ there is a 5% chance that the stock will run out, while with $k = 2.33$ there is a 1% chance. Hence

$$ROL_{5\%} = 45.23 \times 4 + 1.65 \times \sqrt{(4 \times 95.98 + 45.23^2 \times 0.5657)}$$

$$= 180.92 + 1.65 \times \sqrt{1541.20}$$

$$= 180.92 + 1.65 \times 39.26$$

$$= 180.92 + 64.78$$

$$= 245.70$$

$$ROL_{1\%} = 180.92 + 2.33 \times 39.26$$

$$= 180.92 + 91.48$$

$$= 272.40$$

Now we are ready to perform the simulation. Let us use an order quantity of 300 and a re-order level of 245, which will give approximately a 5% chance of the stock running out before the order arrives. Let us also assume that it costs nothing to store material over Saturday and Sunday. Let us assume that the initial stock is 300 units. If we start at the end of the day on a Friday we can compare the stock with the re-order level. It exceeds the ROL and therefore no order will be placed. Next day, Monday, we take a random number as described, 81, and obtain a demand figure of 52. This will leave a stock on hand of $300 - 52 = 248$, which is still above the ROL. On Tuesday the demand will again be 52 so the ending stock will be 196, which is below the ROL so an order for 300 will be placed on the supplier.

We need a value for the lead time for the order, so we pick a uniform random number and apply it to the lead time distribution. From table 3.3 we know that the lead time is either 3, 4 or 5 days with probability 28%, 44% and 28%. The cumulative probabilities are 28%, 72% and 100%. Therefore, if the random number is in the range 00 to 27 use a 3-day lead time, if it is 28 to 71 use 4 days, and if it is 72 to 99 use a 5-day lead time, as laid out in Table 3.5. Let us select lead time random numbers from the bottom of the random number table reading from right to left. The random number is 65 which gives a 4-day lead time. Since the order was placed on Tuesday it will arrive at the end of the day on Monday, and the new material will be available for use the following day, Tuesday.

Table 3.5 Random numbers associated with lead times

Random numbers	Lead time
00–27	3 days
28–71	4 days
72–99	5 days

We continue to draw random numbers for demand, adjust the stock level, place orders when necessary, draw a random number, find the lead time, and so on until we have simulated a sufficient period. Table 3.6 gives the results of this procedure. Then we calculate the costs involved and compare them with alternative ordering policies.

Now we can compute the costs involved:

Cost of holding inventory = $C_i \times \Sigma$ final stock

$= £0.1 \times 10360$

$= £1036$

Cost of orders $= C_o \times$ number of orders

$= £100 \times 8$

$= £800$

Cost of shortages $= £0$

Total cost $= £1836$

Table 3.6   Results for Fixed Quantity System simulation

				Order quantity = 300, re-order level = 245					
D	IS	RN	DD	SH	QR	FS	QF	QO	RN	LT
F						300	300	0	0	0
M	300	81	52	0	0	248	248	0	0	0
T	248	78	52	0	0	196	196	300	65	4
W	196	89	57	0	0	139	439	0	0	0
T	139	59	47	0	0	92	392	0	0	0
F	92	40	42	0	0	50	350	0	0	0
M	50	33	42	0	300	308	308	0	0	0
T	308	60	47	0	0	261	261	0	0	0
W	261	67	47	0	0	214	214	300	67	4
T	214	19	37	0	0	177	477	0	0	0
F	177	21	37	0	0	140	440	0	0	0
M	140	94	62	0	0	78	378	0	0	0
T	78	51	42	0	300	336	336	0	0	0
W	336	37	42	0	0	294	294	0	0	0
T	294	19	37	0	0	257	257	0	0	0
F	257	31	42	0	0	215	215	300	22	3
M	215	89	57	0	0	158	458	0	0	0
T	158	34	42	0	0	116	416	0	0	0
W	116	62	47	0	300	369	369	0	0	0
T	369	23	37	0	0	332	332	0	0	0
F	332	42	42	0	0	290	290	0	0	0
M	290	21	37	0	0	253	253	0	0	0
T	253	75	52	0	0	201	201	300	58	4
W	201	28	42	0	0	159	459	0	0	0
T	159	18	37	0	0	122	422	0	0	0
F	122	20	37	0	0	85	385	0	0	0
M	85	10	32	0	300	353	353	0	0	0
T	353	76	52	0	0	301	301	0	0	0
W	301	43	42	0	0	259	259	0	0	0
T	259	91	57	0	0	202	202	300	27	3
F	202	69	52	0	0	150	450	0	0	0
M	150	35	42	0	0	108	408	0	0	0
T	108	88	57	0	300	351	351	0	0	0
W	351	90	57	0	0	294	294	0	0	0
T	294	88	57	0	0	237	237	300	46	4
F	237	53	47	0	0	190	490	0	0	0
M	190	0	22	0	0	168	468	0	0	0
T	168	80	52	0	0	116	416	0	0	0
W	116	77	52	0	300	364	364	0	0	0
T	364	42	42	0	0	322	322	0	0	0
F	322	96	67	0	0	255	255	0	0	0
M	255	60	47	0	0	208	208	300	41	4
T	208	94	62	0	0	146	446	0	0	0
W	146	49	42	0	0	104	404	0	0	0
T	104	75	52	0	0	52	352	0	0	0
F	52	69	52	0	300	300	300	0	0	0

Table 3.6   (*continued*)

D	IS	RN	DD	SH	QR	FS	QF	QO	RN	LT
Order quantity = 300, re-order level = 245										
M	300	26	42	0	0	258	258	0	0	0
T	258	96	67	0	0	191	191	300	37	4
W	191	24	37	0	0	154	454	0	0	0
T	154	5	32	0	0	122	422	0	0	0
F	122	91	57	0	0	65	365	0	0	0

Key:
D   Day of week
IS Initial stock
RN Random number
DD Day's demand
SH Shortage

QR Quantity received
FS Final stock
QF Quantity on hand plus on order
QO Quantity ordered
LT Lead time

3.5.2   Simulating the Fixed Period System

The decision variables are the period and the maximum stock level. The most economic period will be that closest to the average interval between orders in the economic order quantity rule.

$$\text{Optimum period} = \text{EOQ/Mean demand}$$

$$= 300/45.23$$

$$= 6.65 \text{ days}$$

For organisational reasons we wish the period to be a number of weeks with orders placed on Fridays, ie with a period of 5 or 10 days. Let us use a period of 5 days.

The maximum stock level is less easily determined than the re-order level in the FQ system. In the case when the lead time is fixed the maximum stock level is given by the formula:

$$\text{MSL} = (P + \mu_L) \times (\mu_D + k \times \sigma_D/\sqrt{(P + \mu_L)})$$

where P is the length of the period and k is the safety stock factor, Lewis.[5] In the case when the lead time varies randomly there is no simple formula for setting the level. In fact, Lewis suggests the use of simulation to discover the best level. However for our purposes let us use the formula above, which should slightly underestimate the correct maximum stock level. Inserting 5 for P and 1.65 for k and the other values as previously calculated we obtain:

$$\text{MSL}_{5\%} = (5 + 4) \times (45.23 + 1.65 \times 9.80/\sqrt{(5 + 4)})$$

$$= 9 \times (45.23 + 1.65 \times 3.27)$$

$$= 9 \times (45.23 + 5.39)$$

$$= 9 \times 50.62$$

$$= 455.58$$

Now we can run the simulation again, using exactly the same sequence of random numbers, and placing an order every Friday for sufficient material to bring the stock on hand plus on order back up to 455. Table 3.7 gives the results.

We can calculate the costs of the FPS policy in the same way as before:

Cost of holding inventory = $Ci \times \Sigma$ final stock

$$= £0.1 \times 9090$$

$$= £909$$

Cost of orders $= Co \times$ number of orders

$$= £100 \times 11$$

$$= £1100$$

Cost of shortages $= Cs \times \Sigma$ shortages

$$= £5 \times 35$$

$$= £175$$

Total cost $= £2184$

Table 3.7   Results for Fixed Period System simulation

Review period = 5, maximum stock level = 455

D	IS	RN	DD	SH	QR	FS	QF	QO	RN	LT
F						300	300	155	65	4
M	300	81	52	0	0	248	403	0	0	0
T	248	78	52	0	0	196	351	0	0	0
W	196	89	57	0	0	139	294	0	0	0
T	139	59	47	0	155	247	247	0	0	0
F	247	40	42	0	0	205	205	250	67	4
M	205	33	42	0	0	163	413	0	0	0
T	163	60	47	0	0	116	366	0	0	0
W	116	67	47	0	0	169	319	0	0	0
T	169	19	37	0	250	282	282	0	0	0
F	282	21	37	0	0	245	245	210	22	3
M	245	94	62	0	0	183	393	0	0	0
T	183	51	42	0	0	141	351	0	0	0
W	141	37	42	0	210	309	309	0	0	0
T	309	19	37	0	0	272	272	0	0	0
F	272	31	42	0	0	230	230	225	58	4
M	230	89	57	0	0	173	398	0	0	0
T	173	34	42	0	0	131	356	0	0	0
W	131	62	47	0	0	84	309	0	0	0
T	84	23	37	0	225	272	272	0	0	0
F	272	42	42	0	0	230	230	225	27	3

Table 3.7 (*continued*)

			Review period = 5, maximum stock level = 455							
D	IS	RN	DD	SH	QR	FS	QF	QO	RN	LT
M	230	21	37	0	0	193	418	0	0	0
T	193	75	52	0	0	141	366	0	0	0
W	141	28	42	0	225	324	324	0	0	0
T	324	18	37	0	0	287	287	0	0	0
F	287	20	37	0	0	250	250	205	46	4
M	250	10	32	0	0	218	423	0	0	0
T	218	76	52	0	0	166	371	0	0	0
W	166	43	42	0	0	124	329	0	0	0
T	124	91	57	0	205	272	272	0	0	0
F	272	69	52	0	0	220	220	235	41	4
M	220	35	42	0	0	178	413	0	0	0
T	178	88	57	0	0	121	356	0	0	0
W	121	90	57	0	0	64	299	0	0	0
T	64	88	57	0	235	242	242	0	0	0
F	242	53	47	0	0	195	195	260	37	4
M	195	0	22	0	0	173	433	0	0	0
T	173	80	52	0	0	121	381	0	0	0
W	121	77	52	0	0	69	329	0	0	0
T	69	42	42	0	260	287	287	0	0	0
F	287	96	67	0	0	220	220	235	76	5
M	220	60	47	0	0	173	408	0	0	0
T	173	94	62	0	0	111	346	0	0	0
W	111	49	42	0	0	69	304	0	0	0
T	69	75	52	0	0	17	252	0	0	0
F	17	69	52	35	235	235	235	220	98	5
M	235	26	42	0	0	193	413	0	0	0
T	193	96	67	0	0	126	346	0	0	0
W	126	24	37	0	0	89	309	0	0	0
T	89	5	32	0	0	57	277	0	0	0
F	57	91	57	0	220	220	220	235	8	3

Key:
D Day of week
IS Initial stock
RN Random number
DD Day's demand
SH Shortage

QR Quantity received
FS Final stock
QF Quantity on hand plus on order
QO Quantity ordered
LT Lead time

3.5.3 Discussion on the results

The results suggest that a periodic system of 5 days is some 20% more expensive than a fixed quantity system using the economic order quantity. Theory predicts that the best period would be 7 days so that we would expect using a 5-day period to be slightly more expensive, but probably not as much as 20% more. However, we cannot draw firm conclusions from such a limited experiment. Several factors may be affecting the result. For example, we have included the cost of 11 orders in the fixed period system, even although one of these orders was placed on the first day of the run and one on the last day, with its material not arriving until after the end of the simulation run. If we had calculated the costs over the 10 complete weeks, the difference would have been reduced by $100. The way we started the simulation biased the results against the fixed period system. Had we simulated 20 weeks instead of 10 the effect of the initial conditions would have been less. Or we could have allowed the model to run for a few weeks before starting to assess the results. We started with an initial stock of 300. What would have been the effect of starting with a stock of 200? This would have caused an order to be placed immediately in both systems, and might have yielded fairer results.

We also notice that there were shortages in the fixed period system, but not in the fixed quantity system. Is this a generally valid result? Or, was it caused by particular circumstances in the model? Looking at the demand in the days prior to the shortage, we see that there was a maximum 5-day lead time coupled with slightly above average demand per day during the lead time. But the demand figures are not exceptional by any means, so it looks as though this was just one of these occasions which we predicted would occur with 5% probability, or on average about one week in twenty. When we look at the figures for the fixed quantity system, we note that on the corresponding day the initial stock was 52 and the demand for the day was also 52, so that the closing stock was zero, and it was fortunate that the replenishment order arrived that day. The lead time was 4 days, had it been 5 days then there would have been a shortage the following day. The lead time was 4 days not 5 days, as in the fixed period system, because the spacing of orders was slightly different and so the coincidence of above average demand and long lead time did not occur. Was this just a lucky accident, or was it the sort of thing to be expected in real life? We used only one set of random numbers for demand or lead time in the experiments. A different set might have given a different result. We ought to assess the quality of our random numbers. One simple check would be to compare the average demand of our samples with the average of raw data. The total demand over the 50 days was 2335, giving an average of 46.70. This is slightly above the expected value. Is it close enough? The difference is $46.70 - 45.23 = 1.47$. This is $1.47/9.80 = 0.15$ standard deviations, which is statistically very close. On the basis is this simple test, the random numbers appear to be acceptable, or rather do not appear to be unacceptable.

In running the experiment we took the mid-point values of each cell. This introduces an undesirable lumpiness in the figures, for example the order quantities in the fixed period system were all multiples of 5. Also, as observed above the cumulative frequency curve is not very smooth. If we have no reason to suppose that the demand distribution is not smooth then we might be wise to draw a smooth curve through the cumulative frequency points and use the curve to obtain values which vary continuously rather in increments of 5.

3.6   ACCURACY AND VALIDITY

It is appropriate to bring together some of the points which have been made in the above sections. In general there are four sets of factors which limit the validity of our results. These are

1. simplifying assumptions,
2. transient conditions,
3. bias in the random samples,
4. accuracy of the data used.

We will give a brief note on these points, but can only scratch the surface of what are fundamental points about simulation. They are dealt with at length in the literature, for example, in Tocher,[1] Fishman,[6] Pidd.[7]

3.6.1 Simplifying assumptions

In this model we ignored the fact that in real life there are seven days in the week and it is doubtful whether the assumption that no cost is incurred in holding stock over the weekends is acceptable in practice. We could have corrected for this by counting the stock on Friday three times, ie for Friday, Saturday and Sunday. However, the real point is that most simulations make some assumptions. Often this is a matter of the level of detail in the model, with supposedly minor characteristics overlooked. Thus the results are not exact, and the amount of error introduced by the assumptions is not measured. Consequently simulation is more appropriate for comparing alternatives than for predicting values of parameters. In simulation runs comparing alternatives the same assumptions can be made in each run, as was done in this model.

3.6.2 Transient conditions

In this model we started with one particular value of initial stock. Although the value used was probably acceptable, we cannot be sure that the value we used did not bias the results in favour of one of the decision rules. In general, there are two alternative conditions for starting a model. Either we make a guess at an "average" condition, or we start the model "empty". Starting the model empty in this case is meaningless, but in chapter 1 this is what we did in the machine shop model, and it is the frequently adopted method in manufacturing system simulation. If we start the model empty it will clearly take some time for normal conditions to build up. Until the steady state is reached any results obtained will be misleading. If we start the model from some average condition which we expect is close to the steady state condition it will not take so long to reach steady conditions. However in either case we should have a "run-in" period when we do not collect results, and only collect statistics when the steady state has been reached. Thus in the model above we should perhaps have excluded the first five weeks from the results calculations.

3.6.3 Bias in the random samples

In the model we used one stream of random numbers. As already suggested it is not inconceivable that the samples obtained were not truly representative of the population from which they were drawn. To avoid this there are two simple solutions. One is to run the simulation for long enough so that, provided the random number generator is not itself biassed, a valid series of samples is obtained. This would involve checking as we go how the samples compared with the population. The other method is to repeat the experiment several times with different random number streams and hence different sequences of samples. In this case we could analyse the results and by using analysis of variance techniques could estimate the extent to which the differences between results was due to the different random numbers or to the different decision rules which the simulation was being done to evaluate.

3.6.4   Accuracy of the data used

In most simulation projects some of the data may be approximate only. In this model we observed that the histogram of the daily demand had roughly a normal distribution, but the frequency of some cells was more and of some others less than would be expected in a normal distribution. In this case we could have drawn a smooth curve through the observed frequencies, if in our judgement we considered that a normal distribution was more likely than the lumpy distribution of the observed data. However, by doing so we might be introducing a further error. Whether the data can be smoothed in this way or not, we should always try to assess the sensitivity of our results to the possible error in the data. Generally this means running the model several times with different values of the parameter concerned. In this case, we might have repeated the experiment using normal distributions for demand with means varying between, say, 40 and 50 units per day.

SUMMARY

If a parameter in the model varies stochastically, the Monte Carlo method of drawing random samples from its distribution may be used to provide values for use in the model. When the parameter values are expressed in the form of a histogram, this consists of associating a value of the parameter with the value of uniform random numbers. Computers generate pseudo-random numbers which are statistically indistinguishable from genuinely random ones. Most packages provide functions for generating random numbers according to all of the common probability distributions. If the parameter conforms to one of these distributions random values can be generated automatically, and the graphical technique avoided.

We have used this technique in an inventory control model to compare the relative costs of two ordering policies. During the course of this exercise and the preceding discussion, comments were made concerning the validity of the results obtained. The principal limitations on the validity of the results arise from the bias, if any, in the random numbers generated, which is related to the length of the run and the value of the seed for the random number stream. To eliminate this bias it is normal to repeat the experiment using different seed values or to extend the length of the simulation run. Similarly, to ensure that the results are not affected by the particular initial conditions, it is normal to repeat the experiment from different initial states, such as initial stock.

EXERCISES

Draw a smooth curve for the cumulative probability curve and rerun the experiments:

1. Using several different sets of random numbers by taking them from different columns of the random number table.

2. Over 20 weeks. Compare the results after 10 and 20 weeks.

3. With an initial stock of 200.

4. With the values of ROL and MSL for 1% probability of run out ($MSL_{1\%} = 475$).

5. With a review period of 10 days ($MSL_{5\%} = 694$, $MSL_{1\%} = 719$).

6. Write a computer program to do the work for you, run all these experiments and plot the cost of each experiment on a graph.

7. Check the quality of your computer's random number generator.

8. Repeat the simulation using normal random numbers to generate the sales figures, instead of using uniform numbers which are then applied to the histogram, and compare results.

9. Repeat the experiments using values of 40 and 50 for the mean demand per day.

REFERENCES

1. Tocher, K. D., *The Art of Simulation*, English Universities Press, London, 1963.

2. Poole, T. and Szymankiewicz, J., *Using Simulation to Solve Problems*, McGraw-Hill, London, 1977.

3. Clementson, A. T., *ECSL User's Manual*, page C-110, Cle-Com Ltd, Birmingham, 1985.

4. Naylor, T. H., Balintfy, J. L., Burdick, D. S. and Chu, K., *Computer Simulation Techniques*, John Wiley, New York, 1966.

5. Lewis, C. D., *Scientific Inventory Control*, 2nd edition, Butterworths, London, 1981.

6. Fishman, G. S., *Principles of Discrete Event Simulation*, John Wiley–Interscience, New York, 1978.

7. Pidd, M., *Computer Simulation in Management Science*, John Wiley, London, 1984.

CHAPTER 4

Discrete simulation

4.1   INTRODUCTION

We have already observed that manufacturing systems are modelled by discrete simulation, sometimes combined with continuous aspects. In chapter 1 we began a manual simulation of a small machine shop, in which machines and jobs were involved and the machines performed operations on the jobs, which took periods of time. This was clearly a discrete simulation. In it, we noted the life history of each machine and job. Time moved in jumps from one instant when something happened to another. We will now consider the structure of discrete models in a more formal manner.

4.2   THE ELEMENTS OF DISCRETE SIMULATION

Several terms will recur throughout any work on simulation. Some have already appeared in the earlier chapters, and we shall now explain the meaning of the following terms:

> Entity,
>
> Activity,
>
> Events,
>
> Queues,
>
> Attributes,
>
> Sets,
>
> States.

4.2.1   Entities

Entities are the components of the system, such as the machines, workpieces, handling equipment and so on. In the grammar of simulation, entities are the nouns. Entities are generally of two types:

> permanent entities,
>
> temporary entities.

Permanent entities are those which are in the model for the entire duration of the simulation experiment. Temporary entities are ones which enter the model at some time, pass through it, and leave at some later time. In some simulation systems the permanent and temporary entities are referred to as facilities and transactions. These names follow from the concept that the transactions make use of the facilities as they work their way through the model. In queueing networks they are the service centres and customers.

In the machine shop model, the machines are the permanent entities and the jobs are temporary entities.

4.2.2   Activities

Activities are the things that the entities do or have done to them. In the grammar of simulation, activities are the verbs. In virtually every activity more than one type of entity is involved, and hence an activity could be defined as the coming together of two or more entities for a period of time. It is an essential aspect of activities that we know how long they will last, so that when they start we can specify when they will end. This is essential if the flow of time is to be handled in the manner described earlier. Even if we are uncertain of the duration of the activity we sample a value from a distribution, as in stochastic simulation. Activities are assumed to begin or end instantaneously.

In the machine shop model the only type of activity was the processing of jobs by machines. For example, machine B worked on job 1 for 8 hours. The durations were all given deterministically. This is somewhat unusual, since one of the main objectives in simulation is to assess the effects of possible variation in the durations of the activities.

4.2.3   Events

Events are the instants in time when some change takes place in the model, ie when one or more activities begin or end, Figure 4.1. The state of the model as a whole, and of each of its entities, changes only an event occurs. During an activity the state of an entity is considered constant. Usually more than one activity begins or ends at an event. For example, in the machine shop when job 5 was completed on machine C at time 15 (the end of an activity) two more activities began, namely machine A began processing job 5 and machine C began processing job 4. Later we shall distinguish between "bound" events and "conditional" events. The completion of job 5 by machine C was "bound" to happen once the activity began, but the other two activities were "conditional" on the state of the model at the time.

Events are sometimes classified into two types:

endogenous, or internal, events which are caused by conditions in the model, such as the completion of an operation, and

exogenous, or external, events which are caused from outside the model, such as the arrival of a job from the outside world.

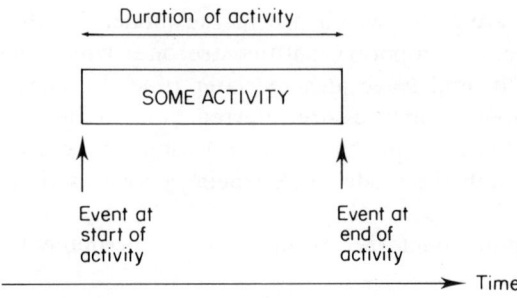

Figure 4.1   Relationship between activities and events.

4.2.4 Queues

Queues are passive states of an entity, while it waits for conditions to change so that another activity can begin.

In the machine shop model, we observed jobs waiting for machines to finish operations on other parts. When a machine became available the next activity could begin. There were also queues of machines waiting for jobs to arrive at them so that they could start an operation. Although the machines do not physically line up behind one another, logically there is no difference between the queues of jobs waiting for machines and the queues of machines waiting for jobs.

4.2.5 Attributes

Attributes are characteristics of entities, such as the type of a machine or the number of operations required on a job. In the grammar of simulation, attributes are the adjectives. Attributes are used to distinguish one entity of a type from another, and the selection of entities from queues frequently depends on the attributes of the entities in the queue. For example we will often wish to select the entity which has been waiting longest, so the time the entity entered the queue would be an attribute of the entity.

In the machine shop model, we did not explicitly recognise the use of attributes in selecting jobs from queues. We said that when job 2 arrived it had to wait for machine B, but we did not consider the queueing arrangement. There might have been a separate queue for each machine, or there could have been one queue for all the machines in which the jobs had a label showing which machine they were waiting for. Physically both methods are possible, and logically both methods are also possible. In one case, when the activity finished we would have to determine which queue to put the job into, according to its attribute value, and in the other we would have to determine which job to select from those in the single queue, according to their attribute values.

4.2.6 Sets

The set is a general concept in simulation. We could talk of the set of jobs in a queue, indeed several simulation languages use the word set to describe the queues in the model. Sets can also be used to group entities in any convenient way. For example, we might use a set named "needed" to hold the list of tools required for an operation on a machining centre. "Present" could be the set of tools in the magazine at any time. Then we could test whether the set "needed" was contained within the set "present" to find out if any tools had to be changed before an operation could be done. Many simulation languages provide powerful facilities for manipulating sets.

4.2.7 States

Generally we use the term "state" to refer to the condition of the model or its entities so that we can test whether some action can or must be performed, or to choose between possible actions. We refer to entities being either in an active state or a passive state, depending on whether they are engaged in an activity or are waiting in a queue.

4.3 ACTIVITY CYCLE DIAGRAMS

Almost all simulation languages use some form of diagram or flow-chart to describe the system being modelled, either to assist in the model-building process, or in implementing the model on a computer, or just to help understanding and communication. One of the most powerful methods is to use activity cycle diagrams,[1] or entity life cycle diagrams as they are sometimes called.

Every entity is considered to perform some cycle of activities in its life, and to pass through queues between one activity and another. For example, in the machine shop the jobs entered, then waited for the first machine which they had to be processed by, were processed by that machine, then waited for the next machine to become available, then were was processed by that machine, waited again and so on. In some cases the job did not have to wait because the machine was already available, but this does not spoil the concept, because we could think of it passing through the queue without a delay. The activity cycle diagram (or ACD) depicts the cycle for each type of entity. There are five conventions in drawing activity cycle diagrams:

1. Each type of entity has an activity cycle.
2. The cycle consists of activities and queues.
3. Activities and queues alternate in the cycle.
4. The cycle is closed.
5. Activities are depicted by rectangles and queues by circles or ellipses.

There are two basic forms of activity cycle, concerned with the case where the entity either

1. May perform one or more different activities in any sequence or is idle,
2. Must perform activities in a definite sequence.

These can be illustrated by a somewhat simplified model of a cafeteria. We will consider only two types of person, counter staff and customers, and ignore other staff such as cash-desk or cleaning staff. The counter staff do two different activities, serve food to customers and put food out on display. These activities can be done in any order, depending on the number of customers waiting and the quantity of food on display, so the activity cycle for these staff is of the first type, as illustrated by Figure 4.2. The customers on the other hand perform activities in a definite sequence, such as enter the cafeteria, get served by the counter staff, pay for food, eat the food, and leave the cafeteria. Between each pair of activities there will be a queue, either in the real sense, such as wait for service queueing at the cash desk, or to separate successive activities, such as paying for food and eating the food. The customer's cycle could be depicted as in Figure 4.3, although other activities such as walk to table could also have been included.

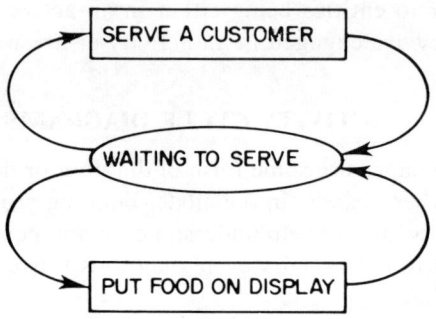

Figure 4.2   Activity cycle for member of counter staff in a cafeteria.

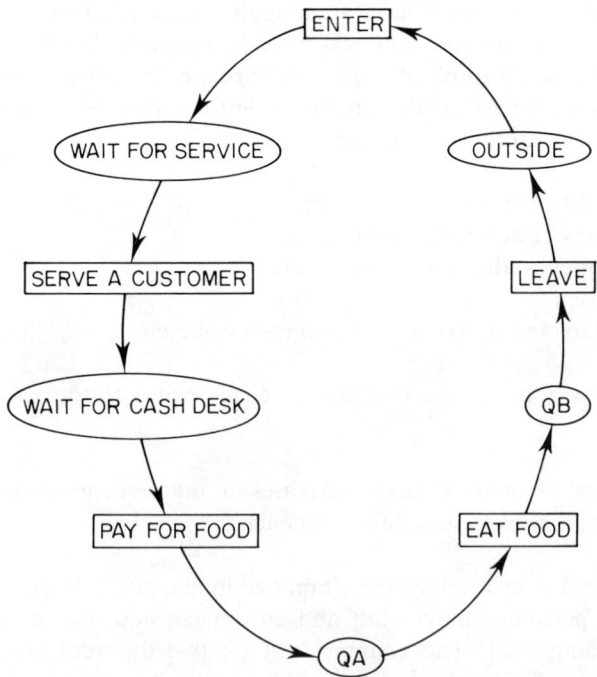

Figure 4.3   Activity cycle for customer in a cafeteria.

Figure 4.2 shows that cycles can have branches in them, since a member of the counter staff who is idle has the choice of whether to serve a new customer or put food on display. In this case the branch is from a queue, but there can also be branches from an activity. For example, Figure 4.4 shows a section of an activity cycle diagram in which entities branch from an inspection operation depending on the results of the inspection. Whenever there is a branch a rule has to be stated to determine which branch should be followed, or which of the possible activities an idle entity should perform in preference to others. These priority rules will be discussed later.

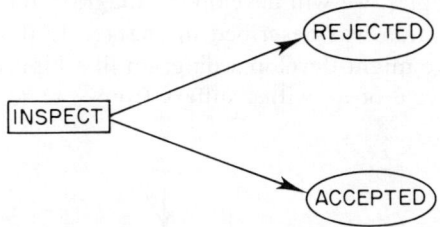

Figure 4.4   Part of activity cycle showing branching from activities.

Each activity draws entities from the queues which lead into it, and supplies entities to the queues which follow it. Starting an activity is conditional on there being suitable entities in each of the queues which lead into it. For example, as illustrated by Figure 4.5 the activity "serve a customer" depends on there being a customer in the queue "waiting for service" and a member of the counter staff in "waiting to serve". If there are these entities at any time the activity can begin. When the activity is complete the customer moves on to "wait for cash desk" and the member of counter staff returns to "waiting to serve". The member of staff can start to serve another customer immediately if there is one waiting, and so can pass through the queue "waiting to serve" instantaneously. Frequently, tests of attribute values are necessary to identify the entity with the required characteristics. Possibly several entities are required before an activity can begin, such as left-hand and right-hand models.

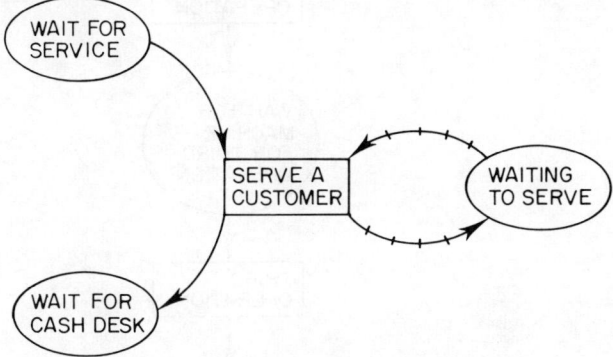

Figure 4.5   Activities draw entities from queues to start an activity and send them to queues on completion of the activity.

Each queue holds only one type of entity, and should have a unique name. Thus we cannot place both customer and staff member in the queue "waiting", but have to give their respective queues different names. Strictly, some packages allow more than one type of entity to be in a single queue. This has certain advantages and complications, which will be mentioned at an appropriate stage in the text. We will adopt the more restrictive definition of a queue throughout this text.

4.3.1 Activity cycle diagram for the simple machine shop

To illustrate these principles, we will develop the diagram for the simple machine shop for which a hand simulation was described in chapter 1. If we examine the sequence of operations of a job, we might develop a diagram like Figure 4.6. Unfortunately, this way of describing the cycle of activities suffers from a serious defect which limits its

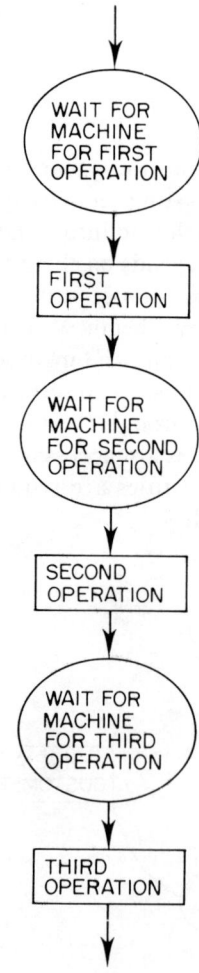

Figure 4.6   An attempt at the activity cycle for a job in simple machine shop model.

generality. Since each job has a different number of operations the cycle for each job would be different. But since the jobs are all the same basic type of entity we need a method which is more general. Fortunately, this is easily solved since the different activities, perform operation one, perform operation two and so on, are all instances of the same basic activity, PROCESS, because they involve the same types of entity, namely a job and a machine. Thus the activity cycle for the jobs could be better described by the simple cycle shown in Figure 4.7. The job will be processed then wait then be processed again, wait again, and so on. To conform to the third rule stated above, even if the job does not have to wait, we don't let it bypass the queue, but we allow it to pass through the queue with no delay. This is perhaps a logical quirk, but is very close to real life. When we go into a Post Office for service we anticipate that we may have to wait in a queue. If we find that there are no people ahead of us we can go straight to the head of the queue. Is there any difference between saying we did not queue and saying that we reached the head of the queue immediately?

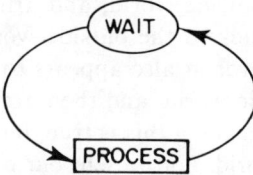

Figure 4.7   Improved activity cycle for job.

This cycle is only a first stage in developing the ACD for the jobs. Since jobs are temporary entities, the cycle must include the job arriving into the job shop, and leaving once all the operations are complete. To show these activities we need a diagram like Figure 4.8.

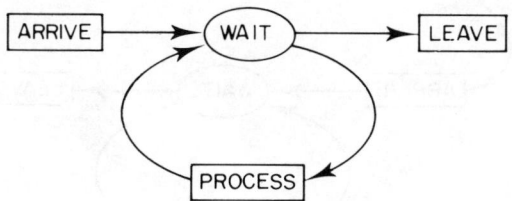

Figure 4.8   Activity cycle for job including ARRIVE and LEAVE activities.

It may seem strange to show ARRIVE and LEAVE as activities, since jobs will arrive in the system and leave it virtually instantaneously, whereas an activity, by definition, takes a period of time. A few words of explanation can resolve the problem.

In the case of the LEAVE activity it may be convenient to have a separate activity in which statistics concerning a temporary entity may be recorded as it leaves the system, such as the time it has been in the system. It may be possible to set the duration of the activity to zero so that no time elapses in leaving the system. Therefore we will include a LEAVE activity in the job's activity cycle. The methods and models presented in later chapters will show that there are other ways to handle the way jobs leave the system. Often, it is largely a matter of convenience which method is selected.

The reason for including ARRIVE in the cycle as an activity is rather more theoretical. It was pointed out that there are two types of event, one of which, exogenous events, are caused from outside the system. The arrival of jobs in the machine shop was given as an example. In most simulation the arrival of temporary entities is controlled by time between arrivals, which is usually expressed in terms of a probability distribution, frequently a negative exponential distribution. The duration of the ARRIVE activity therefore represents the time interval between the arrival of one job and the next, and the instant when a job arrives into the system is the event at the end of the arrival activity.

The diagram is not yet complete for we have not yet closed the cycle, as required by rule 4. In simulation we frequently introduce a queue known as the world set, representing the world outside the system being modelled. If we introduce the OUTSIDE WORLD, the cycle may be completed as shown in Figure 4.9. This shows that jobs arrive into the model from the outside world, and after being processed a number of times leave the system and go back to the outside world. This seems a very accurate summary of what happens. However, it also appears to suggest that jobs may leave the system, join a queue called outside world, and then arrive into the system again, which does not sound correct. Strictly speaking this is true, but we have to distinguish between a computer model and the real world. For present our purpose, we shall regard the cycle as starting and ending at the outside world. In the next chapter, we will enter the model into a computer and will give an explanation of this point then.

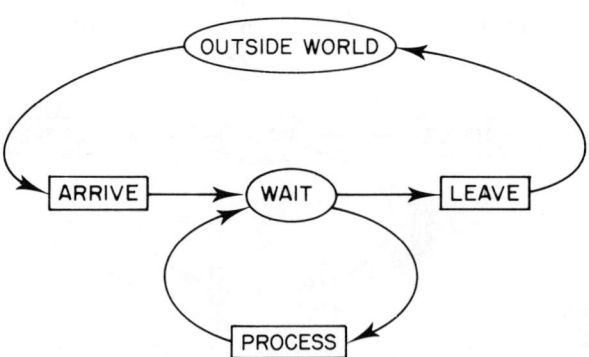

Figure 4.9   Activity cycle for job including the OUTSIDE WORLD.

Now let us consider the machines. They perform only one type of activity, processing jobs, and when not doing so are idle. If we assume that the machines do not break down and there are no other complications then their cycle is very simple and may be drawn as in Figure 4.10.

The cycles for the two types of entity can be put together to form the complete Activity Cycle Diagram, as in Figure 4.11.

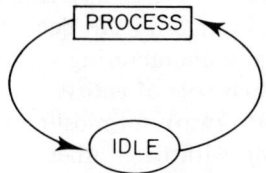

Figure 4.10   Activity cycle for machines in simple machine shop model.

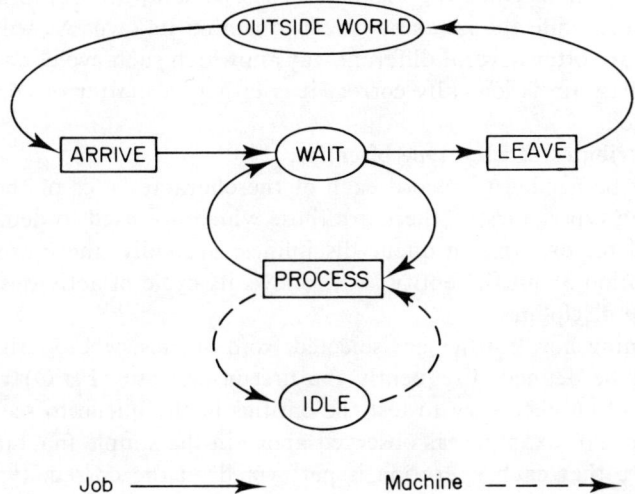

Figure 4.11   Combined activity cycle diagram for simple machine shop model.

4.4   BUILDING A MODEL USING ACTIVITY CYCLE DIAGRAMS

This activity cycle diagram shows the queues and activities for each type of entity, but it is still some way from being a complete model. For example, the cycle for the job has a branch in it. What determines whether the job moves to the PROCESS activity or to the LEAVE activity? Also, when a PROCESS activity is to be done it will not be correct to use just any machine for the job, because we know that each operation requires a particular type of machine. These and various other considerations have to be dealt with in building a simulation model from the ACD. Before we go further with the machine shop model it will be advantageous to review the various stages which have to be gone through:

1. **Define the entities.**
   Usually most of the entities in the model will be obvious from a simple description of the system to be modelled. However for other entities it may not be obvious. For example, if a workpiece is placed in a fixture and placed on a pallet, do all three objects need to be included as entities in the model, or can the combination be considered as a single entity, perhaps called a job? In general if the different items may follow separate existences they should be treated as separate entities; if they

always act as a team they can be considered as a single entity. This is a fundamental aspect of model building, and examples of this decision will be given when we come to discuss models of various real manufacturing systems.

2. **Develop the activity cycle for each type of entity.**
Once again this is fairly straightforward for most entities. However, in order to have queues and activities alternating with each other it may be necessary to introduce "dummy" queues, that is queues which the entity will pass through with no delay. Dummy queues are necessary when one activity may follow immediately on another. For example, a machine may be capable of running on its own, but it must first be set up by a setup man. Once the machine has been setup the setup man goes on to another machine, while the machine is left to run on its own. As will be illustrated later on, there are often several different ways in which such cycles can be described. Provided the diagram is logically correct it is largely a matter of convenience as to which is adopted.

3. **Identify the attributes of each type of entity.**
Attributes will be needed to record each of the characteristics of the entity. These are of two main types. Firstly, there are those which are used to determine how the entity behaves, for example in queue disciplines. Secondly, there are those used to record information about the entity as it follows its cycle of activities.

4. **State the queue disciplines.**
Rules determining how entities are selected from queues, when activities are being initiated, must be defined. Frequently the first-in-first-out (FIFO) rule will apply. However, it is often necessary to test the entities in the queue to see whether some condition holds. For example, as observed above in the simple machine shop model, we must ensure that each operation is performed on the correct type of machine. When a cycle has branches, as in our example machine shop, rules must be stated to specify under which conditions the entity follows each branch. These involve testing the values of one or more attributes of the entities concerned.

5. **Define the durations of the activities.**
Normally in simulation activity durations are randomly sampled using the Monte Carlo method. In manufacturing system models we may prefer to look up a table of operation times, when the times are known exactly and are fixed. This might apply if the operations are done on NC machines. Otherwise the value obtained from a table might be used as a mean value for the Monte Carlo method. When a table is to be looked up, the entry will be found with reference to the value of attributes of the entity.

6. **Define the calculations needed on the values of attributes.**
In order that the attributes should have the correct values when being tested for priority rules, activity durations or collecting results, it is necessary to define the formulae and occasions when the values are to be calculated. For example, in the machine shop model, the number of operations which have been carried out on a job will have to be increased by one every time an operation is performed.

7. **Specify the statistics to be collected.**
We have to arrange for data to be collected for any results which are to be obtained. There is no point building a model and then not obtaining any results from it. The statistics are usually of two kinds: time functions, such as the number of machines busy, and observed values, such as the time each job is in the system, as was discussed

in chapter 2. Attributes of the entity concerned are normally used to collect data pertaining to an individual entity, where a value must be stored for later analysis. For example, to record the time for which an entity is in the system we need to record its time of arrival and then subtract that time from the current clock time when the entity leaves the system. Thus the time of arrival would be held as an attribute of the entity. When the entity leaves the value would normally be added to a variable, statistic or histogram summarising this data for all the entities for printing out at the end of the run. Clearly, it is necessary to define not only which values are to be observed, but where in the model, ie when in the entity's activity cycle, the values are to be recorded.

8. **Specify the initial conditions.**

At the start of the simulation run it is necessary to know the condition of each entity in the model. They will be either in a queue or performing an activity. For each activity in progress, the time when it will be completed must be given, as will the values of any associated attributes of the entities involved. All the remaining entities will have to be placed in queues, and again the values set for attributes as necessary. In manufacturing systems models the initial conditions are often set so that the shop is empty and all the jobs are waiting outside the model, and all the machines and other facilities are idle. This may resemble the conditions at the start of work on a Monday morning, although it is perhaps unreasonable to assume that when a week finishes there is no work in progress in the shop. More important for the validity of the results is that the system would take some time to build up to a "normal" condition. Until a steady state condition is reached results statistics should not be collected.

9. **Specify the initial values of any other variables.**

Every variable must have its initial value specified, even if it is set to zero by default. In particular, each variable whose value will not be reset while the model runs must have its initial value specified. These would include the seed values of random number streams, as in the inventory control model, or attributes of entities. For example, in the machine shop model the type of each machine and job, as well as the operation sequence and time data, must be given.

4.4.1 Completing the simple machine shop model

Now let us apply these steps to the simple machine shop model. We have already defined the entities and their activity cycles. Now we will consider the queue disciplines. First consider the branch in the job's cycle. Jobs can move from the queue WAIT to PROCESS or to LEAVE. We have already stated that the job will leave when all its operations are done. Therefore, we need one attribute to hold the number of operations needed and another to count the number of operations actually performed. This counter will start at one and increase by one every time an operation is done. When it exceeds the value of the other attribute, the job will LEAVE. Another way to achieve the same effect would be to set a counter when the job enters the system, to the number of operations required, and to decrement it each time an operation is completed, then when it reaches zero the job will leave. This method needs only one attribute, which might be important if computer memory is in short supply. However, in this case, since the counter will be used to look up a table for operation times and machine type required, the former method is more convenient. We can call the operation counter OPNO, and the number of operations required NOOPS. The value of NOOPS for each job would be set when it arrives, by looking up a table. The table could be an array OPSREQD(I), giving the number of operations required jobs of type I. The table of values would have to be set as part of the initial values of variables, to the values shown in Table 4.1(a). The rule and the associated arithmetic can now be shown on the diagram as in Figure 4.12.

Table 4.1   Data tables for the simple machine shop model

(a) Number of operations required

Job type, I:	1	2	3	4	5
OPSREQD(I):	4	3	3	4	4

(b) Type of machine required for each operation

Job type, I	Operation number, J			
	1	2	3	4
		MCREQD(I,J)		
1	B	C	A	D
2	B	D	E	0
3	A	E	A	0
4	D	C	E	A
5	C	A	B	E

(c) Duration of each operation

Job type, I	Operation number, J			
	1	2	3	4
		OPTIME(I,J)		
1	18	8	5	6
2	14	2	16	0
3	4	3	2	0
4	3	4	9	3
5	8	5	8	6

(d) Values of machine type attributes

Machine number	1	2	3	4	5
MTYPE	A	B	C	D	E

We also have to ensure that each operation is done on the correct type of machine. To do this we will need an attribute for the type of machine, which we could call MTYPE, and for the type of machine required for the next operation on a job, which we could call NEXT. In selecting machines from queue IDLE we will need to ensure that the MTYPE of the machine is the same as the MCREQD of the job which is in WAIT. We can annotate the diagram to show this, as in Figure 4.13.

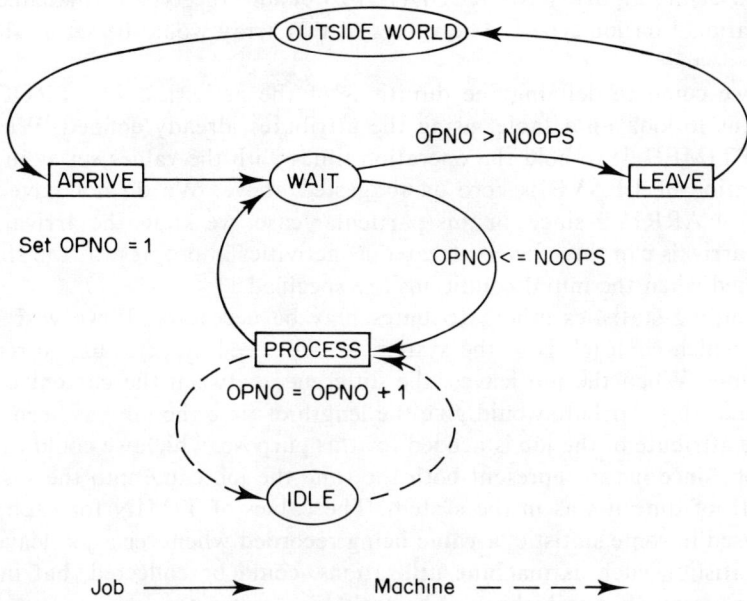

Figure 4.12 Activity cycle diagram including decision rule for the branch.

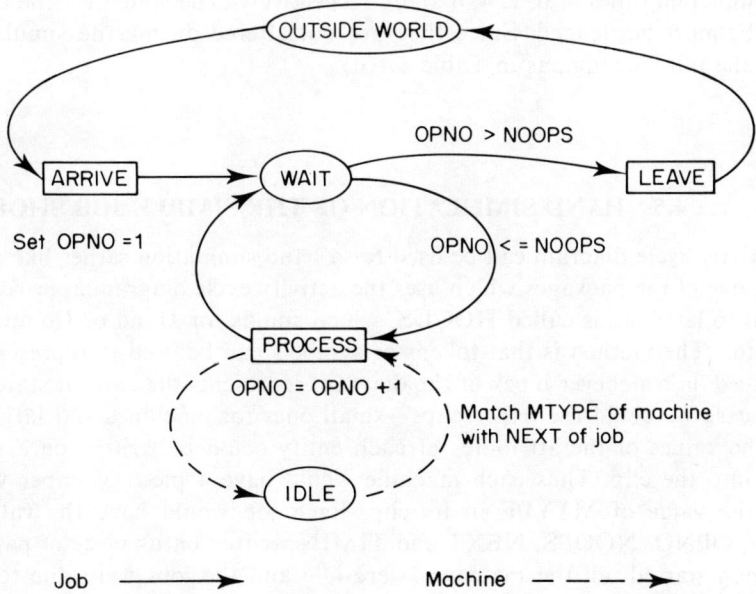

Figure 4.13 Activity cycle diagram showing the queue discipline for matching jobs and machines.

The values of NEXT would have to be obtained from the table of operation sequences. We could define an array MCREQD(I,J) to denote the type of machine required for the operation J on job type I. The values of this array would be set as shown in Table 4.1(b).

Now we come to defining the durations of the activities. The PROCESS activity requires us to look up a table, using the attributes already defined. We could use an array OPTIME(I,J) to hold the operation times with the values set as in Table 4.1(c). The duration of LEAVE is zero as suggested above. We do not have to define the duration of ARRIVE since, in this particular case, we know the arrival time of each job. Job arrivals can therefore be treated as activities in progress at the start of the run, and defined when the initial conditions are specified.

In recording statistics other attributes may be necessary. If we wish to record the time for which each job is in the system we will need an attribute to record the job's arrival time. When the job leaves, the difference between the current clock time and the value of this attribute would give the length of time the job has been in the system. Only one attribute of the job is needed for this purpose, which we could call the TIMIN of the job, since it can represent both the time the job came into the system and later the length of time it was in the system. The values of TIMIN for each job would be summarised in some statistic, a value being recorded whenever a job leaves the system. Other statistics, such as machine utilisations, could be collected, but in order not to introduce too much detail, this will be omitted at present.

The next requirement is to specify the initial conditions. All the machines were idle, and all the jobs outside the system at the start. However, since the arrival time of each job is known, we can consider that all the jobs are engaged in the ARRIVE activity with completion times of 0, 2, 4, 6 and 7 respectively. The values for the machine types, MTYPE, must be defined. Since they are not altered during the simulation they are part of the initialisation, as in Table 4.1(d).

4.5   HAND SIMULATION OF THE SIMPLE JOB SHOP

The activity cycle diagram can be used for a hand simulation rather like a board game. In fact, one of the packages which uses the activity cycle diagram approach, and will be referred to later on, is called HOCUS, which stands for Hand or Computer Universal Simulator. The method is that tokens or symbols can be used to represent each entity, and placed on whichever block of the diagram represents the current state of the entity. In this case we could use paper clips—small ones for machines and large ones for the jobs. The values of the attributes of each entity could be written on a piece of paper placed into the clip. Thus each machine would have a piece of paper with the letter giving the value of MTYPE in its clip. Each job would have the values of its job number, OPNO, NOOPS, NEXT and TIMIN written on its piece of paper. When the simulation started, all the machines were idle and the jobs were due to arrive, so all the large paper clips would be placed on the ARRIVE block and all the small ones in the IDLE block. Because the ARRIVE activities are in progress at the start of the run we would write the arrival time as the value of the TIMIN attribute. We would also set the value of NOOPS to the value in the OPSREQD table, set OPNO to one, and NEXT to the type of machine given in MCREQD(I,1).

As we run the simulation Job 1 arrives at time 0 and its paper clip will be moved to the WAIT block. To start a PROCESS activity we need a job in the WAIT block and a machine in IDLE with its MTYPE the same as the job's MCREQD. Job 1 requires machine B. There is an entity in IDLE whose MTYPE is shown as B on its paper, so we start a PROCESS, moving the clip for Job 1 and the clip for machine B from their present blocks to the PROCESS block, and noting on the PROCESS block when it will be completed, namely at time 18. When the clock has ticked its way round to 18 we will take the clips out of PROCESS and move them into WAIT and IDLE. At time 2 Job 2 arrives and we will move its clip into WAIT, and again check whether we can start a PROCESS for it. We will find that there is no machine in IDLE which can process it because it requires machine B which is in the PROCESS block, so that JOB must WAIT until machine B re-enters the IDLE block. At each point in time the various blocks can be scanned to find when the next event will occur, and the appropriate action taken. When a job leaves the value of TIMIN can be computed, and noted in a table or histogram. This method can be repeated throughout the life of the model, and to assist there is a rule to formalise the procedure.

4.5.1 The three-phase rule

The three-phase rule is the basic rule for handling the flow of time in discrete simulation. The three phases are:

Phase A: Advance the clock to the time of the next event.
Phase B: Terminate any activity due to end at this time.
Phase C: Initiate any activities which the conditions in the model now permit.

Using this rule it is convenient to keep a record of the clock times, the activities terminated and the activities initiated, in the format of Table 4.2. The events which occur when activities are terminated are the "bound" events, in that they are bound to happen once the activity has started. The events which occur when activities are started are "conditional" events, since they depend on conditions in the model. These terms were introduced in the early part of this chapter, and give very convenient words for each phase of the rule:

Phase A: Advance the clock,
Phase B: Bound events,
Phase C: Conditional events.

Table 4.2 Table for applying the three-phase rule

A-PHASE	B-PHASE	C-PHASE
Advance clock	Bound events	Conditional events
CLOCK TIME	ACTIVITIES ENDED	ACTIVITIES BEGUN

In phase B we would move the paper clips from the activity blocks of any activity due to end at that time into the succeeding queue blocks, after updating any attributes which have changed by the activity. For example in the PROCESS block, OPNO would be incremented, and the type of machine needed for the next operation noted in NEXT. Once all the activities ending at that time have been terminated, we start the C phase. In this we examine each activity in turn to see whether it can be initiated. This involves looking at the queues from which arrows lead into the activity and observing whether all these queues contain suitable entities. Because all the terminations occurring at any point in time are processed before any new activities are started, and because the B and C phases are performed at the same simulation clock time, the rule allows entities to enter a queue and then to leave them at the same value of simulation clock. We need to keep track of which entities are involved in each activity. An easy way to do this is to give every entity an attribute called TIME, ie the time of the entity. When an activity starts we set the value of the time of the entities involved to the time when the activity will end. Then as the clock moves on we compare this value with the clock time and those entities for which the values match are due to end their activities then. This TIME attribute has another value. If we do not reset it when the activity ends, it will tell when the entity last ended an activity, in other words when it entered the queue in which currently resides. The value can therefore be used to rank the entities in a queue in the order in which they entered the queue, ie using the FIFO rule.

Applying this to the simple machine shop, Tables 4.3(a) and 4.3(b) show the layout of the information on the piece of paper to be inserted in the paper clip for each entity. Table 4.4 shows the information corresponding to the Gantt charts presented in chapter 1.

Table 4.3   Layout of data for entities

(a) Job

ENTITY TYPE: Job

Attribute values:

TIME:
TIMIN:
OPNO:
NOOPS:
NEXT:

(b) Machine

ENTITY TYPE: Machine

Attribute values:

TIME:
MTYPE:

Table 4.4 Three-phase record for the simple job shop

ACTIVITIES INITIATED BEFORE START OF RUN:

		Arrival of job 1 (due to end at time 0) Arrival of job 2 (due to end at time 2) Arrival of job 3 (due to end at time 4) Arrival of job 4 (due to end at time 6) Arrival of job 5 (due to end at time 7)
A-PHASE	**B-PHASE**	**C-PHASE**
Advance clock	Bound events	Conditional events
CLOCK TIME	**ACTIVITIES ENDED**	**ACTIVITIES BEGUN**
0	arrival of Job 1	1st operation on Job 1 (due to end at time 18)
2	arrival of Job 2	—
4	arrival of Job 3	1st operation on Job 3 (due to end at time 8)
6	arrival of Job 4	1st operation on Job 4 (due to end at time 9)
7	arrival of job 5	1st operation on Job 5 (due to end at time 15)
8	1st op. on Job 3	2nd operation on Job 3 (due to end at time 11)
9	1st op. on Job 4	—
11	2nd op. on Job 3	3rd operation on Job 3 (due to end at time 13)
13	3rd op. on Job 3	—
15	1st op. on Job 5	2nd operation on Job 5 (due to end at time 20) 2nd operation on Job 4 (due to end at time 19)

4.5.2 Computer simulation

Although this section is entitled "hand simulation of the simple job shop", the method presented, with the stages of model development, the attributes and the time flow control rule are handled in the same way if the simulation is being performed on a computer. This will be demonstrated in the next chapter.

SUMMARY

In this chapter the elements of discrete simulation have been introduced, including entities, activities, events, queues, attributes, sets and states. The relationship between activities and events was pointed out. Activity cycle diagrams were presented as a method of defining the logic of a model. The stages in developing a model using activity cycle diagrams were described. The three-phase rule for controlling time flow in discrete simulation was stated. These were applied to the simple machine shop model of chapter 1.

EXERCISE

Draw the activity cycle diagram on a large sheet of paper, make up a sufficient number of cards as per table 4.3(a) and 4.3(b) and perform the simulation manually. Verify the figures given in table 4.4, and at each stage verify that the Gantt charts in chapter 1 are correct.

If you feel confident that you understand the procedure before the run is complete, there is no need to pursue it to the very end.

REFERENCES

1. Hutchinson, G. K., Introduction to the use of activity cycles as a basis for systems decomposition and simulation, *Simuletter*, 7 (1), 15–20, 1975.

2. Poole, T. and Szymankiewicz, J., *Using Simulation to Solve Problems*, McGraw-Hill, 1977.

CHAPTER 5

A job shop model with material handling

5.1 CONVERTING THE SIMPLE MACHINE SHOP
TO A JOB SHOP MODEL

FMS models can be developed from the simple machine shop model of chapters 1 and 4, by adding features to it and changing some details. The most obvious omission from that model is of material handling facilities, and the activities and queues associated with them. In this chapter we will add material handling to the model, change some details and enter it into a computer package and obtain results.

5.1.1 Adding material handling

If material handling is performed by some handling equipment such as fork trucks then we will need to add a new type of entity to the model, which we could call TRUCK. We must also define the activity cycle for this entity. The basic activity of the truck will be MOVE, and while not moving, the truck will be idle. However, since the name IDLE was given in chapter 4 to the queue of idle machines, we need to use another name, such as STOPPED. The basic activity cycle for the truck will be as Figure 5.1.

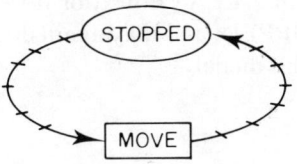

Figure 5.1   Basic activity cycle for TRUCK.

We must also amend the activity cycle for the jobs by including the MOVE activity, which will occur before and after each PROCESS activity. In addition there will be a queue of jobs waiting for trucks as well as the queue of jobs waiting for machines. These queues will have to be given different names. The former name of WAIT cannot be used for both queues. Let us use MWAIT to denote the queue for machines and

HWAIT for jobs waiting for handling. We will assume that OUTSIDE represents a raw material or finished parts store or an assembly shop, so that jobs will need to be moved to the machine for their first operation and away from the machine after their last operation. These changes will alter figure 4.9, giving an activity cycle for a job as in Figure 5.2.

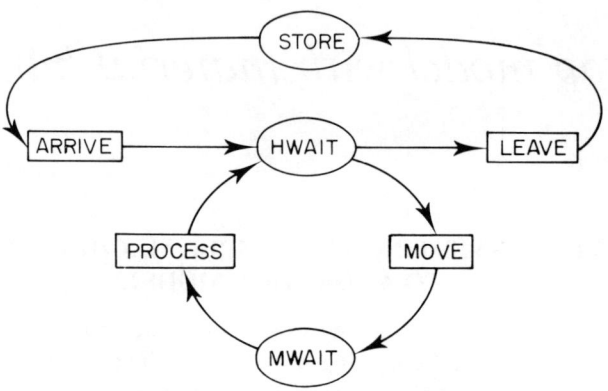

Figure 5.2   Activity cycle for job including MOVE activity.

The LEAVE activity has to be re-thought. In chapter 4 it was a "dummy" activity of zero duration in which statistics were recorded such as the time each job took to pass through the system. Now it can represent the movement of the job from the machine on which its last operation was performed to the outside world. This move will be done by the truck, and could be regarded as an example of the MOVE activity. However, because we will want to collect statistics on the time in the system when a job leaves, this activity will include extra logic compared to a normal movement activity, and it seems wise to preserve the distinction. The truck is used in both the activities, therefore its cycle must include the LEAVE activity, and will become as in Figure 5.3. This is another example of the type of cycle in which there is no particular order of activities. A STOPPED truck may perform a LEAVE next or it may do a MOVE. After a LEAVE (or MOVE) it reverts to STOPPED, and may again do a LEAVE or it may do a MOVE, depending on conditions in the model.

Figure 5.3   Activity cycle for TRUCK including LEAVE activity.

The machines' cycle will be unchanged, remaining as shown in Figure 5.4 and the complete activity cycle diagram will be as in Figure 5.5.

Figure 5.4   Activity cycle for machine.

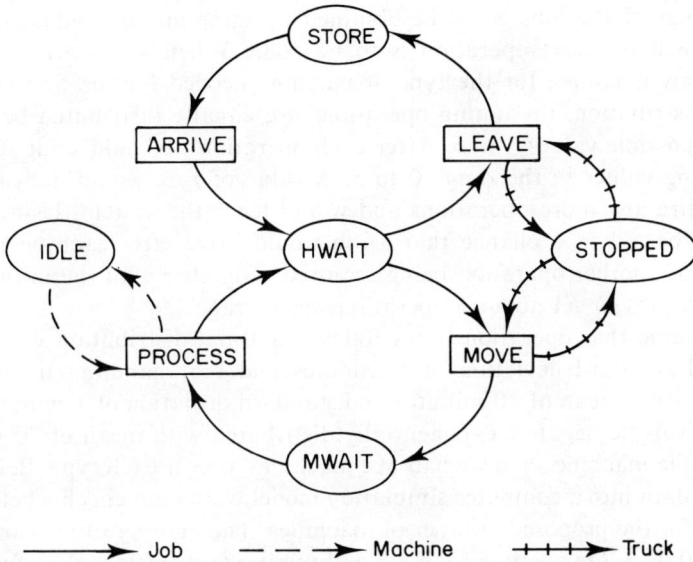

Figure 5.5   Complete activity cycle diagram for job shop model, including job, machine and truck entities.

5.1.2   Operators

So far we do not have any machine operators in our model. This raises the general question of whether every item in the real system needs to be included in the model. If every operator has his own machine, and he never operates any other machine nor does any activity other than operating the machine, and the machine cannot operate without him, then his activity cycle would be similar to the machine's cycle in figure 5.4. The only activity which he would engage in would be the PROCESS activity and the conditions which would make a machine available are exactly those which make an operator available. Since the operator's cycle and the machine's are identical, including either one of them in the model is as effective, and simpler, than including both types of entity. The operator and his machine are effectively inseparable, and can be thought of as a single combined entity. It is as though the operator were chained to his machine, perhaps as a result of a new human relations policy. On the other hand, if the operator may be absent, or has other tasks to perform as well as running the machine, or could operate any of a number of machines then operators will have to be added to the model as a new type of entity. For now, let us assume that our machines and operators are inseparable, or alternatively the machines are fully automatic and can operate without human intervention, so that the operators can be omitted from the model.

5.1.3   Operation sequences and times

Let us also make a further modification to our model. In the earlier model we were concerned with a specific number of jobs which followed defined sequences of operations and had pre-arranged times of arrival. In the classic job shop, the arrival times and the number, sequence and times of the operations of the jobs all vary and are usually modelled by sampling from distributions.

The routings of the jobs could be obtained by sampling at random the type of machine on which the next operation is to be done. When a job arrives in the system we could draw a sample for the type of machine needed for the first operation, using a uniform distribution, (assuming operations are equally distributed between machine types) with possible values 1 to 5. After each operation we could sample again, but this time obtaining values in the range 0 to 5. A value of zero would indicate that the job did not require any more operations and would leave the system. Using this approach there would be a 1 in 6 chance that a job would leave after each operation, and a 5 in 6 chance of another operation being required. Together with the initial operation we would expect jobs to get about 6 operations on average.

Let us assume that operation times follow a normal distribution with a mean of 30 minutes and standard deviation of 5 minutes and that handling times are normally distributed with a mean of 10 minutes and standard deviation of 1 minute. Let the time between arrivals be negative exponentially distributed with mean of 20 minutes.

In the simple machine shop we had five machines, one of each type. Before proceeding to enter this data into a computer simulation model we should check whether these times are sensible for the proposed number of machines. The number of jobs arriving per hour will be 60/20, ie 3. On average each job will require 6 operations therefore in each hour we expect jobs to arrive involving $3 \times 6 = 18$ operations, with a work content of $18 \times 30 = 540$ minutes of work. The capacity of the system with five machines is 5×60 per hour, ie 300 minutes. Since the work load would be 180% of capacity, more than five machines are clearly needed. To bring things into a more realistic balance let us have two machines of each type, ten in total, with an expected utilisation of 80%. Similarly, we can make a rough estimate of how many trucks should be available. If each job requires 6 operations with movements between each and before the first and after the last there will be 7 movements per job, on average. The load on the trucks will be $3 \times 7 \times 10 = 210$ minutes per hour, and the capacity will be only 60 per truck. Four trucks might cope with this load, but their utilisation would be rather high. Five seems more realistic, and their utilisation should be around $210/5 \times 60 = 70\%$.

5.1.4   Attributes and other details

Before converting the model into computer form we need to check whether all the attributes and so on have been considered. As in the simple machine shop we can use NEXT to specify the type of machine for the next operation on each job. Similarly we can use TIMIN to record the time when the job arrives in the system. We do not need a counter for the number of operations to be done before a job leaves, but it might be interesting to record the number of operations actually performed on each job, as a check on the above arithmetic. An attribute will be needed for this, and OPNO can be used. Each machine will have an attribute MTYPE to define its type.

We must also consider the initial conditions in the model. For simplicity, let us start with all machines and trucks idle and all jobs outside the system. The model will take some time to fill up and for its "normal" conditions to develop. We should not start recording statistics until the model has been run for some time. Let us have a "run-in" period of 500 minutes and run the model for 2000 minutes. These figures may turn out to be too short to give valid results, but we can check the results and study them to know whether this will be the case or not. If necessary, the run-in and run length can then be amended.

Another feature of the model requiring consideration are the queue disciplines. The logic discussed in chapter 4 requires a little amendment. Instead of comparing OPNO with NOOPS, the criterion for a job to leave the system is, as observed above, that NEXT is zero, otherwise it will have another operation. This test will be applied to jobs in the HWAIT queue. If NEXT of the job is equal to zero, the job will proceed to LEAVE, and if greater than zero, it would MOVE to another machine. When selecting from the queue MWAIT and machines from IDLE, it is necessary to match the MTYPE of the machine with the type required for the NEXT operation on the job. A slightly different consideration concerns STOPPED trucks. Suppose that when a truck becomes available there are several jobs waiting in HWAIT, and that some want to LEAVE while others want to MOVE to another machine. Should the truck perform a LEAVE in preference to a MOVE, or vice versa? If we are keen to get completed jobs out of the system as soon as possible then the truck should perform a LEAVE activity. If we want to minimise inter-operation delays we would prefer it to do a MOVE first. Let's adopt the first alternative.

5.2 PUTTING THE MODEL INTO A COMPUTER

We are now ready to enter the model into the computer. Figure 5.6 summarises the activity cycle diagram together with queue disciplies, attribute calculations, activity durations and the other details which have been defined.

At this point it might be appropriate to consider the various computer software packages for simulation and how models are developed using them. However, since that would introduce a very large subject we will postpone that discussion until a later stage. Instead we will demonstrate the approach using one particular package. In addition to providing the software to run simulation models, some packages provide interactive model builders which enable those who are not computer programming experts to define the system to be modelled in conversational terms, using words which they might use to describe it to a colleague. One such software package is CAPS/ECSL.[1] ECSL stands for Extended Control and Simulation Language. CAPS (Computer Aided Programming System) is a conversational model builder for ECSL. It develops an ECSL model of a system by asking the modeller questions about the entities, activities, attributes, queue disciplines and so on. These questions are grouped in five sections:

Logic: which asks for details about the entities and their activity cycles,
Priorities: which asks about the various queue disciplines,
Arithmetic: concerning the formulae for activity durations and for calculating attribute values,
Recording: which deals with collecting statistics,
Initial conditions: which enquires about activities in process and the contents of queues at the start of the run.

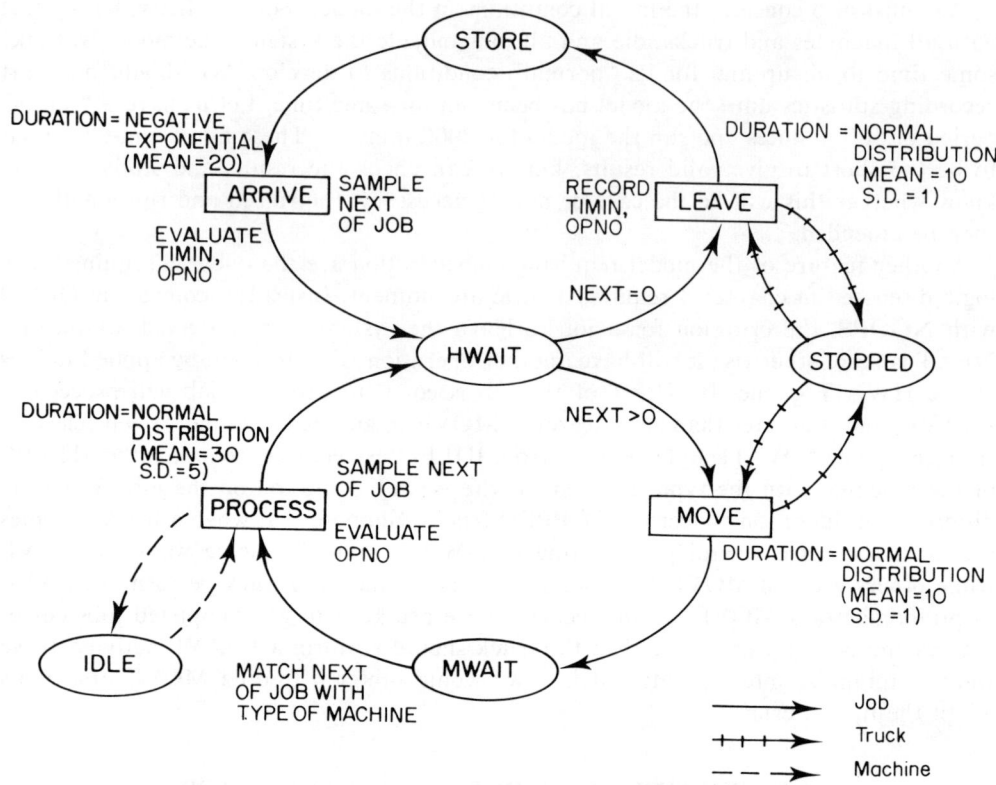

Figure 5.6   Activity cycle diagram including queue disciplines, attribute calculations and activity durations.

5.2.1   The CAPS dialogue

The dialogue between CAPS and the user when building the model we have just described is reproduced in Figure 5.7. Most of the text is produced by CAPS, with the user's responses in upper case characters following the question marks. Names given by the user are also in upper case when reprinted by CAPS. Occasionally, no response is given by the user to indicate that there are no further responses of the kind invited. Some comments on the dialogue may be helpful.

At the start of the dialogue there are some preliminary questions. If the model is to be run with a graphical display of the model the system needs to know what type of terminal is being used, hence the question about the visual display unit.

Instructional comments, offered by CAPS for users unfamiliar with the system, have been included to give a more comprehensive impression of the principal features of the package and its capabilities. Normally, they would not be requested, to save time. A further facility provided by CAPS for speeding up the dialogue, is its "accelerated mode", in which an experienced user can, in certain circumstances, anticipate the next question by appending the next response to the response to the question currently posed.

The system permits a new model to be developed, or an earlier one (perhaps one whose description was not completed at the last session) to be amended. If the response to the question "Do you wish to start a new problem?" is "no", the system will search for a file with the stated problem name and then enquire how the existing model is to be amended.

When we tell the computer how many jobs are in the model we are not giving the number of jobs which will pass through the system, but the maximum number of jobs which we expect will be in the system at one time. We are in effect telling it how much memory must be reserved for storing information about the jobs. In effect a table is constructed in computer memory with one row for each entity and one column for each attribute of that entity type. When a job has passed through the system and its history added to the model's statistics the memory locations which were used to hold information about it can be re-used to hold data about another, later, job. In the previous chapter, when the activity cycle diagram was being developed, it was observed that the cycle appeared to suggest that an job could leave the system, go to outside and then later arrive back in the system. By clearing the memory locations, we effectively create a new entity even though it is using the same memory locations. If we do not reserve enough memory space there will be no space available to record data about new arrivals and the model will not be valid. For entities such as the truck and machines which are either idle or doing some activity, it is only necessary to name the activities concerned.

The system has added a dummy entity, ZZARRI, which is necessary to control the arrive activity. If it did not exist the only requirement for ARRIVE to occur would be the presence of jobs in OUTSIDE. Since we started the model empty with all jobs in OUTSIDE they would all have arrived at time zero. By providing only one entity of type ZZARRI the system ensures that jobs will arrive one at a time with inter-arrival times corresponding to the duration of the ARRIVE activity. When an ARRIVE is complete the ZZARRI goes back into its inactive state, and another ARRIVE can begin immediately. The utilisation of the ZZARRI should therefore be 100%, if there are always jobs in OUTSIDE waiting to ARRIVE. If the utilisation of ZZARRI turns out to be less than 100% it will indicate that on one or more occasions when an ARRIVE activity should have commenced there was no job in OUTSIDE, since that is the only other requirement for ARRIVE to begin. This would in turn indicate that we had not defined a sufficient number of jobs.

At the end of the logic section, the cycles and entities involved in each activity can be displayed. Notice that CAPS has truncated all names to six characters, as with MACHIN, since ECSL treats only six characters as significant although the names may be longer if desired.

In the priorities section, the instructional comments include a list of the available queue disciplines. The responses depend on the discipline specified. In this case two types of discipline are needed. One for branching from queues depending on the value of an attribute, and the other matching one type of entity with another so that they have equal values of specified attributes.

In the arithmetic section, whenever a random sample is to be drawn from a distribution the system asks us to define the stream of random numbers to be used. When we come to experiment with the model we may wish to run the model several times with different initial values of the random number seeds, so that the effects of the particular numbers picked can be separated from the inherent characteristics of the model. This

```
Are you using a visual display unit?N
Extended Control and Simulation Language
-----------------------------------------------
Computer Aided Programming System

Do you wish to have instructional comments?Y
During  this  discussion  you will be asked for  a   number   of
lists.   When  a  list  is complete a blank  line   should   be
entered.    If  when  typing  you  make   errors,   these   may
be  corrected     by  using the delete (erases last  character)
or control  X   (erases  whole  of line) keys.    When  a  line
is complete  press return.     Please   note  — once return  has
been  pressed,     it    may    not    be    possible    to   make
corrections  immediately.    When    you   do   not  understand   a
question,    you  may  press     question    mark  followed    by
return,    to   obtain clarification.
The discussion is in five sections—
Logic— Priorities— Arithmetic— Recording— Initial Conditions
At the end of each section  it   is   possible  to return to the
beginning of any earlier section.
Problem name?JBSHOP
Do you wish to start a new problem?Y
Are you going to use implicit queue mode?N
Logic

This   section discusses the logical structure of the system to
be modelled.   The system is made up of several types of entity
and  is described by listing,  for each type  of  entity,   the
states in which they can be and the orders in which the states
can  occur.    The  computer  will then analyse this  data  and
comment upon it. It may be necessary to ask you some questions
to ensure that the problem statement is complete,   logical and
correct.

Examples of the specification of activity cycles.
1) Simple Cycles       2) Linked Cycles        3) Batch Activities
4) Implicit and Dummy Queues               5) Catalytic Entities
6) Facility Cycles                    7) Extended Facility Cycles
8) Branches from Queues 9) Spliting of Batches from Activities
10) Conditional Branches from Activities   11) Random Branches
Type the number of the example you would like to see?
The set of examples just left may be recalled during the logic
section by responding with a question mark to the request  for
the name of one kind of entity.

Type name of one kind of entity?JOB
How many?50
Type a list of the states through which these entities pass.
This should consist of either—
     a) An alternation  of queues and activities,   starting and
         ending with a queue,
or  b) A  list of activities alone— indicating that  they  can
         occur in any order.
```

Figure 5.7   CAPS dialogue.

```
Precede queues by Q and activities by A.
?QOUTSIDE,AARRIVE,QHWAIT,AMOVE,QMWAIT,APROCESS,QHWAIT,ALEAVE,QOUTSIDE
?
Is this cycle correct?Y
At present the model has the following entity types:-
JOB

Type name of one kind of entity?MACHINE
How many?10
Type a list of states as above.
?APROCESS
?
Is this cycle correct?Y
At present the model has the following entity types:-
JOB      MACHIN

Type name of one kind of entity?TRUCK,5
Type a list of states as above.
?AMOVE,ALEAVE
?
Is this cycle correct?Y

Type name of one kind of entity?
Are  there  any  (other) activities which use  more  than  one
entity of a particular type?N
From  what  you said so far,  the following  are the  maximum
number of simultaneous realisation of the activities.
Activity Number
ARRIVE      50
MOVE         5 Limited by the number of TRUCK
PROCES      10 Limited by the number of MACHIN
LEAVE        5 Limited by the number of TRUCK

Do you wish to apply any limits which are below these?Y
Which activity?ARRIVE
What is the limit?1
This  limitation has been implemented by adding an   artificial
entity ZZARRI.
Which activity?
Not  more than   21 of the    50,  JOB   can be active at one
time.

Do you wish to see a summary of the cycles?Y
JOB 50 ,QOUTSID,AARRIVE,QHWAIT ,AMOVE,QMWAIT ,APROCES,QHWAIT
           ,ALEAVE ,QOUTSID
MACHIN  10  ,APROCES
TRUCK    5  ,AMOVE,ALEAVE
ZZARRI   1  ,AARRIVE

ARRIVE  uses  1 JOB    , 1 ZZARRI
MOVE    uses  1 JOB    , 1 TRUCK
PROCES  uses  1 JOB    , 1 MACHIN
LEAVE   uses  1 JOB    , 1 TRUCK
Do you wish to make any changes in the Logic Section?N
```

Figure 5.7   (continued).

Priorities
This section discusses priorities. It is divided into two
parts.
Part1: Queue Disciplines. This discusses the rules which are
used to select entities from queues for participation in
activities.
Are there any queues whose discipline is not F-I-F-O?Y
The priorities allowed are:-
First-In-First-Out   (code FIFO)
Last-In-First-Out   (code LIFO)
Random   (code ANY)
Entity which MAXimizes an expression   (code MAX)
Entity which MINimizes an expression   (code MIN)
First in the queue, but only if the queue contains at least N
entities   (code FULL)
First attribute equals attribute of an entity already chosen
                                                            (code SUIT)
First for which a matching entity can be found   (code PAIR)
For queues from which an activity uses more than one entity
first and N-1 others with same attribute value   (code SAME)
Any N with same attribute values   (code ALIKE)
First to make expression GT zero   (code GT)
First to make expression EQ zero   (code EQ)
First to make expression GE zero   (code GE)
First to make expression LT zero   (code LT)
First to make expression NE zero   (code NE)
First to make expression LE zero   (code LE)
Which queue?HWAIT
This queue is used in more than one activity.
Which activity?LEAVE
Priority code?LE
Type expression?NEXT OF JOB
Do you wish to use this expression to select JOB   from
HWAIT by MOVE ?Y
Priority code?GT
Which queue?MWAIT
Priority code?PAIR
Which attribute?NEXT
Which entity is to be matched?MACHINE
Which attribute?MTYPE
Are there any further conditions within this match?N
Matching complete.
Which queue?
Do you wish to preview the ECSL coding for the test head of
any activity?Y
Which activity?PROCES
   FIND FIRST MACHIN B WITH TIME OF MACHIN LE 0
      FIND   FIRST   JOB   A FROM   MWAIT
        NEXT   EQ   MTYPE
Which activity?
Do you wish to examine and/or alter Q-disciplines?N
Part2: This discusses the order in which the activities are to
be considered. Most of this can be decided by the computer.
Your help will be required only in marginal cases.
The order of the following activities is unimportant:-

Figure 5.7   (continued).

```
PROCES   ARRIVE
I   propose   putting the remaining activities in the   following
order:-
LEAVE
MOVE
Do you wish to raise the priority of any activity?N
Do you wish to make any changes in the Priority  Section?N

Arithmetic
This section discusses the arithmetic.   It is divided into two
parts.
Part1:   In  this part you are asked to give a formula for each
activity  from  which typical durations  will  be  calculated.
These formulae should be stated in the usual form for computer
arithmetic and may use constants (E.G.  12),  system variables
(E.G.  DEPTH),  attributes of entities (E.G.  SIZE OF JOB   ),
functions   (E.G.   NORMAL),   operators  (+,-,*,/ ,**)  and ,
parentheses.
A full list of the available functions can be found in either
the ECSL Manual (section 13) or the caps manual (section 4).
For example, NORMAL (10,1,S) indicates that the duration of an
activity  is  normally distributed with  mean  10  units  and
standard deviation 1 unit.
After each activity name,  type formula for its  duration, if
the duration might be zero, type 0+....
LEAVE ?NORMAL(10,1,SA)
MOVE  ?NORMAL(10,1,SA)
PROCES?NORMAL(30,5,SB)
ARRIVE?0+NEGEXP(20,SC)
Arithmetic
Part2:  In  this part you will be asked to state when and from
which formulae variables and attributes are to be calculated.
I do not have duration for the following activities:-
If  variable  is not to be evaluated by any activity,  give  a
blank response.
In which activity is NEXT   of JOB   evaluated?ARRIVE,PROCES
Please give formula for its evaluation during ARRIVE.
?1+RANDOM(5,SD)
Please give formula for its evaluation during PROCES
?RANDOM(6,SD)
SA     is a "random  number  stream".
Please give an initial (Odd) random number?12311
SB     is a "random  number  stream".
Please give an initial (Odd) random number?12543
SC     is a "random  number  stream".
Please give an initial (Odd) random number?987
SD     is a "random  number  stream".
Please give an initial (Odd) random number?2345
In which activity is MTYPE  of MACHIN evaluated?
Do you wish to specify constant initial values?Y
Please give the list of  10 initial values?1,1,2,2,3,3,4,4,5,5
Does any attribute have further evaluation points?N
Do you wish to define any more attributes for entities?Y
Which entity type?JOB
Which attribute?OPNO
```

Figure 5.7   (continued).

```
In which activity is OPNO    of JOB      evaluated?ARRIVE,PROCES
Please give formula for its evaluation during ARRIVE.
?O
Please give formula for its evaluation during PROCES.
?OPNO + 1
Does any attribute have further evaluation points?N
Do you wish to define any more attributes for entities?Y
Which entity type?JOB
Which attribute?TIMIN
In which activity is TIMIN  of JOB      evaluated?ARRIVE,LEAVE
Please give formula for its evaluation during ARRIVE.
?CLOCK
Please give formula for its evaluation during LEAVE .
?CLOCK - TIMIN
Does any attribute have further evaluation points?N
Do you wish to define any more attributes for entities?N
Do  you wish to preview the ECSL coding for evaluation section
of any activity?Y
Which activity?PROCES
Evaluation section of PROCES :-
   OPNO      OF       JOB     + 1
   NEXT      OF       JOB     = RANDOM( 6 , SD    )
Which activity?
Do you wish to make any changes in the Arithmetic Section?N

Recording
I have arranged your program to record-
        A) The number of times each activity starts,
   and B) The utilization of those entities whose cycles did not
           involve any queues.
Three other kinds of recording may be included-
  1) Length of queue
        A) Recorded as a time weighted histogram        (Code 1)
     or B) Recorded at regular intervals (time slices) (Code 2)
     or C) Recorded as a time series                    (Code 3)
  2) Length of time an entity is delayed in a queue     (Code 4)
3) Utilisation of entity,  with the named queue being included
     in the "un-utilised" category                       (Code 8)
After the queue name type the code as a single digit.  Type  0
if no recording for  that queue is required.  When two or more
types are required,  type the sum of the codes.
MWAIT ?5
Recording requested so far.
MWAIT     5
HWAIT ?5
OUTSID ?1
For  each queue for which delays are to be  recorded,  specify
the histogram range. (This range will be divided into 10 equal
intervals.)
MWAIT  Range=0 to?100
HWAIT  Range=0 to?100
Do you wish to record any attributes?Y
Which attributes?OPNO
In which activity is it to be recorded?LEAVE
What shall we call the histogram?NOPHIST
```

Figure 5.7   (continued).

```
Range = 0 to?20
Which attributes?TIMIN
In which activity is it to be recorded?LEAVE
What shall we call the histogram?TIHIST
Range = 0 to?500
Which attributes?
Output control section.
The  run-in  period is a time at the start of the  run  during
which  no recording is performed.  It is designed to allow the
model  to recover from the initial state so that  the  results
are unbiased.
What length of Run-In period is required?500
Do you wish to have intermediate results?Y
Give interval between print outs?500
Indicate the required extent of intermediate output:-
    1   Activity counts only.
    2   Activity counts and utilisations.
    3   Activity counts, utilisation and queue histogram.
    4   All recorded data.
Which level do you require?2
Estimated quantity of output per interval is        9 lines.
Indicate required output mode-
    1   All output to terminal.
    2   Intermediate output to terminal, final output to printer.
    3   All output to printer.
Which mode do you require?3
Please give the duration of the simulation?2000
Do you wish to make any changes in the Recording Section?N

Initial conditions.
This section discusses the state that will exist at time zero.
This  will consist of some activities which are thought of  as
having  started  before time zero and are still  in  progress.
This  will  use up some of the entities.  The initial  (queue)
states for the other entities should also be given.
Are there any activities in progress?N
Type  how  many entities should be in each queue listed  after
the queue name.
JOB         50 Entities.
MWAIT ?0
HWAIT ?0
OUTSID ?50
Do  you  wish to make any changes in  the  Initial  Conditions
Section?N
Some entities are apparently suitable for   aggregation   I.e.
they  have no attribute,  no delay recording and use only FIFO
Q-discipline.
Do you wish me to aggregate TRUCK ?Y
Do you wish to use a cell structure?N
Your Program has been written. Have you finished?Y
```

Figure 5.7   (continued).

was discussed in chapter 3. By using a different stream for each type of sample it becomes possible to test these effects one at a time.

While this has been a complete dialogue it does not illustrate all the facilities of the CAPS package, such as the error correction facilities, or features not required for this model which a different one might have called upon. It will be apparent that CAPS is a rather sophisticated package.

5.2.2   ECSL program

At the end of the dialogue CAPS writes a file containing the ECSL program for the model. This can then be compiled and executed by the ECSL system. The program created by the CAPS is shown in Figure 5.8. A full explanation of the ECSL language is beyond the scope of this text; however it will be worthwhile to observe the general structure of the program and some of its features.

An ECSL program consists of five sections:

definitions,

initialisation,

activities,

finalisation,

data.

The definitions section gives the name of each type of entity, with its queues (sets) and attributes, then any variables, arrays, histograms, or functions which the program will need. This section begins at the start of the listing and goes on to the line beginning with the word "FUNCTI".

The initialisation section gives details of any activities in progress at the start of the run, including setting any attributes for entities involved in these activities. In this case no activities were in progress at the start of the run, so the section only consists of the five lines beginning RECYCL which specify the run-in period and the time of the first intermediate report of results. The initial content of the queues are given in the data section.

The activities section begins with the ACTIVITIES statement, and is usually the longest section. It consists of a number of sub-sections, a dynamics sub-section for any continuous processes, a recording sub-section and a sub-section for each activity.

The recording sub-section records the values of certain parameters, such as the number of items in queues or the length of time for which an entity has been inactive, in histograms or variables. Since there are no continuous processes in this model there is no dynamics sub-section, and the recording section immediately follows the ACTIVITIES statement, and ends at PREVCLOCK = CLOCK. CLOCK is the system variable storing the current time in the model. The statements are executed on every time advance. The time value on the previous occasion was stored in the variable PREVCLOCK, a name assigned by CAPS. CAPS has also assigned names to the statistic counters and histogram names, such as ZAMWAIT, ZBMWAIT which are associated with the queue MWAIT. Each entity has an attribute TIME which indicates the time until the activity in which the entity is currently engaged will terminate. TIME is measured relative to CLOCK and is recalculated on every time advance. For entities which complete an activity at the current clock time the value of TIME will be set to zero. Any entity remaining inactive will acquire a negative TIME value when the clock is advanced. Figure 5.9 shows two machines, one of which is working from clock time 92 until 99 and the second is working from 97 to 105. If the current clock time is 100 the TIME of the first machine is $99 - 100 = -1$ and of the second is $105 - 100 = 5$. Hence the statements

DURATION = CLOCK − PREVCLOCK
FOR MACHIN WITH TIME OF MACHIN LT 0
   ADD DURATION TO AZMACHIN

cause the time since the last advance of the clock to be added to a counter AZMACHIN for any idle machine. In the finalisation section this value is used to calculate the utilisation of the machines. Other software packages have different but largely equivalent methods of keeping track of time.

The main part of the activities section describes the activities in the model. Each activity has a block beginning with a BEGIN statement. If two or more instances of an activity could start simultaneously, for example because there are several entities of each type, then the activity's block of statements will normally end with a REPEAT statement. Each activity has a definite structure, with four basic types of statement:

1. condition statements (the "test head" of the activity),
2. attribute setting and calculation of variables,
3. changing the state of entities, ie time and queues,
4. statistic recording, and sometimes printout.

The first of these are the conditions under which the activity can commence and the other three are the actions which will take place if the activity occurs. These actions include computing the duration of the activity, setting the values of attributes of the entities involved, changing the state of entities (usually by removing them from queues which "feed" the activity, and placing them into the queues which follow the activity after the appropriate delay), recording statistics and perhaps printing messages. Figure 5.10 gives the listing of the PROCESS activity annotated to show these elements.

The final activity is OUTPUT, which is called to provide intermediate results. Once a model is debugged and only final results are required this could be moved to the finalisation section, which provides final results. The utilisation of various entities are computed with reference to the variables to which the idle periods of the entities of each type were added in the recording section. The value of that variable divided by the product of the length of the period over which recordings were made and the number of entities of that type gives the proportion of idle time. By subtracting this value from unity the utilisation is obtained.

The finalisation section is introduced by the FINALISATION statement and is called at the end of the run to perform any calculations required and print out final results. A substantial amount of arithmetic calculations may be included, however in this case only simple print statements are required. These use the PICTURE function which prints histograms.

```
 THERE ARE 50 JOB SET MWAIT HWAIT OUTSID WITH TIMIN
+OPNO NEXT
 THERE ARE 10 MACHIN WITH MTYPE
 THERE ARE 1 ZZARRI
 HIST TIHIST( 11 25 50 ) NOPHIS( 11 1 2 )
 FUNCTI PICTUR RANDOM NEGEXP NORMAL
 RECYCL
 RUNINZ= 500 AND PREVCLOCK = RUNINZ
 SWITCH ADD ON AFTER RUNINZ
 TIME OF ZZRPRT= RUNINZ+ 500
 ACTIVITIES 2000
 DURATION= CLOCK - PREVCLOCK
 ADD MWAIT TO HIST ZAMWAIT DURATION
 ADD TIME IN MWAIT TO HIST ZBMWAIT ( 11 5 10 )
 ADD HWAIT TO HIST ZCHWAIT DURATION
 ADD TIME IN HWAIT TO HIST ZDHWAIT ( 11 5 10 )
 ADD OUTSID TO HIST ZEOUTSID DURATION
 FOR MACHIN WITH TIME OF MACHIN LT 0
   ADD DURATION TO AZMACHIN
 ADD DURATION* TRUCK TO BZTRUCK
 FOR ZZARRI WITH TIME OF ZZARRI LT 0
   ADD DURATION TO CZZZARRI
 PREVCLOCK = CLOCK
 BEGIN LEAVE
 TRUCK GE1
 FIND FIRST JOB A IN HWAIT
   NEXT OF JOB A LE 0
 TIMIN OF JOB A= CLOCK - TIMIN OF JOB A
 DURATION= NORMAL( 10, 1 , SA )
 ADD TIMIN OF JOB A TO TIHIST
 ADD OPNO OF JOB A TO NOPHIS
 JOB A FROM HWAIT INTO OUTSID AFTER DURATION
 TRUCK - 1 AND TRUCK + 1 AFTER DURATION
 ADD 1 TO LEAVE
 REPEAT
 BEGIN MOVE
 TRUCK GE1
 FIND FIRST JOB A IN HWAIT
   NEXT OF JOB A GT 0
 DURATION= NORMAL( 10, 1 , SA )
 JOB A FROM HWAIT INTO MWAIT AFTER DURATION
 TRUCK - 1 AND TRUCK + 1 AFTER DURATION
 ADD 1 TO MOVE
 REPEAT
 BEGIN PROCES
 FIND FIRST MACHIN B WITH TIME OF MACHIN LE 0
   FIND FIRST JOB A FROM MWAIT
     NEXT EQ MTYPE
 OPNO OF JOB A+ 1
 NEXT OF JOB A= RANDOM( 6 , SD )
 DURATION= NORMAL( 30, 5 , SB )
 JOB A FROM MWAIT INTO HWAIT AFTER DURATION
 TIME OF MACHIN B= DURATION
 ADD 1 TO PROCES
```

Figure 5.8   ECSL program.

```
REPEAT
BEGIN ARRIVE
TIME OF ZZARRI LE O
FIND FIRST JOB A IN OUTSID
TIMIN OF JOB A= CLOCK
OPNO OF JOB A= O
NEXT OF JOB A= 1 + RANDOM( 5 , SD )
DURATION= NEGEXP( 20, SC )
CHAIN
  DURATION GT O
  OR RECYCL
JOB A FROM OUTSID INTO HWAIT AFTER DURATION
TIME OF ZZARRI = DURATION
ADD 1 TO ARRIVE
BEGIN OUTPUT
CHAIN
  CLOCK EQ 2000
  PRINT **"Final report from simulation JBSHOP    "/
  OR TIME OF ZZRPRT LE O
  TIME OF ZZRPRT= 500
  PRINT **"Report up to time" CLOCK " from simulation JBSHO   "/
PRINT 'LEAVE   was started' LEAVE ' times'
PRINT 'MOVE    was started' MOVE ' times'
PRINT 'PROCES    was started' PROCES ' times'
PRINT 'ARRIVE was started' ARRIVE ' times'
PRINT 'Utilization of MACHIN'+4,(1-AZMACHIN/( 10. *(CLOCK -RUNINZ)))
PRINT 'Utilization of TRUCK '+4,(1-BZTRUCK /( 5. *(CLOCK -RUNINZ)))
PRINT 'Utilization of ARRIVE'+4,(1-CZZZARRI/( 1. *(CLOCK -RUNINZ)))
FINALISATION
PRINT 'Histogram of length of queue MWAIT '/PICTURE(ZAMWAIT)
PRINT 'Histogram of delays at MWAIT '/ PICTURE(ZBMWAIT )
PRINT 'Histogram of length of queue HWAIT '/PICTURE(ZCHWAIT)
PRINT 'Histogram of delays at HWAIT '/ PICTURE(ZDHWAIT )
PRINT 'Histogram of length of queue OUTSID '/PICTURE(ZEOUTSID)
PRINT 'Histogram TIHIST of TIMIN  '/ PICTURE(TIHIST)
PRINT 'Histogram NOPHIS of OPNO   '/ PICTURE(NOPHIS)
DATA
OUTSID 1 TO *
TRUCK 5
MTYPE 1 1 2 2 3 3 4 4 5 5
SD 2345
SC 987
SB 12543
SA 12311
  END
```

Figure 5.8   (continued).

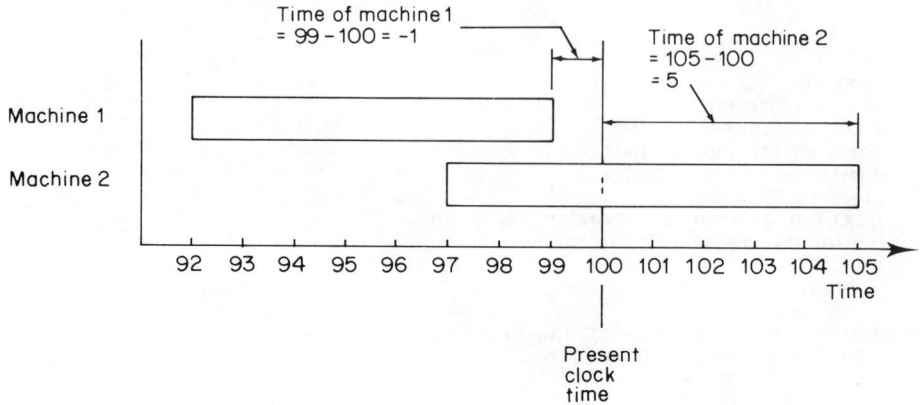

Figure 5.9   Time of entities in ECSL.

The data section begins with the word "DATA" and ends at "END". This section gives the initial values of variables, initial contents of queues and the initial values of random numbers streams.

This has been a very brief overview of ECSL. The language provides many features not involved in this simple model, such as continuous variables and animated display. A major attraction of ECSL in this context is that, being an English-like high level language, it is almost as readable as normal text. Consequently, it will be used to describe some of the FMS models which will be developed in this book. The more complex models will be less easy to read and at that stage other languages will be introduced.

5.2.3 Results

The results which were obtained when this program was run are shown in Figure 5.11.

The utilisation of the trucks at approximately 0.62 is somewhat less than the figure of 70% obtained from the rough calculation earlier in this chapter. The utilisation of the machines is about 80%, which was the value predicted. However whereas the utilisation of the trucks is fairly stable the utilisation of the machines increased at each report. Although the difference between one value and the next is quite small it may suggest that the run-in period was not long enough and that steady conditions had not been reached. The histogram of OPNO is of the shape expected. Its mean can be calculated easily as 5.5. Is this sufficiently close to the expected value of seven? The model could be re-run for a longer period and with different values of the random number stream seeds, or with more frequent intermediate reports, to examine these questions.

A word of explanation may be helpful concerning the histograms. In each histogram the first column is the value of the mid-point of the cell and the second is the number of observations falling in each cell. Note that the histograms deal with two types of statistic, time-dependent variables, such as the length of a queue, and observed statistics, such as the time each entity remained in the queue. In the case of time-dependent variables the total frequency of the cells is equal to the period over which the statistics were recorded, the duration of the run less the run-in period, ie $2000 - 500 = 1500$. The

```
BEGIN PROCES
        This line  is  the  beginning  of a new activity, which you
        have named PROCESS.

FIND FIRST MACHIN B WITH TIME OF MACHIN LE 0
    FIND FIRST JOB A FROM MWAIT
        NEXT EQ MTYPE
        These three lines are the "Test head" of the activity. Find
        the first machine which is idle,  and if there is any refer
        to it as machine B then find  the  first job in queue MWAIT
        for which NEXT of the job is equal to MTYPE of the machine.
        If  a  suitable  job is found refer to it  as  job  A.  The
        indentation of the lines  causes  the  program  to  examine
        every  idle  machine in turn until a suitable job is found.
        If the test  is successful the following statements will be
        executed, otherwise the computer  will  jump to the start of
        the next activity.

OPNO OF JOB A+ 1
        Increment the OPNO attribute of the job by 1.

NEXT OF JOB A= RANDOM( 6 , SD )
        Set the value of NEXt for the job to the value of a uniform
        random number in the range 0   through 5 using a sample from
        random number stream SD.

DURATION= NORMAL( 30, 5 , SB )
        Set DURATION to the value of a  normal  random number using
        stream SB to represent the duration of the proces activity.

JOB A FROM MWAIT INTO HWAIT AFTER DURATION
        Take job A out of queue MWAIT and insert it into queue
        HWAIT after a period equal to DURATION has elapsed.

TIME OF MACHIN B= DURATION
        Set the time of machine B to DURATION.

ADD 1 TO PROCES
        Increase the variable PROCESS which is used to count the
        number of times this activity occurs.

REPEAT
        Repeat this logic to see whether another PROCESS can be
        started.
```

Figure 5.10   Explanation of ECSL statements of the PROCESS activity.

```
Report up to time        1000 from simulation JBSHO

LEAVE   was started          25 times
MOVE    was started         135 times
PROCES  was started         132 times
ARRIVE  was started          31 times
Utilization of MACHIN    .7756
Utilization of TRUCK     .6360
Utilization of ARRIVE   1.0000


Report up to time        1500 from simulation JBSHO

LEAVE   was started          48 times
MOVE    was started         252 times
PROCES  was started         258 times
ARRIVE  was started          44 times
Utilization of MACHIN    .7782
Utilization of TRUCK     .6022
Utilization of ARRIVE   1.0000


Final report from simulation JBSHOP

LEAVE   was started          65 times
MOVE    was started         403 times
PROCES  was started         401 times
ARRIVE  was started          70 times
Utilization of MACHIN    .8073
Utilization of TRUCK     .6257
Utilization of ARRIVE   1.0000

Histogram of length of queue MWAIT
CELL    FREQUENCY
    0      22*****
    1     122*****************************
    2     171*******************************************
    3     270*******************************************************************************
    4     206******************************************************
    5     163******************************************
    6     131*******************************
    7     110****************************
    8      94***********************
    9      45***********
   10      56**************
   11      56**************
   12      39**********
   13      14****
   14       1
```

Figure 5.11   Results obtained from the ECSL program.

```
Histogram of delays at MWAIT
CELL   FREQUENCY
   5     201**********************************************************************
  15      57******************
  25      42*************
  35      31**********
  45      31**********
  55      13*****
  65       9***
  75       1*****
  85       3*
  95       2*

Histogram of length of queue HWAIT
CELL   FREQUENCY
   0    1405**********************************************************************
   1      71***
   2      16*
   3       4
   4       4

Histogram of delays at HWAIT
CELL   FREQUENCY
   5     468**********************************************************************

Histogram of length of queue OUTSID
CELL   FREQUENCY
  23      16*****
  24      16*****
  25      49**************
  26      56****************
  27      18*****
  28      72********************
  29      23*******
  30     196****************************************************************
  31     110********************************
  32      54***************
  33     103******************************
  34     164*****************************************************
  35     197*****************************************************************
  36     225**********************************************************************
  37     128**************************************
  38      73*********************
```

Figure 5.11   (continued).

```
Histogram TIHIST of TIMIN
CELL   FREQUENCY
    25     1*
    75    10**********
   125     9*********
   175    10**********
   225     7*******
   275     6******
   325     4****
   375     1*
   425     3***
   475     4****
   525    10**********

Histogram NOPHIS of OPNO
CELL   FREQUENCY
     1    15***************
     3    16****************
     5    13*************
     7     7*******
     9     4****
    11     4****
    13     0
    15     2**
    17     2**
    19     2**
```

Figure 5.11   (continued).

total frequency of observed data is equal to the number of observations. For example, the total frequency of the TIMIN histogram equals the number of occurrences of the activity LEAVE.

In chapter 3 various aspects of the validity of simulation results were introduced and several of them have been encountered here. We could repeat the run using different run-in periods and random number streams, and we could also record the sampled operation durations so that we could examine directly the "randomness" of the samples obtained. However, at this stage we wish to concentrate on model building, and will omit further experimentation on this model.

This very brief set of results illustrates typical output from the ECSL package, but such a simple model does not use all the facilities of the package. Facilities also exist for a graphical display as the simulation proceeds and for analysing histograms to see whether they conform to mathematical distributions, for plotting graphs and charts and various other forms of output.

SUMMARY

In this chapter we have developed a model of a job shop with materials handling, using activity cycle diagrams. We then used the CAPS package to construct a computer model of the system in the ECSL language, listed the ECSL program, ran the program, and obtained results. We have thus illustrated an interactive model builder and briefly examined one of the leading high level simulation languages.

This model can be expanded to form the basis of various FMS models which will be developed in subsequent chapters. We will also compare the characteristics of several simulation packages and present FMS models written using some of them.

EXERCISE

Analyse the ECSL coding of the model for each activity. Identify the statements which are the conditions for starting the activity, and the function of each. Compare them to the description of the model given in the earlier part of the chapter.

Identify the statements which would need amended if we wished to evaluate the efficiency of different job shop dispatching rules.

REFERENCE

1. Clementson, A. T., *Extended Control and Simulation Language*, Cle-Com Ltd, Birmingham, England, 1985.

CHAPTER 6

Simulation software — an overview

6.1 INTRODUCTION

In the chapters so far we have introduced the concept of modelling, and presented an approach to developing a model of a manufacturing system using activity cycle diagrams. We then used a software package, CAPS, to construct a computer model written in the ECSL language. These two packages are only two among many in the simulation field. In this chapter we intend to explain why so much software has been developed for simulation, to give an overview of the main types of simulation software product, and to relate these to modelling manufacturing systems. The sheer number and variety of simulation packages makes it impossible to give a comprehensive review of these products in a text whose main purpose is of a broader nature. Nor will it be possible to give either a complete description of any one package, or even a brief mention of every package. Instead, we will describe the principal approaches to simulation software and give examples of packages which adopt them. In later chapters models will be presented using several of them.

6.2 ACQUIRING SIMULATION MODELS

Simulation modelling involves creating a model using computer software. So far as obtaining a working model is concerned, there are four basic possibilities, which are:

— write your own program, in a general-purpose computer language such as FOR-
  TRAN,
— acquire a simulation package and write the model in that language,
— acquire and use an already written generic model,
— engage a consultant to carry out the simulation project for you, or to provide you
  with a working model.

Making the selection between them is a fundamental consideration, as the decision may constrain developments for several years. The skills available within the organisation is a major factor. Some companies have staff specialising in Operational Research, some of whom will probably be simulation experts. Most companies, especially the medium and small sized engineering companies will have very few such staff, if any at all. It is mainly with these companies and their present and future staff in mind that this review of software is undertaken. A brief comment on all four of these approaches will now be given, but the chapter as a whole will concentrate on the second and third alternatives.

6.2.1 Using a simulation package

All simulation programs involve similar functions. We have illustrated the model building process using CAPS and ECSL. The development of a package such as CAPS is only possible because the process of developing any simulation model involves defining entities, activities, queues, priority rules and so on. Because of this similarity, packages can be written exploiting these similarities and providing an easy method of model development. Many simulation packages have now been developed to simplify the process of writing models. In many of them the facilities are quite sophisticated. We will review the facilities to be offered by simulation packages below. The more extensive the facilities the longer is the learning time. Simulation packages may be written to use particular hardware, such as high resolution colour graphic terminals, or may run on any common computer. Fortunately, as personal computers improve, more of these are being re-written to run on PCs, perhaps using enhanced graphics or professional graphics boards. The principal advantage of these systems is that, several man-years having generally gone into their development, they provide powerful aids to modelling, and are usually bug-free. They also provide a structure within which the model will be developed and once a programmer is used to this way of thinking, error-free programs can be developed. Their cost may be considerable, up to about $25,000, but this is small in comparison to the benefits which can be obtained from their use. As will be discussed, packages may adopt one of several approaches to model building. Some are easier to learn and quicker to use than others, although there is usually a trade-off between these factors and the power of the system. Because of the need to acquire skill in their use, packages may be more appropriate for use by someone who is going to specialise in simulation rather than the manufacturing engineer who has many responsibilities and regards simulation as just one of his tools. We will give a more extensive discussion of simulation packages later.

Advantages
1. Powerful software tailored to the purpose of simulation.
2. Permits user to concentrate on logic of system to be modelled, not on computer programming as such.
3. Makes available many man-years of experience and programming effort.
4. Permits models to be developed more quickly than in a general-purpose computer language.
5. Most suitable for use by a simulation specialist.

Disadvantages
1. Purchase cost.
2. Training needs.
3. May not be well designed for the special features of user's system.
4. Risk of invalid conclusions until substantial expertise is gained.

6.2.2 Writing the model in a general-purpose computer language

Writing your own program in a language such as FORTRAN involves writing routines for the basic facilities which would be included in any simulation package. This will be time-consuming and most unlikely to be cost-effective unless no package is available or frequent use is foreseen. On the other hand the knowledge involved will be built up and retained within the organisation, which may be beneficial in raising the skills available and in generating awareness of what to look for in commercially available packages.

Advantages
1. Avoids the cost of purchasing a package.
2. Programmer knows all the details of his own program.

Disadvantages
1. Time to develop program.
2. The entire program has to be debugged.
3. Learning curve effects.

6.2.3 Using a generic model

A generic, or generalised, model is a model of a specific type of system written in such a way that certain parameters can be altered by the user through the data. For example, a generic manufacturing system model might ask the user to define the number of machines in the system to be modelled, and then proceed using that information. Several generic models have been written for simulating particular types of systems, such as FMS, automated warehouses, automatic guided vehicle (AGV) systems, transfer lines and so on. Their authors have studied a large number of systems and have included in the program as many as possible of the features which that type of system may exhibit. The configuration of a particular system to be modelled, such as the number of machines in an FMS, can be entered as data and the program run. This avoids completely any computer programming, and only requires a data deck to be "punched-up" by the user. Whether one of these models will be suitable for modelling a particular system will depend on whether the system has special features not included in the simulator. If the system to be modelled is a standard configuration, and is similar to others already installed it is likely that a suitable simulator exists. If it involves special features there may be no ready-written simulator to suit your needs. The weakness of all so-called generic models is that they often are incapable of handling these special features, and unfortunately most real systems possess some special features. Because the input to generic models is data they are usually easily learned by non-specialists, such as manufacturing engineers. In a later stage we will review generic models and present a case study using perhaps the best-known FMS simulator.

Advantages
1. Avoids any programming.
2. Minimises the time to get results.
3. Usually acceptable cost.
4. Usually runs on any "standard" hardware.
5. Suitable for use by non-specialists.

Disadvantages
1. May not be capable of modelling specific features.
2. May include unacceptable simplifications or assumptions.

6.2.4   Using a consultant

When the necessary expertise is not available in an organisation, outside assistance may be obtained as an alternative to training company staff. Consultants provide a major source of assistance. They are likely to have extensive experience in simulation and will provide a thoroughly professional service if given a sensible brief. Many of the simulation software suppliers also offer a consultancy service, based upon their simulation skills and software. Similarly, many management consultants have developed software for simulation, or have specialised in a third party software package and can offer simulation expertise. On the other hand, if the plans for the system are subject to revision then their results will become quickly obsolete, unless the consultant is on an extended contract. Engaging the consultant to revise the model as the plans are changed may involve a cost which is thought too great or long-running. Also, when the project is over the company's own staff may have had little chance to acquire expertise for themselves.

Advantages
1. Speed of obtaining results.
2. Professional expertise.

Disadvantages
1. Cost.
2. Difficulty in modifying model if plans change.
3. Expertise may not be transferred to company's own staff.

6.3   HARDWARE

Before discussing simulation software it may be worthwhile to make some observations concerning the computer hardware on which the model will be run and the peripheral devices which are available. There are three basic possibilities:

— a mainframe or super-mini computer,
— a personal computer,
— a computer dedicated to some other purpose such as a CAD system.

If we are to use a mainframe then we may be constrained by features of the operating system. Simulation programs are usually large and run for a long time. There may be limits on the size of program or length of run which we can run interactively during normal office hours. There may also be restrictions on the peripherals which the system will support. We may want to show a graphic picture of the system and the movement of components around the system. A graphics terminal may have to be acquired specially for this purpose. We will need to be sure that the computer will support its protocol. If the computer serves a large number of users the response of the system may be too slow. If the computer gives each user a small slice of its time at a time, then our graphic picture will freeze while it gives other users attention. This could be frustrating.

Personal computers are becoming increasingly popular for simulation modelling. They avoid the problems which can arise with multi-user hardware. Also their graphical capabilities are more precisely defined and generally easier to learn. On the other hand, PCs may have insufficient memory or may be too slow to permit us to do all that we would like in the model. To obtain the desired resolution for graphic displays special cards or screens may have to be purchased. As the speed and memory capabilities of PCs improves it is likely that virtually all simulation of manufacturing systems will migrate to these machines. Exceptions to this will be those cases where the simulation software is implemented on a specific system for interfacing purposes. For example, the software might be mounted on the FMS control computer so that it can access files defining the current status of the system.

A further factor in the implementation of simulation software on PCs is that a vast range of software is available for such tasks as spreadsheets, word processors and graphics written to operate under the PC-DOS, MS-DOS or other standard operating system, to which the simulation software can be interfaced. More recently operating environments such as GEM have made the PC much more user friendly. Mouse and menu driven software can be used to link modules of the simulation package, or to call up other software to perform associated tasks.

If we have a CAD system, we might consider running the simulation model on that system so that its graphical capabilities can be used. For example, if three-dimensional CAD software is available it might be used to advantage for simulation. Most CAD systems can be used for general-purpose programming, but it may be difficult to find someone with the necessary skills in both simulation and CAD system programming.

6.4   THE HISTORICAL TRENDS OF DEVELOPMENT OF SIMULATION SOFTWARE

The main purpose of this chapter is to put the various software packages, or types of package, in context. Although it is not intended to review specific packages, examples of their approaches will be given. To place the packages in context it is important to appreciate the historical trends in their development, the extent to which they share common origins and have developed along parallel lines. Behind these and subsequent developments have been two desires on the part of the software developers:

1. To make it easier to create a model.
2. To make it easier to understand the output of the model.

Thus to achieve the first objective, many of the software developers added some sort of front-end model-building package to the main package. In addition many have developed a front-end tailored to manufacturing system modelling, in which the words used are meaningful in their context, such as machine and job, rather than simulation jargon such as entity and set. These developments will be reviewed later in the chapter. To achieve the second aim most packages now provide some sort of visual, graphic output, substantially replacing tables of numbers in a computer printout. These developments, too, will be reviewed during this chapter.

In the UK, the pioneering work on simulation was done in the steel industry, by Tocher and others. As a result of accumulating experience in modelling various aspects of steelworks Tocher realised that many models had common features, and set about developing a generic model which included many of the features he had encountered. In fact rather than create a generic model, Tocher's creation was much more general, and was the first simulation package, GSP or General Simulation Package.[61] Virtually all simulation development in the UK followed from that, and was done by former colleagues, or by those influenced by them. Figure 6.1, based loosely on a diagram by Mills,[34] presents a family tree of major UK packages. Thus from GSP stemmed CSL[4] and later ECSL,[7] FORSS, SIMON[20] and HOCUS.[22]

SEE-WHY[52] introduced visual interactive modelling harnessing pioneering work by Hurrion,[24] and graphics were added to ECSL, FORSS, as FORSSIGHT,[22] SIMON to form SIMON/G[33] and HOCUS. Model-building front-ends were developed: CAPS for ECSL, FORGE for FORSSIGHT, DRAFT[30] for SIMON and EXPRESS[10] for SEE-WHY, employing various approaches. In some cases, the new method of model building was embodied in a new package rather than a front end, for example, GENETIK[15] superseding OPTIK.[36] Manufacturing generic models have been developed to make model building significantly easier for projects within that field. Examples are WITNESS,[62] written in SEE-WHY, and GENETIK-GMS. Sometimes packages written for a limited application are found to be of wide use. WITNESS is used in many non-manufacturing applications, and has largely superseded EXPRESS.

In the USA, the trends have been similar, although it is less easy to link the various packages in a family tree. Figure 6.2 shows two groups of packages. GASP was first developed in the steel industry. GASP-II,[46] GASP-IV[43] and GASP-V[5] were developments from GASP, adding facilities and making it easier to use. Q-GERT[45] introduced a network convention as a way of describing the behaviour of elements in a system, which assisted in the conceiving of simulation models, and has some analogy with activity cycle diagrams. SLAM[47] combined the concepts of Q-GERT and GASP. SLAM II[44] emerged from SLAM, TESS[59,60] provided an environment assisting with model development, data management and graphical animation of simulation models. MAP/1[48] is a manufacturing-oriented model using SLAM and TESS. SIMAN[40] was developed after experience gained with SLAM, and incorporates some of the concepts of GPSS.[18,50] It has several modules for facilitating model building, and provides animated graphics through CINEMA.[6] GPSS (General Purpose Simulation System) is one of the earliest simulation languages and was developed by IBM. It has descendents such as GPSS V[16] and GPSS/H.[19] Another of the early American languages is SIMSCRIPT, which evolved through SIMSCRIPT II,[26] SIMSCRIPT II.5,[27,56] PC SIMSCRIPT II.5,[39] PC SIMANIMATION,[25] which added a graphics capability, and SIMFACTORY,[55] which is a manufacturing generic model. PC SIMSCRIPT II.5 and SLAM-PC illustrate another general trend, that of implementing packages on personal computers, so that model building can be easier and the results easier to understand, thanks to the graphics available on PCs. Many other products have appeared from independent origins, often being developed on PC for the first time, rather than as a PC version of an earlier package, for example PC-Model[37] and PC-Model/GAF,[38] which provided more sophisticated animation graphics.

98

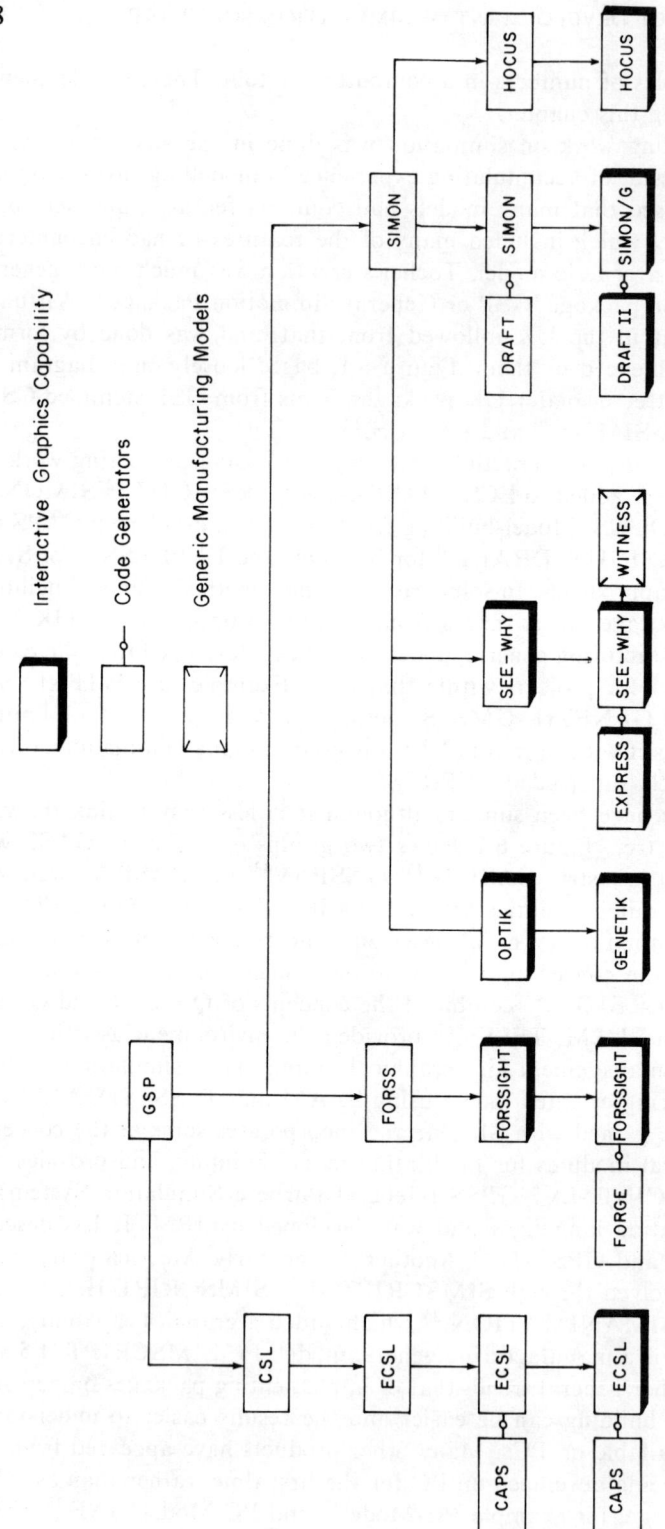

Figure 6.1 *Historical development of simulation packages in the UK, after Mills.*[34]

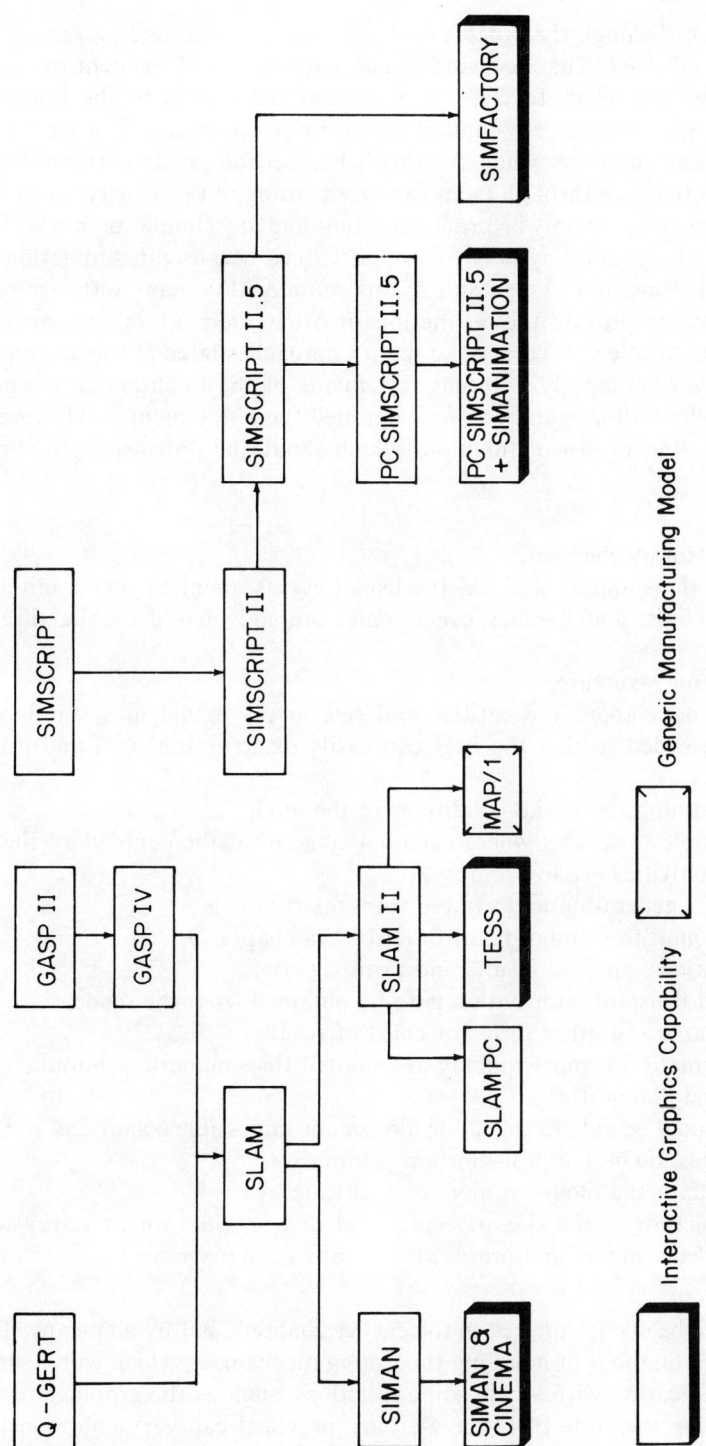

Figure 6.2 Historical development of certain simulation packages in the USA, after Mills.[34]

6.5   FUNCTIONS OF A SIMULATION PACKAGE

In chapter 4 we went through the process of developing a model, and presented a list of the considerations involved. These considerations are largely independent of the nature of the system to be modelled. In chapter 5, we entered details of the model into a computer package which wrote a computer program of the model. It used a standard set of questions. These questions would be asked whatever the nature of the system being modelled, although the path through them varies according to the answers given. Such a standardised approach would only be practical if building any simulation model followed a definite pattern. Model building only follows a pattern because all simulation models involve certain basic functions. As a result of this commonality many software packages have been developed to provide these functions in what their originators consider the most effective manner. We will review the main approaches later in this chapter.

   Different authors give slightly different statements of the functions of a simulation software package, depending mainly on how detailed their statement is. However, most would accept the following list of functions which should be provided by a simulation package:

1. The timing control mechanism.
   This maintains the simulation clock, the list of events about to occur, initiates and terminates activities, and handles event logic, probably based on the three-phase principle.
2. A data base or file structure.
   This holds the data about the entities and sets in the model in a standard form. Routines are provided so that the user can easily place entities into and take them out of queues.
3. A method of defining the initial conditions in the model.
   It must be possible to specify where each entity is, what the contents of the queues are, and what activities are in progress.
4. Random number generation and sample from distributions.
   This is fundamental to simulation, as described in chapter 3.
5. Record observations, analyse results and print reports.
   This is essential if useful information is to be obtained from the model.
6. Display histograms and other forms of chart of results.
   Graphical information is more quickly assimilated than numeric printout.
7. Error checks and diagnostics.
   The package should be able to check for illogical or impossible conditions and provide suitable messages, including post-mortem printouts.
8. Display the state of the model at any point in time.
   An animated picture of the system being modelled provides an easy way to check the operation of the model and present the results to management.

Many of these have been encountered in the earlier chapters, but by no means all. There are several different methods of handling the timing mechanism, which will be described shortly. The sophistication with which some functions, such as the graphical display of results or monitoring the state of the model, are provided can vary enormously, from the extremely rudimentary to the very well developed.

6.6   SIMULATION LANGUAGES AND PACKAGES

We have used the word "package" frequently to refer to a commercially available software product, and we will continue to use it in this sense. We have also used the word "language", for example ECSL was described as a simulation language. There is, however, a fundamental difference between a package and a language. This must now be explained, because it is important in placing the various software products, and their approaches to developing simulation models, into context. We can define them as follows:

Package: a collection of routines written in a standard computer programming language, such as FORTRAN, which provide the user with the basic functions identified above. The user will write segments of program, normally in the same language, to describe the logic of the system to be modelled, calling these routines to handle the standard functions.

Language: a programming language complete with its own vocabulary, grammar, syntax and so on, for performing the basic functions of a simulation program. The user will write his model using the vocabulary and syntax, and does not have to write any code in the underlying language in which the language is written, which might be FORTRAN. Usually the "program" which the user has written is read as a data deck by the language processor and decoded in some way.

Thus there is a subtle difference in the way they are used to write a simulation model, and this will be discussed shortly.

6.7   THE BREADTH AND SOPHISTICATION OF SIMULATION SOFTWARE

So many simulation packages have been developed that they are too numerous for a comprehensive review to be given. However, we can present a discussion of the main types and the approaches to modelling which they adopt. To do this we need a framework within which we can compare them. For our framework we can relate a package to two factors:

1. the breadth of the facilities it provides the user,
2. the level of sophistication with which it provides them.

6.7.1   Breadth of the facilities provided

Obviously, every package must provide the basic core of simulation facilities, and have a method of accepting input and of putting out results. By breadth of facilities we mean the extent to which the package offers facilities over and above this basic core. We have already, in the section on historical trends, observed that there have been additions to many packages to make them easier to use, by for example providing a program generator such as CAPS for ECSL, and easier to understand, by for example adding a graphical display facility. We can consider the facilities offered in addition to a basic core, to be associated with either the developing and entry of the model, or the presentation and analysis of output.

On the input side the main facility which is desirable is some means of avoiding the need to write the code of the model. This could be provided by a conversational program generator such as CAPS, or a menu-driven method of specifying the nature of the system to be modelled by a generic model.

On the output side, the minimum facility would be a printout of summary statistics. In addition we might look for a means of presenting these results in graphical form, for performing a statistical analysis on the results, and a graphically animated display of the model.

6.7.2   Sophistication with which the facilities are provided

By "sophistication", we mean not only the user-friendliness of the package in general, but more specifically the extent to which the package frees the user from the need to write a computer program, but allows him or her to think in "engineering" terms about the practical details of the system to be modelled.

For defining the model, the programmer may need to write a computer program of some sort. We have already observed some of the possibilities in this area, which include:

1. General-purpose computer languages.
2. Simulation packages.
3. Simulation languages.

Although all of these involve some program writing, the second and third represent "higher-level" approaches. In the first of these the user is involved in all the details of computer program writing. With the third, as illustrated by ECSL, the vocabulary and constructs of the language are specially designed to meet the needs of simulation modelling. Within both simulation languages and packages various levels of sophistication can be observed.

However with other tools the user may not need to do any programming as such. This type of package includes:

1. Interactive program generators.
2. Data-driven simulation packages.
3. Broad manufacturing generic models.
4. Generic models for specific types of manufacturing system.

The first of these groups of software generate the code of the simulation program automatically, as demonstrated with CAPS. In the second group the model is defined by entries in data fields for specifying entities, activities, queues and so on. With the third group of package, the terminology is orientated to manufacturing, so that instead of entities the package will refer to machines, parts and so on. They are intended to be suitable for virtually any manufacturing application, whereas the fourth group have narrower terms of references and model specific types of manufacturing system, such as FMS, transport systems and so on. It should be noted that the ease of use of a package improves with its sophistication, with a corresponding decrease in the skill required of the user. However, the flexibility offered to the user to tailor the software to his particular needs and the generality of the package decreases with the level of sophistication.

When considering the facilities offered by simulation packages for the output of results, we should observe the capabilities offered for:

1. printed numerical information,
2. presentation in graphical form.

Tables of figures can present the data, but graphs and charts do so in a much more easily absorbed fashion.

As with defining the logic of the model, we are interested in the extent to which the collection of results is:

1. by instructions programmed by the user, or
2. by the package, automatically.

We are also interested in whether results are provided either:

1. concurrently with the simulation run, or
2. after completing the simulation run.

Similarly, if a graphical presentation of the model is provided we should know whether it is achieved by:

1. post-processed animation,
2. concurrent animation,
3. interactive concurrent animation.

Such classifications are inevitably rather arbitrary. It does, however, provide a framework within which we can review the main types of package and the facilities which they provide to the user.

6.8   THREE APPROACHES TO MODELLING

Before proceeding to consider the facilities offered by any package, or type of package, we must introduce a fundamental characteristic of simulation packages. The "worldview" of a package is its way of thinking about the system to be modelled, and how the logic of the model is expressed. In discussing the ECSL program in chapter 5, it was stated that ECSL[7] is an activity-based language, in that a program consists largely of routines describing the activities in the model. This is only one approach to modelling systems. In fact there are three main approaches to modelling which a package may adopt. Every simulation package has what is known as its "executive" routine, which keeps overall track of the state of the model, in particular the flow of time. The way the executive is written defines the modelling approach of the package. The three approaches are:

— the activity-based approach;
— the event-based approach;
— the process-based approach.

6.8.1   The activity-based approach

This approach has already been illustrated in the discussion of ECSL in chapter 5. In this approach a program segment is written to define every activity which the entities may engage in. The segment includes tests to determine whether the activity may be initiated at any point in time, and defines the actions to be performed if the activity is initiated. The actions involve moving the entities concerned from queues leading into the activity at the start of the activity and placing them in the following queues when the activity is complete, together with associated arithmetic and statistics collection. An activity cycle diagram is used to describe the sequence of queues and activities for each type of entity. This has been illustrated by ECSL in the preceding chapter. A diagram representing the flow of the logic performed by the "executive" routine is given in Figure 6.3. After initialising the model, the executive routine searches for the next time when an activity is due to be completed, and moves its clock to that time. The changes which result from that activity completion are performed, and then a scan of the activities is done to see which, if any, can be started at that time. After scanning all the activities, the clock time is advanced again until the run is complete. The sequence in which the activities are scanned can be important, as described when discussing the priority of activities in chapter 5. The activity approach seems to be more popular in the UK than in the USA. In addition to ECSL, HOCUS[22] uses this approach. Terms vary through common usage, for example OPTIK and GENETIK are described as using the "entity cycle" approach, an equivalent, and perhaps more accurate name for the activity approach.

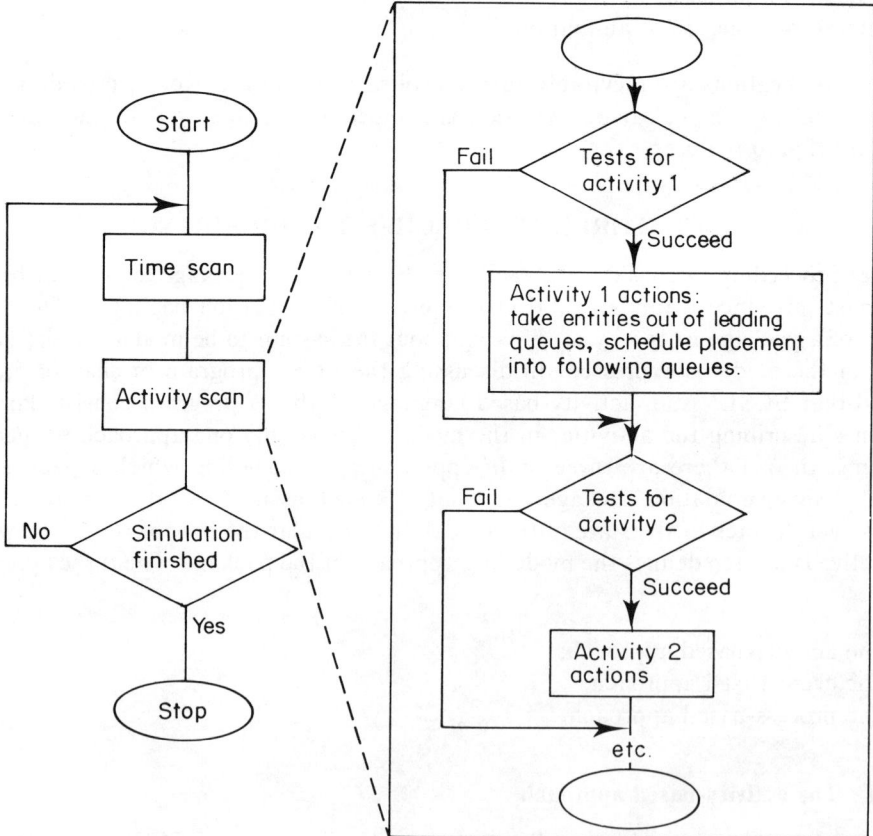

Figure 6.3   Flow diagram for activity-based executive program.

6.8.2 The event-based approach

As stated in chapter 4, events are the instants when a change takes place in the model, and coincide with the start and end of activities. In the event approach a program segment is written to define every event in the model. After initialising the model, the executive routine scans the times for each scheduled bound event and moves the clock to the time of the next event to occur. It then calls the routine to process the changes to be made to the model as a result of that event. There will be a routine for each bound event. Following a bound event, the conditions may allow one or more conditional events to occur. There will normally be a routine for each conditional event, which includes the tests to determine whether the conditions permit that event to occur, and if so the actions to be performed. These actions include making the changes resulting from the event, which corresponds to an activity beginning, and scheduling the bound event corresponding to the end of the activity. The conditional event routines may be called directly from each bound event routine, or, in simple models, the coding may be incorporated into the routine of the bound event itself. Alternatively, the executive may call the conditional event routines. In complex models, correctly tying conditional events to the bound events which might induce them can be tricky. EXPRESS[10,53] provides an event mapping facility to simplify this problem. Figure 6.4 gives a flow-chart for an

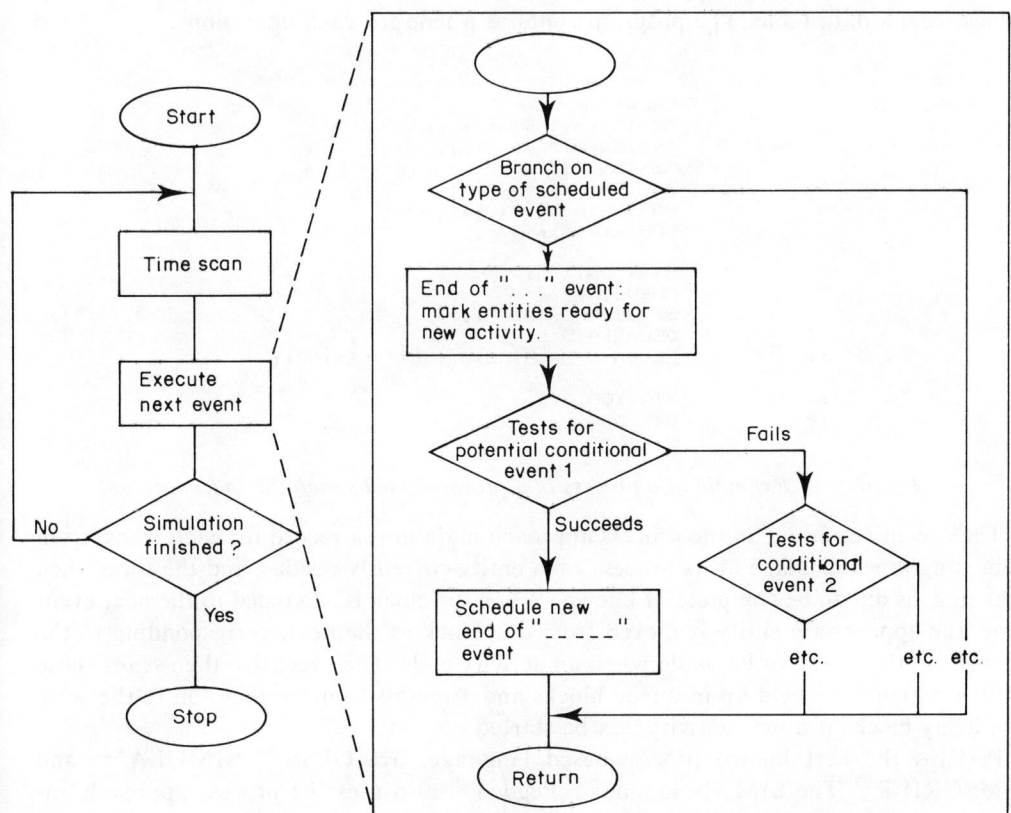

Figure 6.4   Flow diagram for event-based executive program.

event-based executive. This approach is widely used, especially in packages written in FORTRAN. In the UK, SEE-WHY[52] and SIMON[20] use it, while in the USA it was adopted by GASP,[43] is incorporated within GASP's descendants such as SLAM[44] and SIMAN,[40] and is available in SIMSCRIPT II.5.[56]

6.8.3 The process-based approach

In this approach, also known as the process–interaction approach, the entities are classified into transactions or customers and facilities or resources, corresponding roughly to temporary and permanent entities. The terminology varies somewhat between one package and another. Process routines are defined which describe the processes undergone by each transaction. A process routine typically includes the creation of the entity, and then a series of queueing for, seizing and then releasing the facilities it requires in its passage through the system, and finally the destruction or disposal of the entity. The approach frequently uses a keyword for each step in the process, often depicted by a block diagram with one block for each keyword. The process defining the life cycle of a job in a machine shop model might be as shown in Figure 6.5. The purpose of most of these blocks should be obvious. How they behave depends on the parameters modifying each block's keyword. The ASSIGN blocks set up the values of attributes containing the operation number, machine type required and duration for each operation, by obtaining values from a data table. The program contains a loop for each operation.

```
 1                    CREATE:
 2                    ASSIGN:A(1)=job-type
 3      NEXTOP        ASSIGN:A(2)=A(2)+1;
 4                    ASSIGN:A(3)=P(A(1),A(2));
 5                    ASSIGN:A(4)=A(2)+10;
 6                    ASSIGN:A(5)=P(A(1),A(4));
 7                    ROUTE:A(3);

 8                    STATION,1-5;
 9                    QUEUE,M;
10                    SEIZE:MACHINE(M);
11                    DELAY:A(5);
12                    RELEASE:MACHINE(M):NEXT(NEXTOP);

13                    STATION,0;
14                    DISPOSE;
```

Figure 6.5   Example of a process in a process-based simulation package.

The executive routine in the process approach maintains a record for each transaction indicating in which stage of its process each entity currently resides, and the time when that stage is due to be complete, if known. When the clock is advanced to the next event time, the appropriate entity is moved from one block to the next, corresponding to the changes in the model to be made when an activity ends. The executive then scans those entities which were held up in queue blocks and attempts to move them on to the seize and delay blocks if a new activity can be started.

Perhaps the best known process-based languages are GPSS,[50] SIMULA[3,21] and SIMSCRIPT.[27] The SIMAN language, Pegden,[40] also uses the process approach, indeed figure 6.5 is almost a SIMAN program. It will be seen that the process approach can be extremely compact, although it can be difficult to describe complex processes.

SIMAN includes blocks specifically designed for functions involved in simulating manufacturing operations, such as CONVEY and TRANSPORT. In addition to its process
blocks, SIMAN also contains a package of event-based FORTRAN routines for functions which cannot be conveniently modelled using the keywords. Another of SIMAN's
interesting features is its division of the "model" from the "experiment". The model
defines the logic of the model, while the experiment gives the values of the data deck to
be read by the model. The model must be compiled, the experiment is not. This means
that a model can be compiled once and run with many sets of data without the need to
re-compile the model. A SIMAN program will be given in chapter 11.

6.8.4   The approaches compared

The three approaches involve slightly different "world-views". Consequently, they represent different compromises between how well the program models the real world, the
ease of building the model, their computational efficiency, and so on. The process approach, with its block diagramming method, seems attractively similar to a production
process description, but can get complicated when operations involve several types of
facility. For example, one problem inherent in the process approach is that of reserving resources for operations if two independent types of resources are involved. If the
program is written

```
QUEUE (RESOURCE_TYPE_1)
SEIZE (RESOURCE_TYPE_1)
QUEUE (RESOURCE_TYPE_2)
SEIZE (RESOURCE_TYPE_2)
DELAY (OPTIME)
RELEASE (RESOURCE_TYPE_1)
RELEASE (RESOURCE_TYPE_2)
```

then a unit of resource type 1 may be reserved well before a unit of resource type 2
becomes available. In the meantime the unit of resource type 1 is prevented from performing useful work elsewhere. The activity approach avoids this problem by checking
simultaneously for the availability of a unit of both types of resource. Advanced process languages provide keywords to handle this problem; in more basic ones it may be
possible to solve it by additional statements. In some cases it may not be possible to
handle it at all. Some systems provide links to FORTRAN routines to handle situations
which the language cannot handle. On the other hand in the activity approach it may
be less easy to reserve resources in advance. The event approach is probably the most
flexible but usually requires the model builder to become more directly involved in computer programming than either the activity or process approaches. In the experience
of this author, the activity approach is the easiest (excluding generic packages such as
WITNESS) for a non-specialist to learn, and according to Pidd[41] "it is undoubtedly
simpler from the programming point of view to use an activity based approach". However, despite this general assessment, one package may be much more user-friendly than
another, even although they are of the same basic approach. The newer menu-driven
packages virtually eliminate these considerations, although the method which underlies
the package may influence its efficiency, speed and generality.

In terms of computational efficiency, several factors enter the comparison. Much may depend on how a package is implemented rather than on its modelling approach. One major aspect is whether the simulation model is a computer program in the true sense, which can be compiled and executed, or is more of the nature of a data deck, which is read and interpreted by the language computer program, in a way similar to that in which an NC part program in an APT-like programming language is processed. Event-based systems are usually true computer programs, whereas the activity and process approaches are interpreted. This may make these two slower to execute than event-based models, although much depends on how efficiently the interpretation is done. Another factor is how many comparisons are required at each time advance. For example, the activity approach will examine all the activities to see whether they can be started. Most of these tests will fail, so there may be quite a bit of computing involved which does not directly advance the progress of the model. To get around this problem, package writers may include facilities for recognising which activities are fed from the queues which have been affected by the changes and test only those activities. For example HOCUS has this capability, and in ECSL the "cell structure" allows the user to achieve this. With the event approach it may be possible to efficiently map conditional to bound events so that less non-productive testing is done. In the process approach tests have to be done to see whether any held up entities can be advanced due to the changes which have been made at each time step. Thus all three approaches encounter this problem, and package writers employ various devices to maximise the efficiency of these phases of controlling the model, some more effectively than others.

This comparison could be expanded greatly, if it was our purpose to examine how simulation packages should be written. Clearly, that is beyond the scope of this text. The reader who wishes to pursue this topic is referred to the appropriate management science literature, such as Pidd,[41] Fishman,[11–13] Franta,[14] Shannon[54] and many others. Our main concern is how these factors affect the developing of models written with the available software.

6.9   DEFINING THE LOGIC OF THE MODEL

When we come to define the logic of our model various approaches are available to us. They can be subdivided into two groups:

A. approaches which require the model logic to be defined,
B. approaches which do not require defining the logic.

Within the first group there are two subdivisions:

A1. approaches which require a program to be written, and
A2. approaches which do not require any program writing.

In this section we will compare the main approaches to defining the logic of the model, including examples of group A1, namely:

1. FORTRAN,
2. simulation packages of varying levels of sophistication,
3. simulation languages,

examples of group A2:

4. code generators,
5. data-driven packages,

and examples of group B:

6. generic manufacturing models, and
7. generic flexible manufacturing system models.

Perhaps a word of explanation of the distinction between group B, which it was suggested above do not require the logic to be defined, and the rest, which do, is required. In group A1, the logic of the model has to be expressed as a computer program. With group A2, the logic has to be specified, but not as a computer program. Instead it has to be expressed in terms of activity cycle diagrams or equivalent, either in the form of answers to questions or as entries in data fields. With group B, the logic has already been written by the person who wrote the generic model. The user only has to indicate the facilities present and the flow of components through the system. With generic manufacturing models this can usually be done in a rather general way, whereas with with models specifically for flexible manufacturing systems, the user usually has only to select work flow decision rules from a menu presented to him.

These seven approaches therefore illustrate a spectrum of levels of sophistication, from the first in which every detail of the model has to be defined and programmed by the modeller, to the last in which no details of the model logic have to be written and only a few details are left for the user to choose.

We have already dismissed writing the model from scratch in FORTRAN as inefficient and uneconomic, in view of the facilities offered by simulation packages. We will now briefly discuss the nature of the model building task in each of the other alternatives.

6.9.1 GASP: a basic simulation package

One of the key aspects of any simulation software is the way data is held in computer memory, and how the layout is defined. The data includes records for each entity and its attributes. Especially important is how the package keeps track of the active and passive entities. The passive entities are in queues. Associated with each active entity will be the time when its current activity is due to end, ie when its next event is due to occur. A "basic" simulation package would provide subroutines or functions to perform the main functions described in section 6.5, and the framework of a file system to keep track of the active entities and the contents of queues, but leave the programmer to decide for himself how the attributes and queues are to be held in the filing system, using the subroutines provided to manipulate them.

An example of such a simulation package is GASP-IV.[43] We can illustrate it by presenting the statements which might be used in a model of a simple job shop (omitting materials handling between operations). Since GASP is event-based, a routine is written to handle the changes of state when each bound event occurs, such as an operation is completed, and to test to see whether the conditions in the model then permit any conditional event to occur, such as another operation to be started. The FORTRAN statements to accomplish this, with calls to GASP-IV routines are given in Figure 6.6.

```
      SUBROUTINE END OF OPERATION
C
C     Establish which job and machine are involved in the event
      JOBNO=ATRIB(3)
      IOPNO=ATRIB(4)
      MACNO=ATRIB(5)
C
C     Find out if there is a job waiting for machine MACNO.
C     Search for an entity in file 2 with its 5th attribute
C     equal to MACNO.
      K=NFIND(MACNO,5,2,5,0.0)
C
      IF (K.GT.0) THEN
C
C     There is a job waiting for machine MACNO, therefore
C     remove the job from the file 2 and observe its
C     attributes.
          CALL RMOVE(K,2)
          JOB=ATRIB(3)
          JOPNO=ATRIB(4)
C
C     Start a new operation on this job and schedule an end of
C     operation event.
C     Place an entry in file 1.
          ATRIB(1)=TNOW+OPTIME(JOB,JOPNO)
          ATRIB(2)=2
          CALL FILEM(1)
C
      ELSE
C     There is no job waiting for MACNO, set it to idle.
          BUSY(MACNO)=0.0
      ENDIF
C
C     ************************************************************
C
C     Now determine if more operations are required on JOBNO.
      IF(IOPNO.LT.NOOPS(JOBNO)) THEN
C     No more operations are required, therefore the job leaves.
          CALL JOBLEAVES
C
      ELSE
C     More operations are required.
C     Set attributes for the next operation.
          IOPNO=IOPNO+1
          MAC=MCREQD(JOBNO,IOPNO)
          ATRIB(3)=JOBNO
          ATRIB(4)=IOPNO
          ATRIB(5)=MAC
C     Is the machine for its next operation idle or busy?
          IF (BUSY(MAC).EQ.0.0) THEN
C
C     Machine MAC is busy, therefore JOBNO must wait.
C     Place JOBNO into file 2.
              ATRIB(1)=TNOW
              CALL FILEM(2)
C
          ELSE
C     Machine MAC is idle, therefore start a new operation
C     and schedule an end of operation event.
              ATRIB(1)=TNOW+OPTIME(JOBNO,IOPNO)
              ATRIB(2)=2
              CALL FILEM(1)
C     Set machine MAC to busy
              BUSY(MAC)=1.0
          ENDIF
      ENDIF
      RETURN
```

Figure 6.6   *Example of programming in a basic simulation package: a GASP-IV subroutine for the end of operation event.*

The subroutine illustrates the use of three of GASP's functions or subroutines, namely NFIND, RMOVE and FILEM. NFIND locates an entity in a file (or queue) with suitable attributes. RMOVE takes an entity out of a file and places its attribute values into the array ATRIB. FILEM puts an entity into a file by copying the values which have been set in the array ATRIB into the main file system array. Subroutine JOBLEAVES would handle the case where the job is ready to leave the system, and has been omitted for simplicity.

When writing a program in GASP the programmer has to decide how the data in the main files should be laid out. This involves deciding in which files to hold data on entities in which queues, in addition to file 1 which is pre-defined by the system as the list of scheduled future events. In file 1 the first attribute must be the time of the event and the second must be a code number specifying the type of the event. When an event occurs the GASP executive automatically removes the entry from the event file and inserts the values into the array ATRIB and the current clock time into TNOW. In the model file 2 has been used to store jobs waiting for machines. The first field can be used for the time that the entity enters the queue so that the members of the queue can be properly ordered. The third, fourth and fifth fields can hold the job type, the operation number and the machine type on which the operation is done. The machine required and time for each operation is looked up from data tables held in arrays MCREQD and OPTIME. Having decided the layout of the file system, the programmer then has to calculate the total size of the main filing array and DIMENSION it accordingly. The attribute on which each file should be ranked and whether ascending or descending order is to be used have also to be specified. These details are all part of the data deck supplied for the model.

Thus the programmer must be familiar with the housekeeping arrangements of the package, which requires an ability to program in FORTRAN. GASP has been superseded by SLAM and SIMAN, but they retain some of these features of GASP.

6.9.2   EXPRESS: helping with the housekeeping

An enhancement which might be made to a package like GASP would be to add a facility which relieves the programmer of defining for himself how the information about the entities will be held in the memory of the computer. The programmer would define the entities and their attributes, but not have to deal with the file numbers for storage and retrieval. Many packages provide this facility, such as all the SIMON derived packages as indicated in figure 6.1. To illustrate this, and another major advance over GASP, we will describe some aspects of EXPRESS.[10,53] EXPRESS was developed to facilitate model building in SEE-WHY, which is an event-based visual interactive simulation package based on FORTRAN.

The two main stages of EXPRESS are its Define and its Events phases. In the Define phase all the entities and sets are defined, and how they are to be displayed on the screen as the model progresses. Static background features and text are also defined. The events phase adds the logic for each event, and has some of the characteristics of a code generator.

When specifying the event logic the programmer has already defined the entities, sets and some attributes. Other attributes will probably be needed to hold values to be tested in the event logic, and need to be thought out prior to or during the event logic definition.

The logic will be specified by statements not unlike those in the GASP routine above, but by associating attributes directly with the individual entity the need to think about how the data is stored internally is avoided. EXPRESS coding roughly equivalent to the GASP routine above would be as shown in Figure 6.7.

This segment illustrates several of EXPRESS's routines. RATTRB obtains the value of a real attribute of an entity, while IATTRB obtains an integer attribute value. RSETAT and ISETAT are used to set attribute values. ADD inserts an entity into a set. JOBQ is the name given by the programmer to the set (or queue) of waiting jobs. Events are scheduled by the routine SCHEDL whose parameters are the type of event, the time when it will occur and the number of the associated entity. SIZEOF gives the current size of a set. IDNTOF gives the entity identifier for a member of a set. The machine required and duration for each operation have been held, not in look-up tables directly, but in attributes of two dummy entities, MDATA and TDATA, which simulate one-dimensional arrays. The values would be set at the initialisation stage. Another dummy entity, BUSY, is used to indicate whether each machine is busy or idle. It has one attribute for each machine. As with the GASP routine there is a reference to a subroutine, JOBLEAVES, or "procedural event" as it is known in EXPRESS, to handle the case where the job is ready to leave the system.

However, the purpose of presenting this routine was not to illustrate EXPRESS's subroutines, but to illustrate that the programmer no longer has to handle the layout of the data in computer memory, removing one obstacle in the path of a non-expert using the package. In GASP the array ATRIB contains data referring to the most recently considered entity, and there is only one ATRIB array. The values must therefore be stored in "non-GASP" variables, such as IOPNO, if two or more entities are being considered in parallel. This happens in this case, since two jobs, JOBNO and JOB, are being considered. In EXPRESS the system maintains records of the attributes of each entity. The programmer is not relieved of the need to correctly define the model logic.

A further feature of EXPRESS is that the programmer does not actually write FORTRAN, but a simplified version of that language. EXPRESS checks each line as it is entered to ensure that it can be correctly converted into FORTRAN, and will flag errors. The programmer is also relieved of the need to define COMMON blocks and so on, these are inserted automatically by the system. Once EXPRESS has converted the model logic into FORTRAN, there being a SEE-WHY routine in FORTRAN for each of the function words in EXPRESS, the model is compiled into executable code as would any FORTRAN program.

EXPRESS has also been superseded. WITNESS[62] has replaced it for modelling manufacturing systems, and many other types of system, by those who need a programming-free simulation package, and those who wished to work in FORTRAN remained loyal to SEE-WHY.

6.9.3   GENETIK: avoiding compilation

Another step in relieving the model builder from the chores of programming is provided by GENETIK.[15] Like SEE-WHY, EXPRESS and its own predecessor OPTIK, GENETIK is a visual interactive package. The central characteristic of model building in this package is that the model is built up in a number of units. There are several types of unit, such as picture units, data units, logic units, statistic units. Each unit is

```
* Establish which job and machine are involved in the event
JOBNO=IDENTM
IOPNO=IATTRB(JOBNO,4)
MACNO=IATTRB(JOBNO,5)
*
* Find out if there is a job waiting for machine MACNO.
* Scan queue of jobs waiting for machines to find if there
* is one waiting for machine MACNO.
ISIZE=SIZEOF(JOBQ)
K=0
FOR I=1,ISIZE
    JOB=IDNTOF(JOBQ,I)
    MAC=IATTRB(JOB,5)
    IF (MAC.EQ.MACNO) THEN
        K=JOB
        GOTO 10
    ENDIF
NEXT I
*
LABEL 10
IF (K.GT.0) THEN
* There is a job waiting for machine MACNO, therefore start
* an operation and schedule an end of operation event
    T=STIME(1)
    JOPNO=IATTRB(JOB,4)
    K=JOB*10+JOPNO
    OT=RATTRB(TDATA,K)
    T=T+OT
    SCHEDL(2,T,JOB)
*
ELSE
* There is no job waiting for MACNO, set it to idle.
    ISETAT(BUSY,MACNO,0)
ENDIF
*
* ***************************************************************
*
LABEL 1
* Now determine if more operations are required on JOBNO.
NOOPS=IATTRB(JOBNO,6)
IF (IOPNO.EQ.NOOPS) THEN
* No more operations are required therefore the job leaves.
    CALL JOBLEAVES
*
ELSE
* More operations are required.
* Set attributes for the next operation.
    IOPNO=IOPNO+1
    K=JOBNO*10+IOPNO
    MAC=IATRIB(MDATA,K)
    OT=RATTRB(TDATA,K)
    ISETAT(JOBNO,4,IOPNO)
    ISETAT(JOBNO,5,MAC)
* Is the machine for its next operation idle or busy?
    K=IATTRB(BUSY,MAC)
    IF(K.GT.0)THEN
* Machine MAC is busy, therefore JOBNO must wait.
* Add JOBNO to JOBQ
        ADD(JOBNO,JOBQ,TAIL)
    ELSE
* Machine MAC is idle, therefore start a new operation
* and schedule an end of operation event.
        T=STIME(1)
        T=T+OT
        SCHEDL(2,T,JOBNO)
* Set machine MAC to busy.
        ISETAT(BUSY,MAC,1)
    ENDIF
ENDIF
RETURN
```

Figure 6.7   *Example of programming in an enhanced simulation package: an*
EXPRESS segment for the end of operation event.

created by means of on-screen menus, and is checked as it is created and converted into executable code directly. This means that there is no lengthy delay while a model is compiled. It also means that more substantial types of interaction with the model are possible during a run, ie changes which would otherwise need the FORTRAN to be amended and the model re-compiled. In this respect GENETIK is rather like a generic model, but has the advantage that the user is not restricted to the logic (eg decision rules) built into the generic model by its writer.

6.9.4 Simulation languages

We have already illustrated a simulation language, namely ECSL. Since ECSL is activity-based whereas GASP and EXPRESS are event-based, no direct statement by statement comparison can be made. However, the programming style can be compared. The ECSL coding for the activity of performing an operation roughly corresponding to the GASP and EXPRESS segments above would be as shown in Figure 6.8. Since we have already discussed the structure of an ECSL program in chapter 5, no explanation of these statements should be necessary. Look-up tables can be used directly. The case where the job is ready to leave the system would be handled by a separate activity.

```
BEGIN OPERATION
FIND FIRST MACHINE B WITH TIME OF MACHINE LE 0
    FIND FIRST JOB A FROM JOBQ
        NEXT EQ MACNO
DURATION = OPTIME ( JTYPE OF JOB A , OPNO OF JOB A )
OPNO OF JOB A +1
NEXT OF JOB A = MCREQD ( JTYPE OF JOB A , OPNO OF JOB A )
JOB A FROM JOBQ INTO JOBQ AFTER DURATION
TIME OF MACHINE B = DURATION
REPEAT
```

Figure 6.8 Example of programming in a simulation language: an ECSL segment for the operation activity.

It will be evident that a "high-level" simulation language of this type has several major advantages, such as compactness, readability and referring to entities, attributes and sets by name.

A high level language is compact, since one statement represents many FORTRAN instructions. In ECSL the "entity FROM queue INTO queue AFTER duration" statement is particularly powerful.

A simulation language is readable, because the statements are usually English-like. A non-expert could read an ECSL program and make some sense of it. Obviously, the more complex the model the less easy this would be. There is a trade-off between readability and compactness. The more compact the less readable. For example, SIMSCRIPT, another widely used language, is rather less compact because its designers sought to emphasise readability.

For the programmer, the facility to refer to entities, queues, attributes, variables and so on, by meaningful names is a tremendous advantage. He is therefore freed from the need to think in FORTRAN and to continually refer to elements of arrays, such as ATRIB in the GASP segment above, but can think in terms of the entities being modelled.

There is a potential disadvantage with high level languages, however. The strength of a language depends on the richness of its vocabulary. If a language is insufficiently rich there may not be keywords for functions desired by some users. In that case the user must resort to elaborate logical constructions to achieve his aim, or may find it impossible altogether. The example has already been given of simultaneously seizing two types of resource in a process-based language. There is therefore the risk of inflexibility in the language, because the user does not have access to the underlying logic of the language, and its data structure. The data concerning each entity is held in the system somewhere, but the user has no means of "peeking" or "poking" directly into the data structure. To overcome this problem a language might offer the facility to call externally written subroutines.

6.10   CODE GENERATORS

Just as all simulations involve similar concepts and functions, so the process of developing models tend to be rather similar whatever the problem being tackled. As a result, several programs have been developed which accept a simple description of a model and produce a program to run the model in the model generator's target language. They are usually interactive and accept a description of an activity cycle diagram. Examples are CAPS for ECSL[7] and DRAFT for SIMON.[30]

We have already illustrated the use of CAPS. It is an interactive program which asks for data about the system to be modelled under five headings:

Logic
This section asks for the names of each type of entity, their quantity and activity cycle. The validity of each cycle is checked and so is the combined model. Several illogical conditions can be detected and correction invited.

Priorities
This section requests data about the priority rules to be used when entities are selected from queues, or when there are branches in the cycle of an entity. It also enquires about the relative priority of activities in the case of entities which may be assigned to any one of a number of activities from an idle state.

Arithmetic
This section asks for the formulae to be used for calculating the duration of activities and for calculating the values of attributes. It also asks for the initial values of attributes or variables which are not changed during the model.

Recording
In this section the user specifies what statistics are to be recorded, such as length of queues, delays in queues and attributes of entities. It also asks for the length of the run-in period.

Initial conditions
In this section the user specifies the contents of each queue at the start of the run and gives details of any activities in progress. Various checks on the validity and consistency of the specified conditions are performed. The duration of the simulation is also given.

As with many software packages the power of a code generator lies in the richness and variety of situations with which it can deal. For example, the user may is presented with a menu of queue disciplines, from which to make a selection. Although the menu covers many possibilities it is conceivable that there would be situations not covered by them. Similarly, CAPS requires that activity durations are computed by sampling from a distribution, and offers a menu of the available rules. Often in FMS modelling it is desired to look up a table to obtain operation times or the next workstation in a sequence of operations. CAPS does not cater for these possibilities, and the user has to omit some detail from his model and then refine it later on, once the ECSL code has been written. This is a common feature of "front-ends", that they are "narrower" than the package they serve. As with high level languages, the higher the level of sophistication, the easier they are to use but possibly the less the generality.

Code generators have two basic tasks to perform. First, they have to obtain details of the system to be modelled, and second they must write a program for the model in the target language. In a sense these are quite separate functions. In principle, once the details of the system have been obtained the model could be expressed in any desired language, just as an NC post-processor produces a control program for a specific machine tool from data in a standard format cutter location file. In fact DRAFT exhibits this because it has demonstrated code writer modules for SIMON, GASP-II, SIMULA and SIMSCRIPT.[33]

Code generators for graphic animation simulation packages have additional functions to perform. They must obtain data for each element in the model of where, when and how it is to be displayed, and write the necessary code. DRAFT is available in both non-graphic and graphic versions.

EXPRESS[10] is an example of a code generator for a graphic simulation package. It generates SEE-WHY models. It has two main phases, the Define stage and the Events stage. In the Define stage each element (entity, set, time series, and so on) and how it is to be displayed on the screen is defined. Static background features are also defined. This stage is quite a powerful facility. The Events phase, however, is more rudimentary. Whereas CAPS obtains details of the activity cycles and generates a working program in the target language, EXPRESS does not relieve the user of the need to write model logic code. Instead, it aids this process by permitting him to write the logic in "pidgin-FORTRAN", checks the syntax of each statement interactively, indicates references to any undefined elements and so on, then converts the input to a working FORTRAN program. The logic itself is not checked.

6.11   DATA-DRIVEN PACKAGES

At the end of CAPS's logic section a table giving a summary of the entity cycles is printed if requested by the user. Part of this table lists the entities taking part in each activity. HOCUS[22] uses the data in this form to specify the model logic. A standard pro-forma table is presented on the screen into each field of which the user inserts data. Figure 6.9 illustrates this format. For each entity involved in an activity a row in the table is completed, giving the queue from which it is drawn, the queue to which it will be sent on, and data concerning the rule for selecting the entity from the queue. There are also data fields to specify how the activity should be displayed on the screen. Similarly the location on the screen for displaying the content of the queues in the model can be

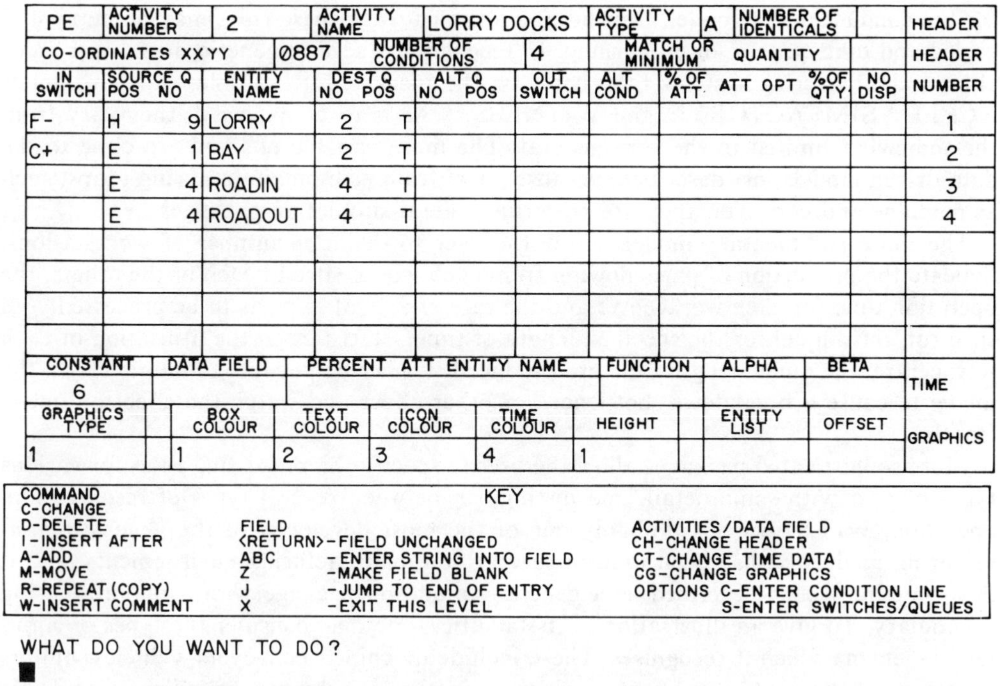

PE	ACTIVITY NUMBER	2	ACTIVITY NAME	LORRY DOCKS	ACTIVITY TYPE	A	NUMBER OF IDENTICALS		HEADER
CO-ORDINATES XX YY		0887	NUMBER OF CONDITIONS	4	MATCH OR MINIMUM		QUANTITY		HEADER

IN SWITCH	SOURCE Q POS NO	ENTITY NAME	DEST Q NO POS	ALT Q NO POS	OUT SWITCH	ALT COND	% OF ATT	ATT OPT	%OF QTY	NO DISP	NUMBER
F -	H 9	LORRY	2 T								1
C+	E 1	BAY	2 T								2
	E 4	ROADIN	4 T								3
	E 4	ROADOUT	4 T								4

CONSTANT	DATA FIELD	PERCENT ATT ENTITY NAME	FUNCTION	ALPHA	BETA	TIME
6						

GRAPHICS TYPE		BOX COLOUR	TEXT COLOUR	ICON COLOUR	TIME COLOUR	HEIGHT	ENTITY LIST	OFFSET	GRAPHICS
1		1	2	3	4	1			

```
COMMAND                              KEY
C-CHANGE
D-DELETE          FIELD                              ACTIVITIES/DATA FIELD
I-INSERT AFTER    <RETURN>-FIELD UNCHANGED           CH-CHANGE HEADER
A-ADD             ABC      -ENTER STRING INTO FIELD   CT-CHANGE TIME DATA
M-MOVE            Z        -MAKE FIELD BLANK          CG-CHANGE GRAPHICS
R-REPEAT (COPY)   J        -JUMP TO END OF ENTRY      OPTIONS C-ENTER CONDITION LINE
W-INSERT COMMENT  X        -EXIT THIS LEVEL                   S-ENTER SWITCHES/QUEUES
```

WHAT DO YOU WANT TO DO?

■

Figure 6.9   HOCUS activity definition screen layout (reproduced by permission of P-E Consulting Services, Ltd).

specified. HOCUS therefore accepts information of a similar nature to that given to CAPS, but runs the model directly. It is a data-driven package in the sense that the model logic is expressed in data, and no code writing is required of the user.

HOCUS therefore has some of the characteristics of both a code generator and a generic model. It is not a code generator in that no code is generated. Since no programming is required it is rather like a very broad generic model. It is also constrained by the limitations of both code generators and generic models, namely it cannot handle factors or possibilities not thought of by its writers, and therefore may not be as general as a language or package. The decision rules available in HOCUS are more extensive than in CAPS, and can involve quite complex functions, using data fields other than the source and destination queues. Details are available in Poole and Szymankiewicz[42] and the HOCUS Manual. Like code generators and generic models HOCUS provides the user with a quick and easy-to-learn method of creating models.

6.12   GENERIC MANUFACTURING MODELS

A generic model is a package written to model a particular type of system, in which the details of the specific system to be modelled can be specified in the data supplied to the package. The user is therefore freed from any need to write a computer program, and has merely to learn the input specifications for the package, and any optional rules for the conduct of the model. Depending on the package these rules may be many and varied and take quite a lot of learning. In recent years several generic models for modelling manufacturing systems have come on the market, for example, Modelmaster,[35] XCELL,[8] SIMFACTORY[55] and WITNESS.[62] As is to be expected, they vary from the somewhat limited to the very general. The more capable of them are close to the data-driven models just described, because apart from using manufacturing terms, such as machine and conveyor, they are powerful general simulation packages.

The more rudimentary models allow the user to define a number of workstations, indicate the proportion of parts flowing from each workcentre to each of the others, the operation time for each workcentre and the rate of arrival of parts to be processed, and then run this model for the specified length of time. Statistics on the utilisation of each workcentre, the queue lengths, etc are produced. It may be possible to interact with the model to cause a breakdown, bottleneck or other effect and watch the effect on queue lengths.

More sophisticated packages allow the user to specify the processing rules for various types of part with some detail, and distinguish between several types of resource and operation. WITNESS[62] is probably one of the most developed of the generic manufacturing models. It has been used in several non-manufacturing environments, and it has sometimes been described as a general simulation package with a manufacturing vocabulary. To give an illustration of its facilities, one may consider the types of simulation element which it recognises. These include machines, conveyors, vehicles, buffers or stores, labour, and parts, most of which have various subtypes, as indicated in Table 6.1 The types of operation peformed by the machines are known as single, batch, assembly or production (ie making a number of discrete items from a single input). Each element has attributes. The three main stages of developing a model, which is entirely menu-driven, are the define, display and detail stages of WITNESS. In the define stage all the elements in the system to be modelled are defined. In the display stage a display is set up, which may extend over four screens. The machines, conveyors, buffers and vehicle paths are positioned on the screen. A library of icons to represent machines is available and they can be made to change colour depending on the status of the machine, ie running, waiting, blocked, broken down, or being set up. In the detail stage the rules for work flow, the times of operations, frequency of arrivals and so on, are specified. Ten decision rules are available for defining the input and output rules for work flow between machines, conveyors and buffers. These include Push or Pull, Most, Least, Percentage, Sequence, Select, If and Wait. A variety of parameters can be used with these rules so that quite complex decisions can be modelled. WITNESS is written in SEE-WHY, and it is possible to link SEE-WHY segments to it for any features which cannot be achieved with the standard facilities. For the same reason WITNESS provides powerful interactions when a model is being run, as described in sub-section 6.20.2 on Visual Interactive Simulation below. One of the interesting effects of the way WITNESS is implemented is that it is no longer necessary for the user to complete the model before running it. The user can develop his or her model a bit at a time, defining each element through the define, display and detail stages before beginning on a new element. Normally in simulation it is necessary to handle all the elements through each stage of model development. As with most generic models the results are printed in a standard format, the user only having to indicate that statistics are to be collected for the element of interest.

Table 6.1 Simulation elements and their subtypes in the WITNESS manufacturing simulation package

Element type	Subtypes	Description
Parts	—	—
Machines	Single	processes a single part
	Batch	processing cannot start until a batch of parts is present
	Assembly	where a number of parts are combined into a single part
	Production	where a single part is converted into a number of new parts
Conveyors	Fixed	where spacing between parts is fixed, and which stops if when a part reaches the end it cannot move off
	Queuing	which allows parts to accumulate in a line at the front of the conveyor until the conveyor is full
Buffers	—	—
Labour	—	—
Tracks	—	—
Vehicles	—	—

6.13 GENERIC FLEXIBLE MANUFACTURING SYSTEM MODELS

Another class of generic models is that of models handling specific types of system. In manufacturing systems, there are generic models for material handling systems, guided vehicle modelling, warehousing,[9] production line models, and flexible manufacturing system models.[29] As with the broader generic manufacturing models their capabilities vary widely in their sophistication. In this section we will consider only FMS generic models, of which a review has been given by Bevans.[2] In addition to those available commercially, most of the FMS suppliers have some sort of model, probably written by themselves using one of the general simulation packages, capable of modelling the sort of equipment which they supply.

Generic FMS models may be fairly simple models, intended for modelling materials handling and sizing system proposals to potential customers, or they may be quite sophisticated, including the detailed work flow control decision rules of the real system control software. Perhaps the best-known and most highly developed is MAST.[28,29] The capability of a generic model can be assessed quickly and approximately by considering the number and type of data items it requires or will accept and the number of decision rules which it allows the user. For example MAST provides six groups of decision rules for:

1. part introduction,
2. selection of parts from queues,
3. selection of next operation,
4. station selection,
5. transporter selection,
6. transporter scheduling.

Within each group, a number of different decisions rules are available, including calling functions written by the user. The possibilities include assembly and disassembly operations, cart, conveyor and mixed mode transport systems, various types of machine buffer, and alternative operations. The reliability of each element within the system may be easily modelled. MAST provides for the definition of the time between breakdowns and repair times using any of nine distributions.

As with virtually all generic models results are collected automatically. MAST provides three main reports, each comprising a number of tables:

1. System description, from the input data, comprising:
    part types description,
    station description,
    operation description,
    pallet description,
    cart type description,
    track layout description.
2. System status:
    initial status,
    parts status,
    station status,
    cart status,
    pallet status,
    track status.
3. Performance summary report:
    parts completed,
    part types performance,
    pallet statistics,
    operations performed,
    station performance,
    station shuttle performance,
    cart performance,
    track performance.

In addition, it will provide an trace of events either on printout or as a file. This file can be read by MAST's companion module, BEAM (Background and Enhanced Animation for MAST),[1] referred to again in section 6.19 on post-processed graphics below. BEAM also provides graphical output in the form of charts and graphs. Another companion module, SPAR (System Planning of Aggregate Requirements)[58] provides an FMS sizing tool by which the static performance of a system design can be assessed to ensure that on a rough cut plan the design makes sense. SPAR also creates an input data file for MAST. This initial calculation is extremely valuable, since there is little point in performing a lengthy and expensive simulation run if a quick approximate calculation will indicate that a fundamental problem exists. Mention may also be made here of queueing network models, such as CAN-Q,[57] which can assess some of the probabilistic or dynamic aspects of a system design, such as queue lengths and throughput times, again without the effort of a full-blown simulation. Lenz[28] suggests the use of the three packages, SPAR, MAST and BEAM, as a suite for the design and evaluation of FMS, using the information flow concept described in Figure 6.10. TOOL, a module for assessing tool assignments is also available.

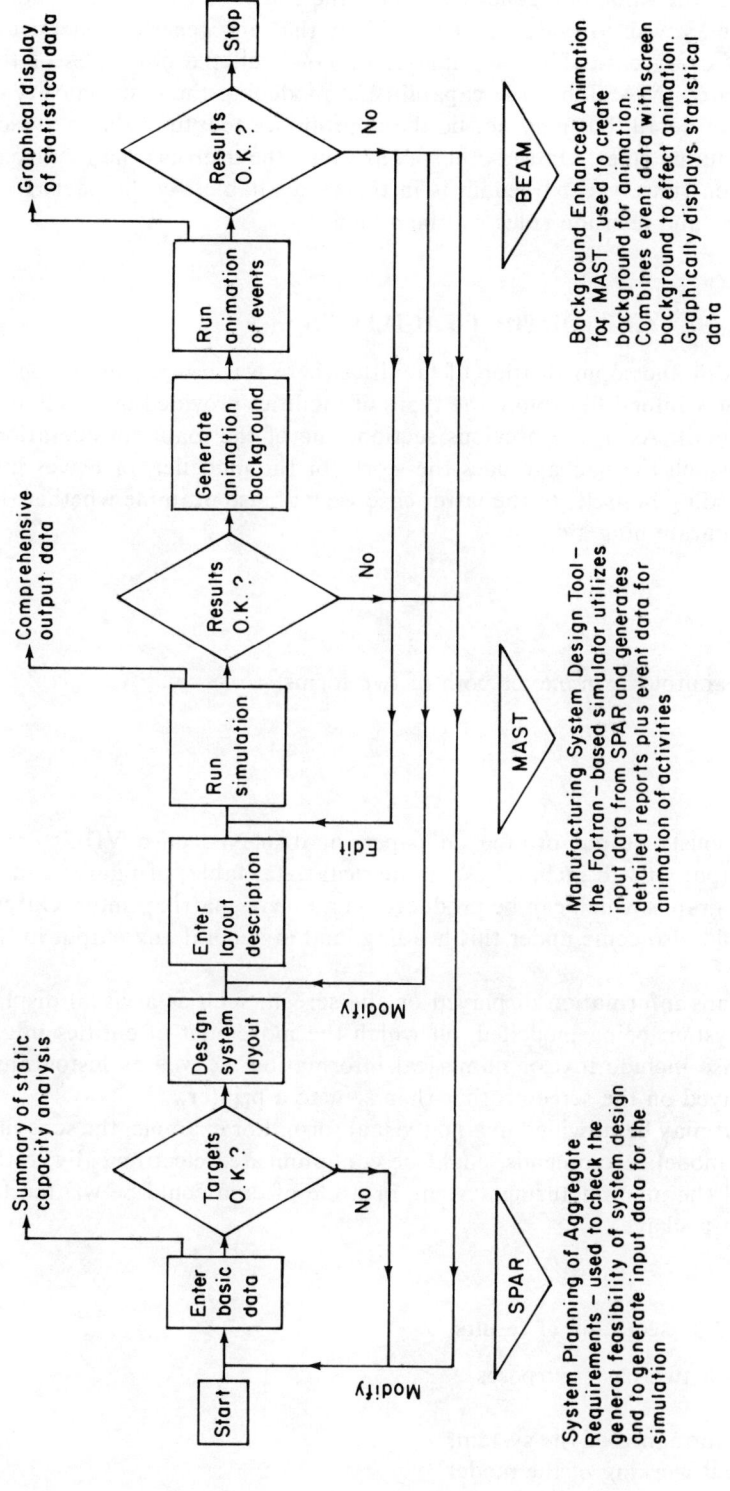

Figure 6.10   Information flow in FMS design using SPAR, MAST and BEAM (reproduced by permission of Citroen Industrie UK Ltd and CMS Research Inc.).

As already suggested, the value of a generic model is the extent to which it can model any system which we may wish to evaluate. It is unlikely that any generic model could handle every detail of every case. This is a matter of how well the objectives of the model and the user match. MAST has the capability of modelling the vast majority of systems to some level of detail, but may not be the appropriate tool to evaluate a wide range of possible scheduling rules. On the other hand, since the user has the facility of writing his own decision rules, great freedom is in theory available. As the package is developed new facilities and decision rules will be added.

6.14 OUTPUT FACILITIES

In section 6.7 on breadth and sophistication of facilities there was a brief discussion on output facilities. We now intend to explore the types of facilities provided by simulation packages in greater detail. As in the previous section, one of the main considerations will be the extent to which the package does the work for the modeller, or leaves him to write the detailed coding himself. In the latter case we will also examine whether the package provides a programming aid.

6.14.1 Output media

With a few exceptions output is in one or both of two forms:

— printed output,
— graphic output.

Printed output is obviously output printed on paper, or displayed on a VDU screen simulating printed output. It will include text, numerical data, tables of figures and so on, and also plots and graphs which can be produced on a line or matrix printer. Output on a graph plotter could also come under this heading, and in general any output in the form of a hard copy.

Graphic output means information displayed on the screen, such as a visual display of the layout of the system being modelled, on which the movement of entities might be shown. It would also include text or numerical information, as well as histograms, graphs or plots, displayed on the screen rather than sent to a printer.

In some cases output may be provided in a non-visual form. For example, the schedule which the simulation model recommends, could be communicated electronically to the executive computer of the manufacturing system, or a file of data could be written for entry to a spreadsheet package.

6.14.2 Collection and presentation of results

Results are required for two basic purposes:

1. to measure the performance of the system,
2. to ensure the correct working of the model.

While the first is clearly the object of the simulation exercise, the second category is just as important. Unless we ensure that the model is correct no results will be valid. Those with little experience of simulation are always amazed how long it takes and how much effort is expended in ensuring that a model is valid. Frequently the results produced in ensuring the validity of the model are much more voluminous than those finally presented on the performance of the system.

It is sometimes useful to distinguish two type of results:

1. observed data, such as the time that an entity took to pass through the system, or the value of a variable at some point in time, and
2. generated data, such as statistical summary or analysis of observed data.

For each piece of data two times are of interest:

1. the time when it is observed or generated, and
2. the time when it is presented to the modeller.

These times can be either:

1. during the simulation run,
2. after the completion of the simulation run.

Combining these we observe that we could have:

1. data generated and presented during the run,
2. data generated during the run, presented at the end of the run,
3. data generated and presented after the completion of the run.

In the early days of computing, when models were run as batch jobs on mainframe computers results could only be obtained after the run was complete. Even although data could be sent to the printer as the model was run, it would become available to the modeller after the run was ended. Sometimes the model would run to completion, which could take several hours, and then terminate without producing any output, because some error had occurred. This was, of course, very frustrating to the modeller. "Trace" facilities were added to packages, whereby, for example, the status of variables at every advance of the clock could be sent to a file and printed, even if no other output was provided by the model. Trace facilities can be vital to ensuring that the model is valid. They represent an example of output generated during the run and presented after the run is complete.

The advent of interactive computers, especially personal computers, made the presentation of data during the run a practical possibility. It is no longer necessary to wait until the end of a run to discover if it worked correctly. As soon as it becomes apparent that something is amiss the run can be terminated. This saves much time for both the computer and the modeller. Visual graphical output is particularly valuable for this purpose.

Not until the run is complete can final results be presented. That is obvious. Perhaps less obvious is that there is an important class of results which cannot be generated until the run is complete. This includes the general class of statistical analysis of results. For example, it may be desired to examine whether there is some statistical correlation between variables, or to discover if any of the results, such as the shape of a distribution, conforms to a standard distribution. Such analysis may be performed to assess the validity of the model, such as to ensure the independence of the results from the random numbers sampled during the run, and other aspects mentioned briefly in chapter 3. We do not wish to expand on that topic now, but the reader should be aware that it is a very large one, with which many authors have dealt, eg Fishman and Kiviat.[13] Several packages, especially those produced by operational research scientists rather than those developing packages for specific applications, such as generic models, provide powerful facilities for analysis of results.

Analysis facilities may be either:

1. within the main simulation package, or
2. a self-contained module of the package, which can be run as a post-processor.

It is perhaps of little significance to the user which form exists in the package being used. For the package writer, however, there are important considerations. The main advantage of separating the simulation model proper from the analysis of results is that the main simulation package can be smaller, will require less computer memory and so a larger model can be accommodated in a given computer, and the model may run faster. Data will probably be written to file on disc in a standard format, which can then be analysed. This tends to simplify the program, and avoids the need to manage internal memory. Since no analysis is done of the data, the speed of execution may be improved, so long as the time taken to write data to disc is not excessive. The speed of execution and size of model which can be built are of course important considerations to the user. In a later section we will review the facilities for analysis of data provided by some of the major packages.

6.14.3 Graphic output

When the simulation model is run on a personal computer or interactively on a mainframe with a graphics terminal, the package may support the presentation of data to the user in graphical form immediately it is observed in the model. Thus time series could be displayed and histograms gradually built up as observed data is collected.

The most significant possibility with graphic output capability is of course the provision of an animated representation of the model, so that entities can be seen moving as the model progresses. The facilities for graphic output of certain major packages will be discussed in section 6.20 below.

6.15   FACILITIES FOR COLLECTING STATISTICS

As suggested in the opening chapters there are two basic forms of statistic which can be observed:

1. Simple variables, such as the time in the system of an entity. These individual observations are usually summarised in mean and variance statistics and/or presented

as histograms. The term used to describe them varies from package to package, and include "observed data", "tally variables", "generated data".

2. Time-dependent variables, such as the number of jobs in a system. These may be stored in the form of histograms and summarised by their mean, variance or other statistics. These methods of summarising the data destroy the time base of the data. To preserve this they can be collected as times series, either sampled at regular intervals, or as a set of pairs of values and time intervals since the last observation or change of value. The names used to describe them vary from package to package, and include "time-dependent", "time series", "time-persistent" and "time-generated data".

As with the methods of expressing the event logic and controlling the simulation model as a whole, so there are various levels of sophistication in commercial packages. These can roughly be classified as follows:

1. programming all the data collection and analysis by the user himself,
2. using subroutines to perform the main functions, which the user calls in his program,
3. in which there is a measure of automation of the collection of statistics, although the user has to define which statistics are to be collected,
4. in which the package will decide for itself which statistics should be collected and do the collection without requiring any programming by the user.

The first of these corresponds roughly to what would be required if the model was being written in FORTRAN by the user; the second to the facilities provided by a basic simulation package; the third is provided by enhanced packages and languages, although there is some variation within these groups; and the fourth is to be found with some model-builders and menu-driven packages.

We will now discuss these approaches and illustrate them with reference to some packages. Our viewpoint will principally be that of asking what they do for the user, how much of the work they leave to him, and how easy they make the user's task.

6.15.1   Programming data collection entirely by the user

It is relatively straightforward to compute the values of simple and time-dependent variables, and to compute mean and standard deviations and other simple statistics, using a standard programming language such as FORTRAN. For example, suppose that we wish to summarise the time in the system of an entity, and that TIMIN is the time when it entered the system. Then all we need are a few statements to initialise some counters, a few more to increment them, and some more to compute the statistics when the run is complete.

Initial statements would be

```
XS=0.
XSS=0.
XN=0.
```

Whenever an entity leaves the system we would write:

```
X=TNOW-TIMIN
XS=XS+X
XSS=XSS+X*X
XN=XN+1.
```

At the end of the run we can compute mean and variance as follows:

```
AVG=XS/XN
VAR=(XN*XSS-XS*XS)/(XN*(XN-1.0))
```

To present the results we need a statement of the form:

```
WRITE (N) AVG,VAR
```

In the case of time-dependent variables we need to know for how long the current value of the variable has existed, so that the counters may be time-weighted. For example, if we are recording the work in progress in a system, then each time a job enters or leaves the system the number would be increased or decreased by one. Thus whenever the value of the variable changes, statements would be written such as:

```
XISYS=XISYS+1.
TINT=TNOW-TLAST
STS=STS+XISYS*TINT
STSS=STSS+XISYS*XISYS*TINT
ST=TNOW
```

At the end of the run we would write:

```
AVG=STS/ST
VAR=STSS/ST-AVG*AVG
WRITE (N) AVG,VAR
```

Although these are technically simple statements they would be very numerous in a serious model. They are also highly repetitive and therefore subroutines should be written to perform their functions as succinctly as possible. This would require that arrays are declared for recording the various counters for frequency, sum and sum-squared of each variable. Reserving computer memory for, say, 20 statistical variables of each type can be done by a statement such as:

```
DIMENSION XS(20),XSS(20),XN(20),ST(20),STS(20),STSS(20)
```

6.15.2   Automatic computation of summary statistics

Virtually all simulation packages provide a facility for computing these summary statistics. One of the parameters of a package is the number of independent statistics which can be collected. If there is a limit on the size of program which can be accommodated within the computer, the package may not provide what the user needs. Fortunately, this is not often a problem today since the memory capacity of mini and personal computers has been expanding greatly in recent years.

A package can therefore be expected to include statements of the form of those given above. Indeed GASP includes almost exactly the statements given above. The user is freed from the responsibility of keeping track of the statistics and is concerned only with how and where the data should be collected. For example, GASP includes the routines COLCT and TMST for recording observations of simple and time-dependent variables, and HISTO for constructing histograms. The user calls these routines in his event logic routines. Thus the statements above would be replaced by:

    X=TNOW-TIMIN
    CALL COLCT(X,N)

and

    XISYS=XISYS+1.
    CALL TMST(XISYS,TNOW,N)

where N is the statistic number to be used for collecting the variable. Summary reports at the end of the run or at intervals during the run are produced by a routine SUMRY called directly by the GASP executive routine. The user has to define in the data deck the statistics to be collected, and must remember which code number he is using for each statistic. Having done so the GASP system initialises all counters, updates statistic counters and prints reports automatically. Figure 6.11 illustrates the format of these reports.

This method makes things easy for the user, and in the absence of carelessness on his part will be error-free. However, it would be nice if the process of collecting the observations could be automated as well.

6.15.3   Semi-automatic collection of statistics

Several packages have the capability to collect the observations semi-automatically. The method is that whenever an event occurs, or in some cases at specified intervals, a call is made to a routine which adds the current value of all requested variables to the appropriate statistical counters. This routine would be called directly by the executive routine.

The EXPRESS package provides some degree of automating the data collection of time-dependent variables. It provides two routines EVALHS and EVALTS which add observations to histograms and time series (ie graphs) respectively. These are called at intervals specified by the user for each histogram or time series. The system automatically calls these routines once for each statistic to be collected. Although the call is made automatically, the user still has to write the code to be inserted into them. In EXPRESS, variables are local to each routine, and consequently attributes of a dummy entity are used to pass the values to be collected to the subroutine. Thus in an arrival event there might be the statements:

    XISYS=XISYS+1.
    RSETAT(STATS,XISYS,4)

```
                        **GASP SUMMARY REPORT**

         SIMULATION PROJECT NUMBER   2  BY  A.  S.  CARRIE

              DATE 10/  9/ 1976      RUN NUMBER   1 OF   1

              CURRENT TIME =  0.5600E+02

         **STATISTICS FOR VARIABLES BASED ON OBSERVATION**
                 MEAN      STD DEV   SD OF MEAN      CV       MINIMUM      MAXIMUM     OBS

RATIO        0.7765E+00  0.2095E+00  0.9369E-01  0.2698E+00  0.5510E+00  0.1000E+01    5

         **STATISTICS FOR TIME-PERSISTENT VARIABLES**
                 MEAN      STD DEV    MINIMUM     MAXIMUM   TIME INTERVAL  CUR. VALUE

XISYS        0.3071E+01  0.1266E+01  0.0000E+00  0.5000E+01  0.5600E+02  0.0000E+00
XBUSY        0.2214E+01  0.1047E+01  0.0000E+00  0.5000E+01  0.5600E+02  0.0000E+00
```

Figure 6.11 Layout of GASP-IV reports.

where STATS is a dummy entity whose attributes are used to store statistical values. In this case attribute 4 holds the current value of XISYS. Then in EVALHS and EVALTS, respectively, the user would insert:

IF(HISTPN.EQ.1) THEN VAL=RATTRB(STATS,4)

and

IF(TMSSPN.EQ.1) THEN VAL=RATTRB(STATS,4)

HISTPN, TMSSPN and VAL are system variables. RATTRB is a function giving the value of a real attribute of an entity. HISTPN and TMSSPN are the values of the statistic which the system wishes to record and the value of VAL is automatically added to the statistical record. The routines can include any arithmetic which may be necessary. Each histogram or time series has to be declared in initalising statements defining the variable, the sampling interval and various other parameters.

EXPRESS therefore provides a framework for automating to some extent the collection of data, but still demands rather a lot of program writing on the part of the user. Simple variables have to be recorded by statements very similar to the GASP ones given above, using routine RECDHS, which records a value in a histogram.

Languages generally provide a greater degree of automation of data collection, and provide keywords for collecting statistics. This was illustrated in the ECSL model in chapter 5. The principal keyword in ECSL is ADD, which adds an observation to a statistic. For recording simple variables, statements equivalent to the GASP ones above would be placed (as was done in the model in chapter 5) in the appropriate activity:

TIMIN OF JOB A = CLOCK - TIMIN OF JOB A
ADD TIMIN OF JOB A TO TIHIST

The resulting histogram would be printed, normally in the finalisation section, using the PICTURE routine, by the statement:

PRINT 'Histogram of Time in System'/ PICTURE (TIHIST)

More powerful statements are available for collecting time-dependent statistics. The activity approach is well adapted to this purpose. In ECSL statements are placed in the recording block and executed every time the simulation clock is advanced. For example to record the length of a queue and the time for which entities are held up in the queue the special recording block would contain the statements:

DURATION = CLOCK - PREVCLOCK
ADD MWAIT TO HIST ZBMWAIT (11 5 10)
ADD TIME IN MWAIT TO ZAMWAIT DURATION
PREVCLOCK = CLOCK

An attractive feature of the ADD command is that it is only active once it has been turned on by a SWITCH statement, so that values can be ignored until the prescribed run-in time has elapsed:

SWITCH ADD ON AFTER RUNINTIME

It will be observed that, although the user must write the statements to compute the values to be recorded of some simple variables, ECSL requires little programming other than to give the actual collection statements themselves.

Languages based on the process approach usually provide still more powerful facilities for collecting statistics. SIMSCRIPT will perform the data collection automatically once the variables to be observed have been defined. It provides TALLY and ACCU-MULATE statements to specify that simple or time-dependent variables, respectively, are to be recorded. Suppose we wish to record the time spent in the system by each of the jobs, and the number of jobs in the system. In the preamble we would include statements to define the global variables and the statistics to be recorded, using statements of the form:

PREAMBLE

  TEMPORARY ENTITIES
    EVERY JOB HAS A TIME.OF.ARRIVAL AND A TIME.IN.SYSTEM
    DEFINE TIME.OF.ARRIVAL AND TIME.IN.SYSTEM AS REAL
                                     VARIABLES
  DEFINE MEAN.TIME.IN.SYSTEM, VAR.TIME.IN.SYSTEM,
    AVG.NUM AND VAR.NUM AS REAL VARIABLES

    DEFINE NUMBER.IN.SYSTEM AS AN INTEGER VARIABLE

    TALLY MEAN.TIME.IN.SYSTEM AS MEAN OF TIME.IN.SYSTEM
    TALLY VAR.TIME.IN.SYSTEM AS VARIANCE OF TIME.IN.SYSTEM
    ACCUMULATE AVG.NUM AS THE MEAN OF NUMBER.IN.SYSTEM
    ACCUMULATE VAR.NUM AS THE VARIANCE OF NUMBER.IN.
                                     SYSTEM
  END "PREAMBLE

The job's process routine would include statements to set the values of the statistics to be observed, such as:

PROCESS JOB

  LET TIME.OF.ARRIVAL = TIME.V
  ADD 1 TO NUMBER.IN.SYSTEM
    .
    .
    .
    .
    .
    .
    .
  LET TIME.IN.SYSTEM = TIME.V - TIME.OF.ARRIVAL
  SUBTRACT 1 FROM NUMBER.IN.SYSTEM

  END "JOB

This illustrates that the process routines can be free from data collection statements. The addition of the variable to the statistical registers is automatic once the TALLY and ACCUMULATE variables have been stated. MEAN and VARIANCE are built-in functions for statistical data collection. Several others, such as MINIMUM, MAXI- MUM, are also available, and the TALLY or ACCUMULATE variable can be defined as a HISTOGRAM.

SIMAN is even more compact in its statistical data collection. It uses the MARK and TALLY keywords to record time

CREATE:EX(2,1):MARK(6);
.

.
TALLY:1,INT(6):DISPOSE;

The MARK modifier causes the time the entity enters the CREATE block to be noted in attribute 6. The TALLY block causes the interval (INT) between TNOW and the time recorded in attribute 6 to be added to tally statistic number 1 before the system disposes with the entity. The average, standard deviation, minimum, maximum and number of observations are automatically recorded. In the Experimental Frame for the model there is a corresponding TALLIES element defining for each tally register the descriptive name to be printed, and whether the observations are to be written to a file for later analysis by SIMAN's output processor:

TALLIES:1,TIME IN SYSTEM;

For collecting time-dependent variables, SIMAN includes a DSTAT element, such as:

DSTAT:1,NQ(1),QUEUE FOR MACHINE 1;

This single entry in the experimental frame causes the specified variable (in this case a system variable recording the length of a queue) to be recorded automatically, summary statistics computed and the observations written to file if desired.

GENETIK has an interesting method of defining the statistics which are to be col- lected. Special types of unit are provided for statistics collection and reporting. Areas of the system are declared as being covered by a statistic unit. This area can be as large as the entire factory or system, or as small as a part of a machine. Data is then collected on these areas and displayed in graphs or histograms. The passage of entities through the area is observed so that time to pass through and the number of entities within the area are recorded.

6.15.4 Automatic collection of statistics

It is conceivable that data collection could be fully automated, by the package automat- ically inserting the necessary statements to collect every variable, to record the length of every queue and so on into the program; or to act upon these statements without ac- tually formulating them. General-purpose packages generally do not do this since there are several alternative ways of collecting statistics (eg as a histogram or numerical summary statistics) and also, perhaps especially, because an enormous amount of computer memory would be required.

   Only with data-driven packages has this been developed to any great extent. For example, HOCUS automatically records the data for some 24 different reports, summarised in Table 6.2. These are of four types:

— trace reports
— snapshot reports
— summary reports
— user-defined reports

Table 6.2   HOCUS standard reports

Report numbers	Report description
1	Activities started
2	Activities ended
3	Activity run count
4,5,6	Activity detail reports, data on each activity in progress, in three levels of detail
7,8,9	Queue detail reports, data on the current contents of queues in three levels of detail
10,11	Entity trace report, giving data on specified entities, in two levels of detail
12,13	Entity location reports, giving the location of each entity, in two levels of detail
14	Histograms
15	Display
16	Attribute trace
17	Attribute trace, when monitor switch is on
18	Queue logs, giving statistics about queues
19,20,21	Queue trace, giving data on specified queues in three levels of detail
22	Data fields
23	Activities failing to start
24	User-defined reports

Many are of a diagnostic nature rather than user's summary statistics, and would be switched off once a model was debugged. Within report number 24, user-defined reports, many different reports can be created and produced, and the user can perform arithmetic and print explanatory text to format his own reports.

With generic models, however, automatic data collection is the norm, with reports being formatted and expressed in system parameters, such as the number of jobs in the system, the manufacturing lead time, or machine utilisations, rather than in terms of queue contents or other simulation parameters. For example WITNESS, produces a report for each type of element, such as, parts, labour, machines, buffers, conveyors and vehicles. Table 6.3 lists the parameters recorded and printed by WITNESS.

Table 6.3   Statistics reported by WITNESS (reproduced by permission of Istel Ltd)

Element type	Statistics reported	Element type	Statistics reported
Parts	Number created Number shipped Number scrapped Number assembled Number rejected (on creation) Number currently in the system Average number in the system Average time in the system	Conveyors	Number of parts that have entered Number of parts currently on Maximum number of parts on Average number of parts on Average time spent on conveyor Percentage of time empty Percentage of time moving Percentage of time blocked Percentage of time queueing Percentage of time broken

Machines	Number of cycles completed
	Percentage of time idle
	  (waiting for parts)
	Percentage of time busy
	  (working on parts)
	Percentage of time blocked
	  (unable to output)
	Percentage of time setting-up
	Percentage of time broken or
	  being repaired
	Percentage of time waiting for
	  labour
		Labour	Number of labour free
			  (Now, Max, Min, Ave)
			Number of jobs waiting
			  (Now, Max, Min, Ave)
			Average waiting time for jobs
			Number of jobs completed
			Percentage of time busy
			Percentage of time idle

Buffers	Number of parts that have
	  entered
	Number of parts that have left
	Number of parts currently in
	Maximum number of parts in
	Minimum number of parts in
	Average number of parts in
	Average time spent in buffer
		Tracks	Number of vehicles that have
			  entered
			Current vehicle on (if any)
			Percentage of time empty
			Percentage of time busy
			Percentage of time blocked

		Vehicles	Total distance travelled Number of loads carried Percentage of time idle Percentage of time demanded Percentage of time loading/unloading Percentage of time loaded Percentage of time stopped Percentage of time blocked

6.15.5   User's requirements

These remarks have given an overview of the types of facility offered by simulation packages for data collection and computing statistics. Any package other than the ones mentioned may be expected to conform to the pattern at one or other level of sophistication. It is recommended that the package should be able to record both simple and time-dependent variables with little work on the part of the programmer, and to be able to output the individual observations to file for subsequent analysis. However, remember that the less work required of the programmer the more must be done by the package itself, which increases its complexity, size, cost and execution times.

6.16   PRESENTATION OF STATISTICAL RESULTS

Since most results are numerical, numerical tabular printed output, or the equivalent of printed data on a VDU, predominates. However, much statistical data, especially time series and histograms, also lends itself to some form of graphical output. We have already given an example, namely ECSL's facility for printing histograms, which can give the output either on a screen or on a printer. Many other packages also use a printer to provide a graphic output. For example, GASP-IV plots frequency histograms and cumulative histograms. By expressing the values as a frequency, scaling the plot is simplified somewhat. GASP-IV also provides a graph plot facility whereby the value of any specified time-dependent variable can be recorded in a table and then plotted at the end of the run. Figures 6.12 and 6.13, from a GASP model of a simple machine shop in chapter 1, illustrate these outputs. The statements required to collect the observations for graph plotting, which must be included in each event routine whenever a change is made to a variable to be plotted, are of the form:

```
PLOT(1)=XISYS
PLOT(2)=XBUSY
CALL GPLOT(PLOT,TNOW,1)
```

where PLOT is an array used to store the observed values, TNOW is the current clock and the values are to be recorded in plot number 1.

ECSL can provide similar plots, at the end of a run, of variables whose values were saved during the run, using its GRAPH command. Like many of the early languages, ECSL was originally written with the line printer in mind as the primary output device. It, like others, has been re-written to take advantage of the developments in personal computers, so that, if the model is being run interactively, the plot can be presented on the VDU.

The facilities offered by simulation packages for graphical output have improved as computer hardware has improved. More recent packages were written with VDUs in mind from the start. Consequently, more sophisticated displays may be provided. For example, EXPRESS can display histograms and time series continuously as the simulation progresses, providing a valuable check on the functioning of the model. Another feature of modern computers which may be exploited is the multi-windowing facility, so that the histograms can be displayed on a different window (or on a different screen) from the main display.

With some packages graphical output is only produced by a post-run analysis module of the package, and we will discuss these in the next section. For end-results it does

```
                              **HISTOGRAM NUMBER  1**

                                       RATIO

 OBSV    RELA     CUML       UPPER
 FREQ    FREQ     FREQ     CELL LIMIT    0         20        40        60        80        100
                                         +    +    +    +    +    +    +    +    +    +    +
   0     0.000    0.000    0.5000E-01    +                                                 +
   0     0.000    0.000    0.1000E+00    +                                                 +
   0     0.000    0.000    0.1500E+00    +                                                 +
   0     0.000    0.000    0.2000E+00    +                                                 +
   0     0.000    0.000    0.2500E+00    +                                                 +
   0     0.000    0.000    0.3000E+00    +                                                 +
   0     0.000    0.000    0.3500E+00    +                                                 +
   0     0.000    0.000    0.4000E+00    +                                                 +
   0     0.000    0.000    0.4500E+00    +                                                 +
   0     0.000    0.000    0.5000E+00    +                                                 +
   0     0.000    0.000    0.5500E+00    +                                                 +
   1     0.200    0.200    0.6000E+00    +**********                                       +
   0     0.000    0.200    0.6500E+00    +         C                                       +
   2     0.400    0.600    0.7000E+00    +********************         C                   +
   0     0.000    0.600    0.7500E+00    +                             C                   +
   0     0.000    0.600    0.8000E+00    +                             C                   +
   0     0.000    0.600    0.8500E+00    +                             C                   +
   0     0.000    0.600    0.9000E+00    +                             C                   +
   0     0.000    0.600    0.9500E+00    +                             C                   +
   0     0.000    0.600    0.1000E+01    +                             C                   +
   2     0.400    1.000    0.1050E+01    +********************                             C
   0     0.000    1.000       INF        +                                                 C
 ---                                     +    +    +    +    +    +    +    +    +    +    +
   5                                     0         20        40        60        80        100
```

Figure 6.12 GASP-IV histogram plot.

```
                                                      **PLOT NUMBER  1**
                                                        RUN  NUMBER  1

                                                        SCALES OF PLOT
J=XISYS     0.0000E+00               0.1250E+01               0.2500E+01               0.3750E+01               0.5000E+01
B=XBUSY     0.0000E+00               0.1250E+01               0.2500E+01               0.3750E+01               0.5000E+01

 TIME       0    5   10   15   20   25   30   35   40   45   50   55   60   65   70   75   80   85   90   95  100 DUPLICATES

0.0000E+00  J                        +                        +                        +                        + JB
0.0000E+00  B                   J    +                        +                        +                        +
0.0000E+00  +                   J    +                        +                        +                        + JB
0.1000E+01  +                   B    +              J         +                        +                        +
0.4000E+01  +                   B    +                        +         J              +                        +
0.4000E+01  +                        +              B         +         J              +                        +
0.6000E+01  +                        +              B         +                        +    J                   +
0.6000E+01  +                        +                        +         B              +    J                   +
0.7000E+01  +                        +                        +         B              +                        J
0.7000E+01  +                        +                        +                        +    B                   J
0.8000E+01  +                        +                        +                        +                        J JB
0.8000E+01  +                        +                        +                        +    B                   J
0.9000E+01  +                        +                        +         B              +                        J
0.1100E+02  +                        +                        +                        +    B                   J
0.1100E+02  +                        +                        +         B              +                        J
0.1300E+02  +                        +                        +         B              +    J                   +
0.1300E+02  +                        +              B         +                        +    J                   +
0.1500E+02  +                        +                        +         B              +    J                   +
0.1800E+02  +                        +                        +                        +    J                   + JB
0.1900E+02  +                        +                        +                        +    J                   B
0.1900E+02  +                        +                        +                        +    J                   + JB
0.2000E+02  +                        +                        +         B              +    J                   +
0.2600E+02  +                        +                        +                        +    J                   + JB
0.2600E+02  +                        +                        +         B              +    J                   +
0.2800E+02  +                        +              B         +                        +    J                   +
0.3100E+02  +                        +                        +         B              +    J                   +
0.3200E+02  +                        +                        +                        +    J                   + JB
0.3400E+02  +                        +                        +         J              +    B                   +
0.3400E+02  +                        +                        +         J              +                        + JB
0.3400E+02  +                        +                        +         J              +    B                   +
0.3400E+02  +                        +                        +         J              +                        + JB
0.3700E+02  +                        +              J         +         B              +                        +
0.3700E+02  +                        +              J         +                        +                        + JB
0.4000E+02  +                   B    +              J         +                        +                        +
0.5000E+02  +                   J    +                        +                        +                        + JB
0.5600E+02  J                   B    +                        +                        +                        +
0.5600E+02  J                        +                        +                        +                        + JB

 TIME       0    5   10   15   20   25   30   35   40   45   50   55   60   65   70   75   80   85   90   95  100 DUPLICATES


OUTPUT  CONSISTS  OF      37 POINT SETS (     74 POINTS)
STORAGE ALLOCATED FOR    140 POINT SETS (    420 WORDS )
STORAGE  NEEDED   FOR     37 POINT SETS (    111 WORDS )
```

Figure 6.13 GASP-IV time-dependent variable plot

not matter whether a display is produced by the main package or a separate module, but clearly no opportunity exists to examine displays for checking the operation of the model in a diagnostic mode, if they are not available until after the run is complete.

Business graphics packages, which can be used for bar charts, pie charts, graphs and so on, are now commonly available on personal computers. Consequently, there is a trend towards not including facilities for graphical output within the simulation package itself, but to provide the data in a standard form for entry to one of these packages. This is commented on further in the next section.

6.17   FACILITIES FOR ANALYSIS OF STATISTICAL DATA

In manufacturing system simulation, perhaps because deterministic data is frequently used for operation times, there is a tendency to overlook the importance of proper statistical analysis of simulation-generated data. Some packages omit the provision of facilities for this type of analysis, presumably on the assumption that a separate statistical package is available. Other packages provide comprehensive analysis facilities.

Post-run analysis can include the presentation of "normal" results, of the form discussed in many preceding sections, including histograms, graphs and bar charts. With many packages specifically aimed at personal computers there has been a tendency to postpone much of the analysis and presentation of observed data until after the run is complete, in order to improve speed and reduce memory requirements. However, in this section we wish to consider more specific data processing functions.

A facility which should be provided in any package is the ability to direct observed data to a file, which can be analysed later. ECSL's GRAPH command is an example of the analysis of data saved by the main package. In addition to that command, ECSL includes an ANALYSE command to test for autocorrelation of a time series or cross-correlation between two time series, and to provide various statistical measures of them, including graph and histogram plots. ECSL also provides a FIT command for testing whether a histogram or time series conforms to any of the standard statistical distributions. This latter facility is useful if later use is to be made of the distribution. Sampling from a standard distribution is generally much simpler and more compact to program and quicker to execute, than is sampling from a histogram.

Facilities for analysis of generated data are probably most highly developed in SIMAN's "Output Processor". Table 6.4 summarises SIMAN's facilities, which include statistical analysis, graphic and data transfer facilities. The graphic capabilities for bar chart, graph plot and so on support either line printer output or to a screen. Presentation quality, bit-mapped graphics are available for selected graphic boards. This is a developing trend, and IBM's enhanced graphics adapter is perhaps the one most frequently supported. Another feature of SIMAN's output processor is its set of data transfer commands, which offer output files formatted for spreadsheets, for Tektronix graph plotting or in ASCII format. The theory behind the statistical commands is beyond the scope of this text, and the reader is referred to specialist material, such as Fishman,[11,12] Schmeiser[49] and Schruben.[51]

Table 6.4   SIMAN output processor facilities (reproduced by permission of Systems Modelling
Corporation)

Command	Description
BEGIN	Directs a copy of the output results to a file
Graphic Capabilities Commands:	
BARCHART	Plots data in a bar-chart format for a specified observational variable
HISTOGRAM	Calculates frequency information from data in a specified file
PLOT	Plots data from specified files on $X-Y$ axes
TABLE	Generates a table of values of observational or time-persistent variables
Statistics Commands:	
CORRELOGRAM	calculates correlation between observations from a specified data file
FILTER	Generates a filtered data set for a specified file containing either observational data, discrete time-persistent data or continuous time-persistent data using truncation and batching
INTERVALS	Constructs confidence intervals of data from specified files
MEANTEST	Performs a comparison of means between two data sets from specified files
MOVAVERAGES	Performs exponentially weighted moving averages to a data set from a specified file
SDINTERVALS	Constructs standard deviation confidence intervals of data from specified files
STDINTERVALS	Calculates standardised time series confidence intervals of data from specified files
ONEWAY	Performs a one-way analysis of variance of data from a specified file
VARTEST	Performs a comparison (ratio) of estimated variances from specified files
Data Transfer Commands:	
DIFFILE	Outputs the data in a text file DIF format which can be read by other micro DIF-format supporting programs
EXPORT	Outputs SIMAN-formatted files as ASCII-formatted files
EZPREP	Outputs SIMAN-formatted files as Tektronix PLOT 10 Easy Graphing Software formatted files
IMPORT	Outputs ASCII-formatted files as SIMAN-formatted files
LOAD	Outputs non-SIMAN ASCII formatted files as SIMAN formatted files

6.18 GRAPHICAL ANIMATION OF SIMULATION MODELS

Thanks to the improvements in computers, especially in personal computers, in recent years, virtually every simulation package now provides some facility for graphical output, permitting the animation of simulation models. With these facilities a visual picture of the system being modelled can be displayed on the screen. As entities move through the system a character or icon can be moved on the screen to mimic the status of the model. Similarly, the colour of resources can be changed as their status alters from "idle" to "running" to "under repair" and so on. Animation of models can have very important benefits in model debugging and validation, in presenting results to management and in education of personnel associated with the system. As stated by Johnson[25] "... the complexity of the system can make it difficult for users and management to appreciate the interactions between system elements. Numerical output isn't always enough. ... : The best way to describe a system often is with pictures. Moving pictures show the dynamic operation of a real system. Animated results are easy to understand"., A brief review of graphic animation packages has been given by Grant and Weiner.[17]

6.18.1 Sophistication of graphic animation

As with most other aspects of simulation software, several levels of sophistication in the graphic animation facilities are offered by simulation packages. A major criterion is the stage at which the display is provided, which may be either:

1. concurrently with the simulation, or
2. as a post-processor, run after the completion of the simulation run.

If the display is concurrent then a further criterion is whether the system is either:

1. interactive, or
2. non-interactive.

Another criterion is the quality of the graphics. At least four levels of sophistication may be discerned:

1. character graphics,
2. bit-mapped graphics for static background features with character graphics for dynamic features,
3. bit-mapped graphics for both static and dynamic features, or
4. three-dimensional graphics.

This categorisation is somewhat arbitrary, since characters may be produced using bit-mapped graphics, and there is normally alphanumeric text on the simulation display. The resolution of graphic driver also affects the quality of the display. A system handling 1280 by 1024 pixels can obviously produce more sophisticated graphics than one supporting only 320 by 256. The number of colours available varies also, often in association with the screen resolution. 3-D graphics may be provided by systems implemented on CAD or other systems which use a similar quality of terminal. Other features which might be offered are several pages of different screens, windowing or pan and zoom facilities.

As with so many aspects of simulation there is a trade-off between the ease and speed of generating a display and the value obtained from the result. The law of diminishing returns applies. A fairly simple animated display may provide 80% of the information for a relatively small effort. In some situations the quality of the presentation may be as important as the results which are displayed. For example, in certain marketing situations it may be vital to give the highest possible quality of presentation. The effort involved will be considerable and goes well beyond that necessary for evaluation of the system. The effort will not normally be justified unless the techniques involved can be used many times.

Several examples of graphics screens will be presented. It should be remembered that much detail and impact is lost in black and white reproductions of colour screens.

6.19   POST-PROCESSED GRAPHICS

In the case of post-processed animation, details of every event in the simulation are sent to an output file. This event-trace is later read by the post-processor program which identifies the movements of entities associated with each event, relates these to screen locations specified in the input data, and moves the appropriate icon (characters or bit-mapped images) to the appropriate location as each event occurs. These images are usually presented over a static background diagram, which would be drawn to match the layout of the system being modelled.

Even if the package being used does not provide graphic animation, it is fairly easy to write a graphics post-processor. Even if the package only supports printed output, the output can usually be directed to a file which can be saved and then analysed. Figure 6.14 shows a display derived by a postgraduate student for animation of the output of an ECSL progam, using an inexpensive colour character-graphics terminal with a VAX computer.

Post-processed graphics is an attractive entry into animated simulation for packages not originally written with a graphics facility. An example of such a package is MAST. To the original package BEAM (Background and Enhanced Animation for MAST)[1] was added. Figure 6.15 shows a printout of a BEAM display. This package runs on a PC with the base colour display in low resolution graphics, or on a PC/AT with enhanced graphics in high resolution graphics. An example of a MAST model is given in chapter 11.

Although the trend is definitely towards concurrent graphics, post-processed graphics has some important advantages for package and model writers. Firstly, the simulation package and the graphics package are separate modules run in series, so that they can be smaller programs than one combined package. This permits any given hardware to run a larger model than might be possible if the combined package had to fit within a limited computer memory. Secondly, because the graphics is separated from the simulation, it should be possible to define a standard format for the event trace so that a single graphics post-processor can operate with any model. Figure 6.16 gives a structure diagram for a graphics post-processor. With a combined package the model writer has to think about both event logic and the necessary display commands when developing and debugging the model. However, with post-processed graphics, the main advantage

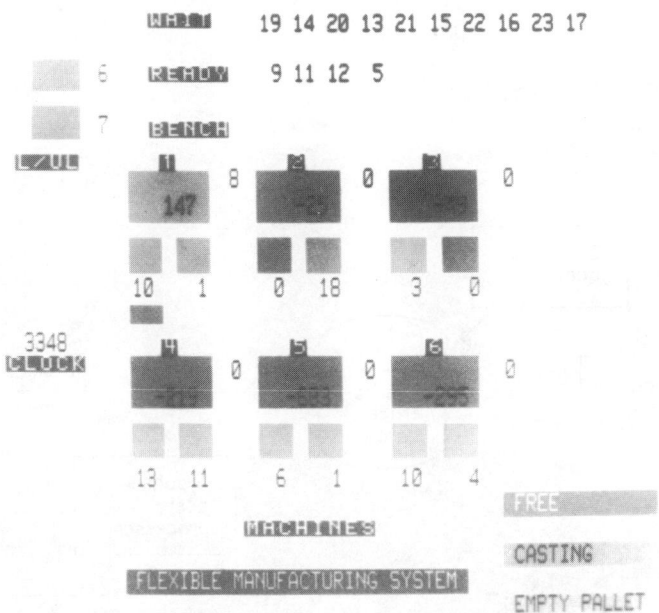

Figure 6.14   Display from a write-it-yourself post-processor graphics program.

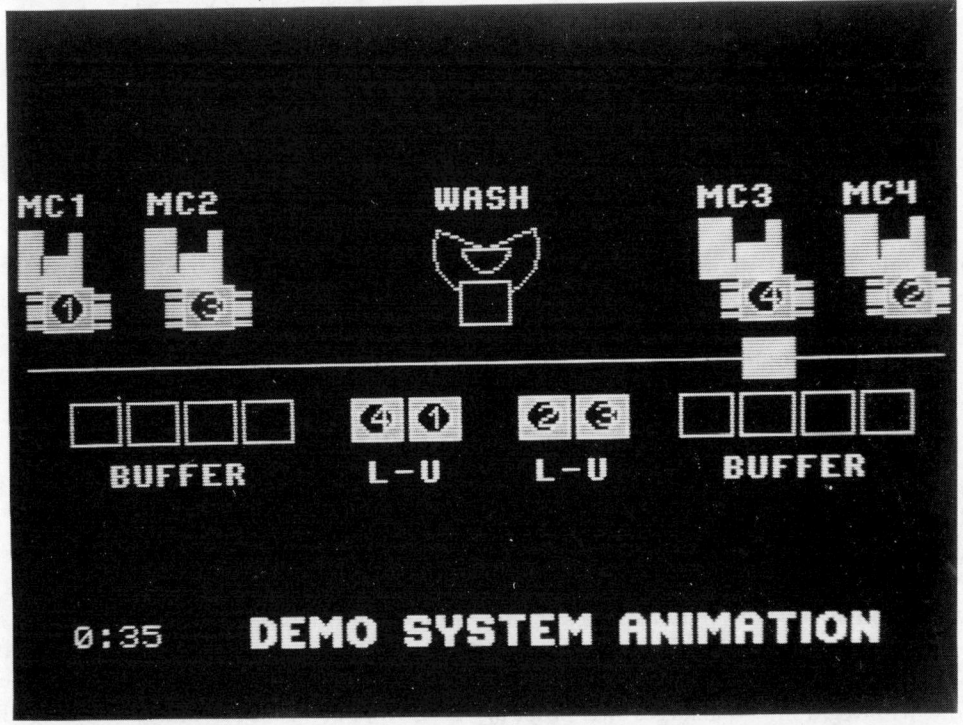

Figure 6.15   Display from BEAM, the Background and Enhanced Animation for MAST package (reproduced by permission of Citroen Industrie UK and CMS Research Inc.).

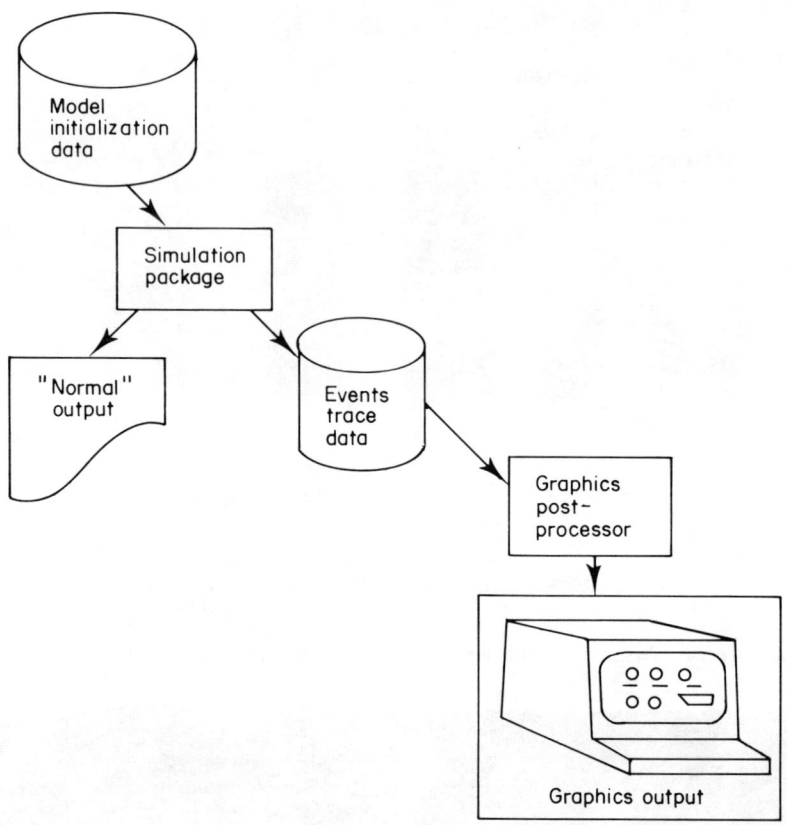

Figure 6.16   Information flow with a post-processor for graphic animation of simulation models.

of graphics is lost, namely that the state of the display and model are synchronised, perhaps permitting more rapid detection of errors in the model. Although post-processed graphics does greatly help in detecting the cause of errors, it cannot have the immediacy of concurrent graphics.

6.20   CONCURRENT GRAPHICS

Most of the packages now available support concurrent graphics. In this method the display is updated as each event occurs and workpieces move from one machine or location to another, or as the state of each resource changes. This technique is obviously more demanding of the software and hardware, creating a larger program which takes longer to run. It may require a faster processor or additional memory. It is obviously a great advantage to be able to see the display as the model progresses, and not have to wait until the end before viewing the display. In the UK at least, the best-known concurrent graphics packages are probably SEE-WHY, HOCUS and CINEMA. By coincidence these are examples of the first three levels of sophistication quoted above, respectively, although developments are continuing in all three cases.

SEE-WHY, and its relations EXPRESS and WITNESS, provides a colour display of characters, lines and block graphics, although since SEE-WHY itself is a FORTRAN package it should in principle be possible to create any effect which the hardware being used and that language can support. In the Define stage of EXPRESS, as each "simulation object" (entity, set and so on) is being defined, the user gives its name, pen (ie colour of foreground and background, height and flash status), x and y co-ordinates where it is to be displayed, description and the number of characters (up to four) of the description which are to be displayed. In the case of sets the user gives the increments in the x and y directions between the points at which successive members of the set are to be displayed. The package includes many routines to vary the parameters during the run of the model. Thus by changing the pen the colour of a machine can be altered; by changing the description the characters displayed on the screen can be varied, perhaps to reflect attribute values. The Beautify phase can be entered to create static background, comprising points, lines, text and block graphics. In the Events phase, routines are provided for manipulating the display as well as the normal logic facilities. For example, the function DXYMOVE will remove an entity from one set and add it to another set, in the process moving the characters associated with it between the locations of the two sets on the screen, first moving in the x and then in the y direction. Figure 6.17 shows an EXPRESS screen display. SEE-WHY provides a similar quality of display. In WITNESS the screen development process is accelerated through the use

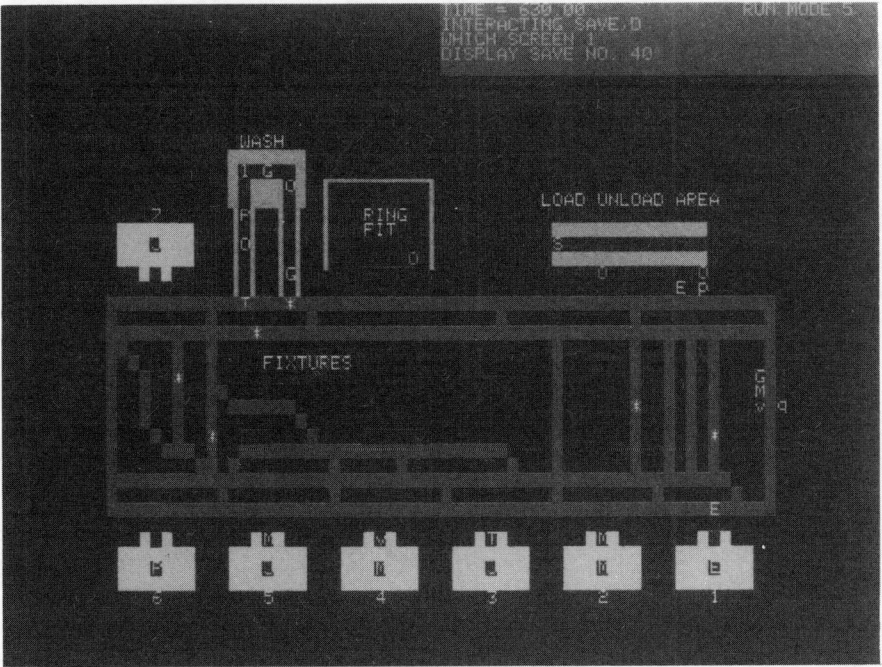

Figure 6.17   EXPRESS screen display (reproduced by permission of Istel Ltd and KTM Ltd, Brighton, England).

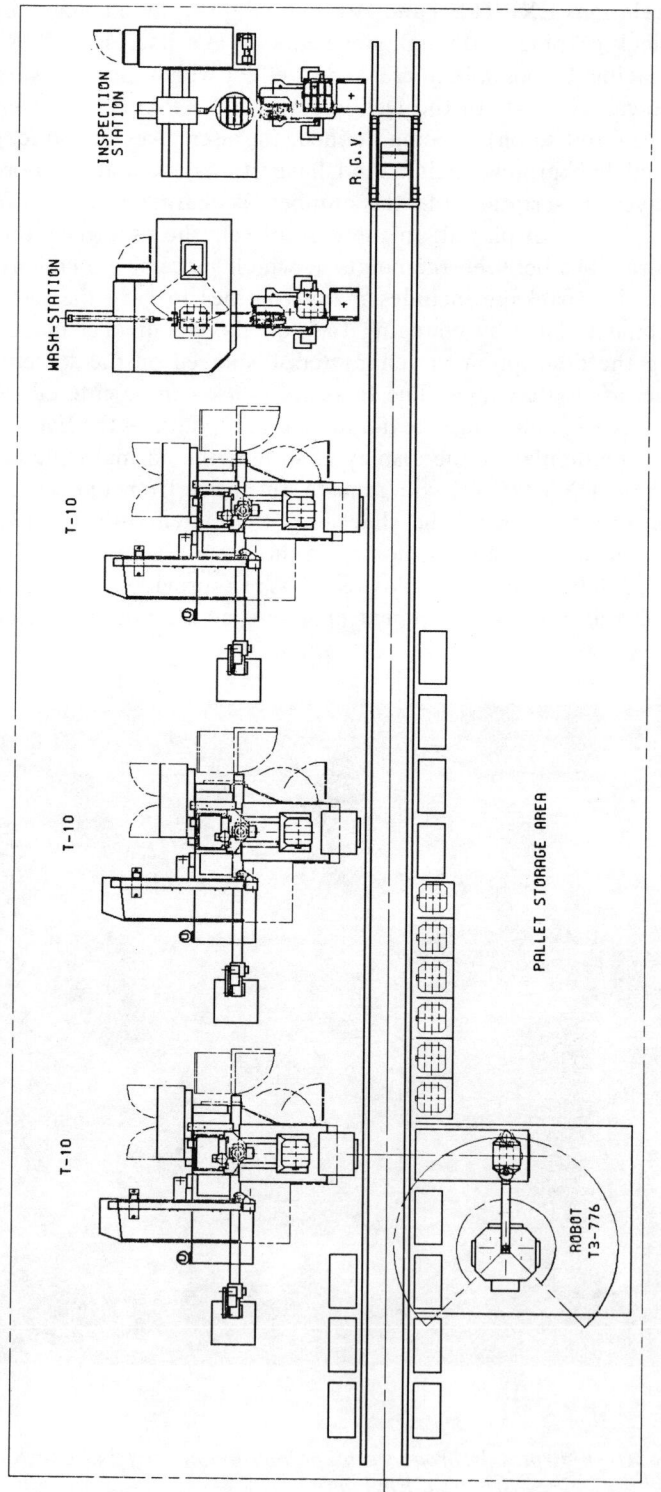

Figure 6.18   Layout of FMS as drawn on CAD system (reproduced by permission of Cincinnati Milacron Ltd and the Institution of Mechanical Engineers).

of a mouse, icons, and more extensive menus. WITNESS illustrates another interesting trend. Since there are now sophisticated packages for developing graphics screens, such as GEM-PAINT, the simulation package could omit the more elaborate facilities for doing so, and instead provide a link to the graphic package. This has the advantage that bit-mapped pixel graphics can be provided more easily.

HOCUS affords a somewhat similar capability of graphics, with some bit-mapped background features. Entities in queues or activities may be displayed either as a block of up to three characters, or as a bit-mapped icon of up to about one character in size. The display can be in either of two formats. In both formats the identifying characters for each entity in a queue or activity can be displayed in a horizontal or vertical line. In one format, a diagram like the layout of the system can be created. For example, Figure 6.18 gives a diagram of a flexible manufacturing system and Figure 6.19 the corresponding HOCUS simulation display. In the other format a diagram similar to an activity cycle diagram is presented. Queues are displayed as a circle bearing the queue's identifying number, with the number of items in the queue and, if requested, the list of entities in the queue. Activities are displayed as rectangles containing the name of the activity, its identifying number, and the names of entities engaged in the activity.

CINEMA, the graphic animation companion to SIMAN, provides a more sophisticated animation capability. Both foreground and background features can be bit-mapped. Entities can have icons associated with them. For resource entities up to four different icons can be defined for each resource. Thus different pictures can be displayed for busy or idle operators, or for working or broken down machines. The icons can be stored in a library and re-used from one model to another. Zooming and panning with up to 16 levels of magnification are available. CINEMA was originally written for a high quality graphics screen, but a version CINEMA/EGA for the IBM PC/AT with enhanced graphics is available.

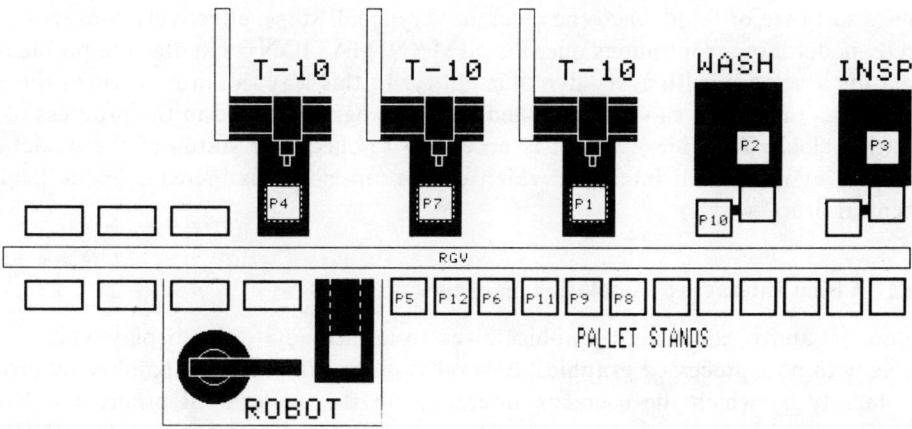

Figure 6.19   HOCUS screen layout for FMS in figure 6.18 (reproduced by permission of Cincinnati Milacron Ltd and the Institution of Mechanical Engineers).

Every package represents a different compromise between cost and capability, ease of use and effect created. The trend is, as has been observed already, to make the model easier to build and easier to understand. This trend will lead more packages to offer high resolution graphics, but the techniques for easily creating and manipulating a complex display have to be developed in parallel. Each user must make his own decision concerning the correct balance between conflicting criteria on his projects.

6.20.1 The problem of display synchronisation

There is a fundamental difficulty in animating a discrete simulation model. We have already described the three-phase method of advancing the simulation clock and ending and starting activities. The simulation clock moves instantaneously from the time of one event to the next, and changes are made to the model only at these discrete event times. This means that the instruction to move an icon from one location to another is issued by the processor at an instant in simulated time, and when the command has been issued the simulation clock jumps onto the next event time. Thus the execution of the instruction to move a pallet from one location to another can be issued at either the event at the start of the move or the event at the end of the move. Generally, there is no control over the time taken by the hardware to execute the actual movement on the screen, it is not synchronised to take a time proportional to the duration of the associated activity. Furthermore, since the computer can only execute one statement at a time it can only process the logic for one event at a time and therefore cannot issue more than one command to the display simultaneously, even if in simulated time two or more movements are going on simultaneously. Suppose for example that a certain operation is the disassembly of an object into two parts, and that after the operation one part moves away from the workstation going in a leftward direction, and the other part moves in a rightward direction. The computer has to issue one command before the other so that on the display the parts would be seen to move one before the other, even although in the model they move simultaneously.

To overcome this problem, it would be necessary to introduce either very sophisticated graphics software or to advance the clock in very small steps, effectively converting the discrete model into a continuous one. PC-SIMANIMATION[25] handles the problem by associating a velocity with every dynamic entity. In this way the rate at which the icon moves across the screen can be controlled and its progress related to the progress of the simulation clock. To achieve this it is necessary to check the status of the model and the display at very small intervals, which causes an enormous increase in the load on the central processor.

6.20.2 Visual interactive simulation packages

As observed above, concurrent graphics gives an immediacy to the display which is not possible with post-processed graphics. Several systems exploit this capability by providing a facility by which the user can interact with the model as it progresses. Visual interactive simulation was pioneered by Hurrion.[24] Both SEE-WHY and HOCUS provide this capability. Not only does this help with diagnosis of faults, but it provides a unique means of tuning a system without having to perform multiple full-length simulation runs. For example, if queues are seen to build up at a certain workstation, the modeller could alter the speed of the machine or add another machine to find out the best balance. Thus the required speed of operation might be found. For example, several models of machine capable of different speeds might be available for purchase. The speed of the machine would be held as an attribute. By interacting with the display and altering the value of the attribute, the speeds of the several machines could be experimented with and the most suitable machine found.

Typical facilities offered by interactive systems are illustrated by the following summary of the categories of interaction in SEE-WHY and EXPRESS:

Run control:
    slow down or speed up the simulation
    step from one event to the next
    switch off the display for fast execution
    stop or restart the execution
Display changes and output:
    reform the display
    display a different screen
    obtain printed copy output
    change the pen used to draw elements
Element inspection and manipulation:
    examine the contents of sets
    de-schedule or re-schedule events
    change the values of attributes
    move an entity from one set to another
Saving and restoring model:
    saving the state of a model
    saving the current display
    restoring a previously stored model
User's own interactions:
    invoking interactive routines written by the user.

Many of these facilities are designed to ease the task of finding the cause of errors. Their power in tuning the system is largely dependent on the skill of the user to conceive of desirable interactions and inserting the appropriate code into the model.

One word of warning is necessary in connection with interactive simulation. For reasons such as those discussed in chapter 3, simulation must be regarded as an inexact science. It is important that the results are statistically valid, especially where sampling is involved. If we run the model and interact with it we may upset an otherwise valid experiment, and render our results worthless. The purpose of any interactions and their effect on the validity of the results must be considered. The importance of validating a simulation model cannot be overlooked.

SUMMARY

In this chapter a review of the main types of package has been given. It has been emphasised several times that it was not intended to give a comprehensive review of any specific package, but to illustrate the kind of facility provided, by reference to specific packages. Inevitably it has not been possible to refer to every package or every facility of any one package.

We have tried to show that the trends in developing simulation packages have been towards:

1. making the package easier to use, by providing, for example, interactive model generators or manufacturing-oriented front ends,
2. making the output of the package easier to understand, for example, by providing graphical animation of the simulation model and results in the form of graphs and charts.

In addition, to the range of facilities provided by a package or group of packages, there is also the important aspect of the level of sophistication of the package, in the sense that some types of package demand more programming effort and understanding of the internal workings of the package than others.

We also observed that in general the "higher" the level of the package, the less easily may it be adapted for specific purposes not envisaged by its creators, and that, as in so many managerial decisions, selecting the most appropriate tool for the job is a matter of compromise among many factors.

REFERENCES

1. *BEAM User Manual*, CMS Research Inc., Oshkosh, Wisconsin, USA.

2. Bevans, J.P., First choose your FMS simulator, *American Machinist*, May, 143–145, 1982.

3. Birtwhistle, G.M., *Discrete Event Modelling on SIMULA*, Macmillan, London, 1979.

4. Buxton, J.N. and Laski, J.G., Control and Simulation Language, *The Computer Journal*, 5, 1962.

5. Cellier, F. and Blitz, A.E., GASP-V: A Universal Simulation Package, Proc. IFAC Conference, 1976.

6. *CINEMA User's Manual*, Systems Modeling Corporation, State College, Pennsylvania, USA.

7. Clementson, A.T., *ECSL User's Manual*, Cle-Com Ltd, Birmingham, 1985.

8. Conway, R., Maxwell, W.L., McClain, J.O. and Worona, S.L., *User's Guide to XCELL+ Factory Modelling System*, Cornell University, 1987.

9. Dangelmaier, W. and Bachers, R., SIMULAP—a simulation system for material flow and warehouse design, *Proc. SIM-1, 1st Int Conference on Simulation in Manufacturing*, Stratford-upon-Avon, IFS (Publications) Ltd., Bedford, England, 1985.

10. *EXPRESS User's Manual*, Istel Ltd, Redditch, England, 1985.

11. Fishman, G.S., Digital computer simulation: The allocation of computer time in comparing simulation experiments, *Op. Res.*, 16, 280–295, 1968.

12. Fishman, G.S., Grouping observations in digital simulation *Mgmt.Sc.*, 24, 510–521, 1978.

13. Fishman, G.S. and Kiviat, P.J., The analysis of simulation generated time series, *Mgmt.Sc.*, 13, 7, 1967.

14. Franta, W.R., *The Process View of Simulation*, North-Holland, Amsterdam, 1977.

15. *GENETIK User Manual* and *GENETIK Simulation Extension User Manual*, Insight International Ltd, Woodstock, Oxfordshire, England, 1987.

16. Gordon, G., *Application of GPSS V to Discrete systems Simulation*, Prentice-Hall, Englewood Cliffs, New Jersey, 1975.

17. Grant, J.W. and Weiner, S.A., Factors to consider in choosing a Graphically Animated Simulation System, *Industrial Engineering*, August 1986, 36.

18. Greenberg, S., *GPSS Primer*, John Wiley, New York, 1972.

19. Henriksen, J.O. and Crain, R.C., *GPSS/H User's Manual*, 2nd edition, Wolverine Software Corporation, Annandale, Virginia, 1983.

20. Hills, P.R., SIMON—A simulation Language in ALGOL, In Hollingdale, S.M., editor, *Simulation in Operational Research*, English Universities Press, London, 1965.

21. Hills, P.R., *An Introduction to Simulation using SIMULA*, Publication No. S55, Norwegian Computing Center, Oslo, 1973.

22. *HOCUS User Manual*, P-E Information Systems, Egham, England.

23. Hollocks, B.W., Simulation and the micro, *Journal of the Operational Research Society*, 34, 4, 331–343, 1983

24. Hurrion, R. D., An investigation of visual interactive simulation methods using the job shop scheduling problem, *Journal of the Operational Research Society*, 29, 1085–1093, 1978.

25. Johnson, G.D., *PC SIMANIMATION User's Guide and Casebook*, CACI Inc., Los Angeles, 1986.

26. Kiviat, P.J., Villaneuva, R. and Markowitz, H., *The SIMSCRIPT II Programming Language*, Prentice-Hall, Englewood Cliffs, New Jersey, 1975.

27. Law, A.M. and Larmey, C.S., *An Introduction to Simulation using SIMSCRIPT II.5*, CACI Inc., Los Angeles, 1984.

28. Lenz, J.E., MAST: A simulation as advanced as the FMS it studies, *Proc. SIM-1, 1st Int. Conference on Simulation in Manufacturing*, Stratford-upon Avon, IFS (Publications) Ltd., Bedford, England, 1985.

29. *MAST User Manual*, CMS Research Inc., Oshkosh, Wisconsin, USA.

30. Matthewson, S.C., *DRAFT II/SIMON Manual*, Department of Management Science, Imperial College, London, 1982.

31. Matthewson, S.C., *A Programming Manual for SIMON simulation in FORTRAN*, Imperial College, London, 1977.

32. Matthewson, S.C., Simulation Program Generators, *Simulation*, 23, 6, 81–189, 1975.

33. Matthewson, S.C., Simulation program generators: code and animation on a PC, *Journal of the Operational Research Society*, 36, 7, 583–589, 1985.

34. Mills, R.I., Simulation for manufacturing systems—a critical review, *Proc FMS-5, 5th Int. Conf. on Flexible Manufacturing Systems*, IFS (Publications) Ltd, Bedford, England, 1986.

35. *Modelmaster, Graphically Enhanced Factory Modeling System*, General Electric Automation Europe, Daventry, England.

36. *OPTIK User's Manual*, Insight International Ltd, Woodstock, Oxfordshire, England.

37. *PCModel*, Simulation Software Systems, San Jose, California.

38. *PCModel/GAF*, Simulation Software Systems, San Jose, California.

39. *PC SIMSCRIPT II.5, Introduction and User's Manual*, Release 2, CACI Inc., Los Angeles, 1986.

40. Pegden, C. D., *Introduction to SIMAN, version 3.0*, Systems Modelling Corporation, State College, Pennsylvania, 1985.

41. Pidd, M., *Computer Simulation in Management Science*, John Wiley, New York, 1984.

42. Poole, T.G. and Szymankiewicz, J., *Using Simulation to Solve Problems*, McGraw-Hill, London, 1977.

43. Pritsker, A.A.B., *The GASP-IV Simulation Language*, John Wiley, New York, 1974.

44. Pritsker, A.A.B., *Introduction to Simulation and SLAM II*, Halsted Press and Systems Publishing Corporation, 1984.

45. Pritsker, A.A.B., *Modeling and Analysis using Q-GERT Networks*, Halsted Press and Pritsker & Associates, 1977.

46. Pritsker, A.A.B. and Kiviat P.J., *Simulation with GASP-II*, Prentice-Hall, Englewood Cliffs, New Jersey, 1969.

47. Pritsker, A.A.B. and Pegden, C.D., *Introduction to Simulation and SLAM*, John Wiley, New York, 1979.

48. Rolston, L.J., Modeling flexible manufacturing systems with MAP/1, *Annals of Operations Research*, 3, 189–204, 1985.

49. Schmeiser, G.S., Batch size effects in the analysis of simulation output, Research Memorandum 80-05, School of Industrial Engineering, Purdue University, W. Lafayette, Indiana, 1980.

50. Schriber, T., *Simulation using GPSS*, John Wiley, New York, 1974.

51. Schruben, L., Confidence intervals estimation using standardised time series, *Op. Res.*, 31, 6, 1090–1108, 1983.

52. *SEE-WHY System Description*, Istel Ltd, Redditch, England.

53. Shanehchi, J., EXPRESS: A man–machine interface for simulation, *Proc. SIM-1, 1st Int.*

Conference on Simulation in Manufacturing, Stratford-upon-Avon, IFS (Publications) Ltd, Bedford, 1985.

54. Shannon, R.E., *Systems Simulation: The Art and The Science*, Prentice-Hall, Englewood Cliffs, New Jersey, 1975.

55. *SIMFACTORY*, CACI Inc., Los Angeles, 1986.

56. *SIMSCRIPT II.5 Reference Handbook*, 2nd edition, CACI Inc., Los Angeles, 1983.

57. Solberg, J.J., *CAN-Q User's Manual*, Report No. 9, The Optimal Planning of Computerised Manufacturing Systems, Purdue University, 1980.

58. *SPAR User Manual*, CMS Research Inc., Oshkosh, Wisconsin, USA.

59. Standridge, C.R., Performing simulation projects with The Extended Simulation System (TESS), *Simulation*, 45, 6, 283–291, 1985.

60. Standridge, C.R., Animating simulations using TESS, *Comput. & Indus. Engng.*, 10, 2, 121–134, 1986.

61. Tocher, K.D. and Owen, D.G., The automatic programming of simulation, Proc. 2nd Conference of the International Federation of the OR Societies, Aix-en-Provence, 50–68, 1960.

62. *WITNESS User's Manual*, Istel Ltd, Redditch, England, 1986.

CHAPTER 7

Flexible manufacturing systems

7.1   THE FLEXIBLE MANUFACTURING PHILOSOPHY

In chapter 1 we gave a definition of a flexible manufacturing system, as follows:

Two or more computer numerically controlled units interconnected with automated workhandling equipment and supervised by an executive computer having random scheduling capabilities.[16]

Many consider this definition too mechanical and restrictive, rather like defining chess as a game played on a board with chessmen which can move in specified patterns. This omits the fundamentals of the game, its strategy and tactics and all those aspects which make the game one of the most compulsive ever devised by man. Similarly, the FMS definition focusses too much on the physical constituents of an FMS, without considering the strategic reasons why we should adopt flexible manufacturing. Many experts now believe that any system which seeks to achieve certain objectives by a variety of means is an FMS, and that instead we should concentrate on these objectives. This is summed up in the "Flexible Manufacturing Philosophy", which has been defined by the Institution of Production Engineers[2] as:

The Flexible Manufacturing Philosophy: The proven response of a *total* facility which can serve a *volatile* market with *minimum response time* from order input to saleable product using the *minimum of working capital*.

This definition can be applied to the complete manufacturing organisation of a company or to each of its departments. In applying this concept to a particular department within an organisation, the various keywords in the definition have to be expressed in terms of the function of the department and the "market" which it supplies. For example, the Flexible Manufacturing Philosophy for a departmental facility, such as a machine shop, might be:

$$\text{The provision of a} \left\{ \begin{array}{c} \text{Departmental Facility} \\ \text{Machine Shop} \end{array} \right\} \text{which can serve}$$

$$\left\{ \begin{array}{c} \text{Customer} \\ \text{Assembly Shop} \end{array} \right\} \text{within} \left\{ \begin{array}{c} \text{Response Time} \\ \text{24 Hours} \end{array} \right\} \text{from receipt of}$$

$$\left\{ \begin{array}{c} \text{Raw Material} \\ \text{Fabrications} \\ \text{Casting} \\ \text{Forging} \\ \text{Bar Stock} \end{array} \right\} \text{ to } \left\{ \begin{array}{c} \text{Saleable Product} \\ \text{Finished Component} \end{array} \right\} \text{ using the minimum}$$

of working capital.[2]

There could be many ways of organising the machine shop to achieve the objectives. Indeed, the diversity of systems using the term Flexible Manufacturing System is so great that it is impossible to classify them on a scheme based purely on their physical components. See, for example, the surveys of FMS published by Dupont-Gatelmand[15] or Cresswell and Edghill.[13] To get over this problem Browne et al.[6] in their paper on a classification of flexible manufacturing systems examined the various forms of flexibility required of manufacturing systems. They gave examples of the type of physical system providing the various kinds of flexibility.

In approaching the subject of simulation modelling of flexible manufacturing systems we have to bear these considerations in mind. Just as important as what the system is composed of, is how it is used to process the parts concerned. Consequently, in this chapter a brief review of typical elements and forms of FMS will be given together with a discussion of the problems and methods of operating them. It is addressed mainly to those who are not manufacturing engineers. Manufacturing engineers will be well aware of these matters already. The examples quoted will be mainly drawn from the field of metal cutting systems, since these systems are most numerous and highly developed; however, much of what will be said is generally applicable.

7.2 ELEMENTS OF FLEXIBLE MANUFACTURING SYSTEMS

This section reviews the main types of facility which may be included in a flexible manufacturing system. The next section will give examples of how they are integrated to form systems, and the following section will discuss the operation of the systems.

7.2.1 Workstations

The workstations vary according to the type of part being produced. In metal cutting systems the machines are usually computer numerically controlled (CNC) horizontal spindle machining centres, if prismatic workpieces are to be produced, or turning centres if rotational workpieces. Some systems consist of both types of machines, when workpieces involving both types of operation are required. Other systems include single-purpose machines, as opposed to machining centres which are designed to perform a range of processes. In addition to metal working machines there may also be washing machines and gauging machines or other types of inspection machines.

There are systems for sheet-metal operations, printed circuit board manufacture and assembly operations. Assembly operations and fabrication operations may be performed within a single system.

7.2.2   Load and unload stations

Parts have to be introduced into the system at some point and there are usually load–unload stations, where parts are placed on pallets, usually by human operators. In some cases parts may be supplied by an orienting device and loaded by robot. Unloading is usually done at the same stations, but there may be separate unload stations. In some systems load–unload stations are dedicated to one (or more) type of workpiece, while in others they may be used by any type of workpiece.

7.2.3   Workpiece transport equipment

Workpieces must be transported from the load positions to the productive equipment, and back for unloading. Three types of equipment are in common use, namely conveyors, vehicles and robots. In conveyor systems there are usually spurs or loop lines serving each workstation, and some form of addressing system is necessary to direct workpieces to the correct station. Conveyor systems seem to be less popular than formerly, probably due to advances in the other methods. There are several types of vehicle. Rail-cars, which run on rails connecting the workstations, usually follow linear tracks and are therefore to be found in the smaller type of FMS. Automated guided vehicles (AGVs) follow inductive signals from electric wires buried in the floor, and can have quite complex track layouts. Some vehicles can carry only one load at a time, while others have two load positions. Some, more specialised, vehicles can handle several loads simultaneously. There are several types of mobile floor-mounted robots, which can be used for workpiece transport. However, overhead gantry-mounted robots are popular for both workpiece and tool handling.

7.2.4   Pallets

Workpieces are normally held in pallets of some sort for transport and locating on machine tables. Two types are common. One type of pallet serves just as a carrier for a batch of small parts, to facilitate and reduce the frequency of movements. They serve as trays into which the parts may be placed at the load stations, and from which they may be lifted and placed into the machines, perhaps by a robot. This type is common in systems which use conveyors and gantry robots, but are also used in AGV systems. The other type of pallet is one on which one or more parts are accurately located and which is itself moved on to the machine table and held in position while machining operations are performed on the parts.

7.2.5   Fixtures

Fixtures are used to locate parts precisely on pallets. They are usually specific to one type of part, so that each part requires a different fixture. In some cases however, several types of part may be sufficiently similar to make use of the same fixture. The fixtures may be permanently bolted on the pallets, or they may be removed from the pallet when a part requiring a different fixture is to be introduced into the system and placed on the pallet. If the parts are small, several parts, either all of the same type or of different types, may be held on a pallet.

7.2.6 Tools

Most operations require some form of tooling specific to the particular operation being performed, typically cutting tools in machining centres. Machining centres have tool magazines in which a set of tools can be held so that any operation on a range of work-pieces can be performed. Tools have to be changed, either because they have exhausted their useful life, or because the part to be worked requires tools which are not currently in the tool magazine.

7.2.7 Tool transport

Tools must obviously be supplied to the machine. This may be done manually, but more recent systems include a tool transport system as well as a workpiece transport system. Robots are frequently used as part of the tool supply operation.

7.2.8 Robots

As observed above, robots are frequently used for both workpiece and tool handling. Robots may be either stationary or mobile. There are several types of mobile robot, principally mounted either on rails on the floor or on overhead gantries. In some systems robots are mounted on AGVs which carry tool pallets to the machines, the robot handling the tools into and out of the tool magazine. Robots may also be used to perform the processes themselves, such as assembly, welding, deburring or light machining operations. The grippers on the robot may have to be changed between handling one type of part and another. Obviously, this reduces the flexibility of the robot and the whole system, imposing batch operation.

7.2.9 Buffer storage at workstations

Most systems include some form of buffer storage at workstations. This permits work-pieces to be brought to a machine while it is operating on the previous part, and the machine to work on the next part while the previous one is waiting to be transported away. There are several types of buffer storage. Often it takes the form of two pallet stands in front of the machine. One provides a queueing position for work waiting to go on to the machine, and the other a queueing position for work waiting to be taken away from it. In another type the machine has a rotary pallet shuttle, with one queueing position and one working position. Some systems provide an automatic work changer, or chain of pallets, giving six or eight positions for queueing.

7.2.10 Other storage facilities

In addition to buffer storage at workstations, many systems have storage facilities for work-in-progress, pallets, fixtures or raw material. There are many different arrangements. Work-in-progress buffer storage points may be distributed throughout the system to smooth out any surges in work flow. In most systems there are storage positions for the pallets. There may be a "home" position for every pallet, or there may just be sufficient to cater for those pallets not currently at a workstation or load–unload station. In some systems the load–unload stations also serve as the home location. Some systems have a central storage facility, which might be used to hold a build-up of work prior to an unmanned shift. Others include an automatic storage and retrieval system (AS/RS) for raw material, which may serve the factory as a whole and not just the FMS.

7.2.11   Human operators

Despite the advances towards unmanned factories, most, if not all, FMSs require human operators for a variety of tasks, although many systems operate unmanned for one shift, or part of a shift, per day. Operators may be required for loading and unloading parts, tool preparation, clearing faults, such as broken tools or faulty location of parts, and system monitoring to cope with other exception conditions, and for generally managing the system.

7.2.12   Control computer

The tasks of the "executive" computer are many and varied. The basic functions are to give instructions to the machines and transport devices so that the parts are moved and processed as required. Monitoring the status of the system so that this can be done is a vital function. Most executive computers also give instructions to the operators at the tool preparation areas and load–unload stations, and make scheduling decisions. In order to schedule the work, the computer must be supplied with production requirements and material availability data. This may be achieved through a link to the main factory production control system computer. The computer will also provide management reports on the work done, system productivity, diagnostics and so on. Programmable logic controllers may be used to control small systems or sections of systems. There will normally be a separate controller for the AGVs in systems with several vehicles.

Perhaps surprisingly, the control computer seldom features in a simulation model of an FMS. Instead the decision rules which it employs are built into the logic of the model. Only if the capacity of the computer to handle the input and output needed to control the system is in doubt would it be part of the model. In that case the model would probably be a special one to model the communications network.

7.3   SOME EXAMPLES OF FLEXIBLE MANUFACTURING SYSTEMS

These and other elements can be combined in a variety of ways to form flexible manufacturing systems. In this section some examples of systems will be presented, concentrating on their physical structure. Operational planning aspects of the systems will be discussed in the following section. A large number of systems could be described, but it is not intended to give a comprehensive review of FMSs. Instead, only a sufficient number will be mentioned to illustrate the types of systems which exist, and some of the planning problems involved in them.

As previously observed, it is difficult to present a classification system for FMSs, due to their enormous variety. However, in order to give this discussion some structure, systems of three types will be discussed:

1. Systems with linear tracks.
2. Systems with AGV networks.
3. Systems comprising a number of distinct cells.

7.3.1 Systems with linear tracks

Systems with linear tracks usually have only one vehicle. The vehicle is frequently a rail-guided vehicle, but an AGV could be used. One of the simplest systems, so far as physical layout is concerned, is the FMS at Anderson Strathclyde plc,[12] in Motherwell, near Glasgow in Scotland. A diagram of the system is given in Figure 7.1. There are six machines, each with a tool magazine of 100 tools capacity. One of the machines has a facing head, required for certain operations, the other five are identical. Each machine has two pallet stands providing on-queue and off-queue buffers. There are also two pallet stands at the load–unload area. At the load–unload area there some work stands where the part is put into the fixture before it is put on the pallet at one of the pallet stands. There are 13 pallets, all identical. Fixtures are generally specific to one part. There is one vehicle, an AGV, which can handle one pallet at a time, and follows a track which has no branches. Although the system is rather simple in layout, the workpieces are complex and many planning problems arise. These are discussed in the next section.

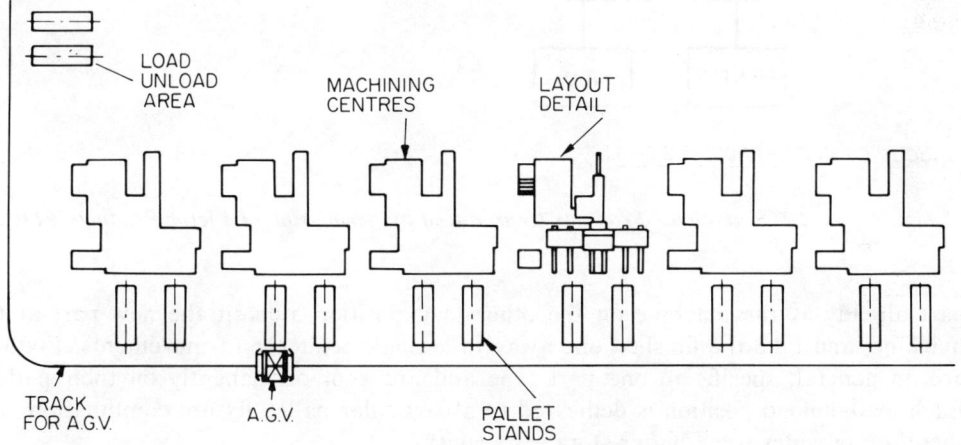

Figure 7.1   Anderson Strathclyde FMS.

A potential problem in the Anderson system is that apart from the pallet stands at the machines and load–unload area, there are no places where pallets with part-machined components may be parked between operations. Many systems include several pallet storage points, such as the small FMS at Victor Products,[5,23] Figure 7.2. It consists of two machining centres with rotary pallet shuttle, a rail-car, two load–unload positions and several pallet storage locations.

Figure 7.3 shows a system which performs vertical turning operations in one of the Caterpillar plants.[21] It consists of three vertical turning centres, a vehicle and 15 load–unload stations. Due to the geometry of vertical turning machines there are no buffer positions at the machines. However, the vehicle has two load positions. This allows it to bring a part occupying one load-carrying position to a machine, pick up the finished

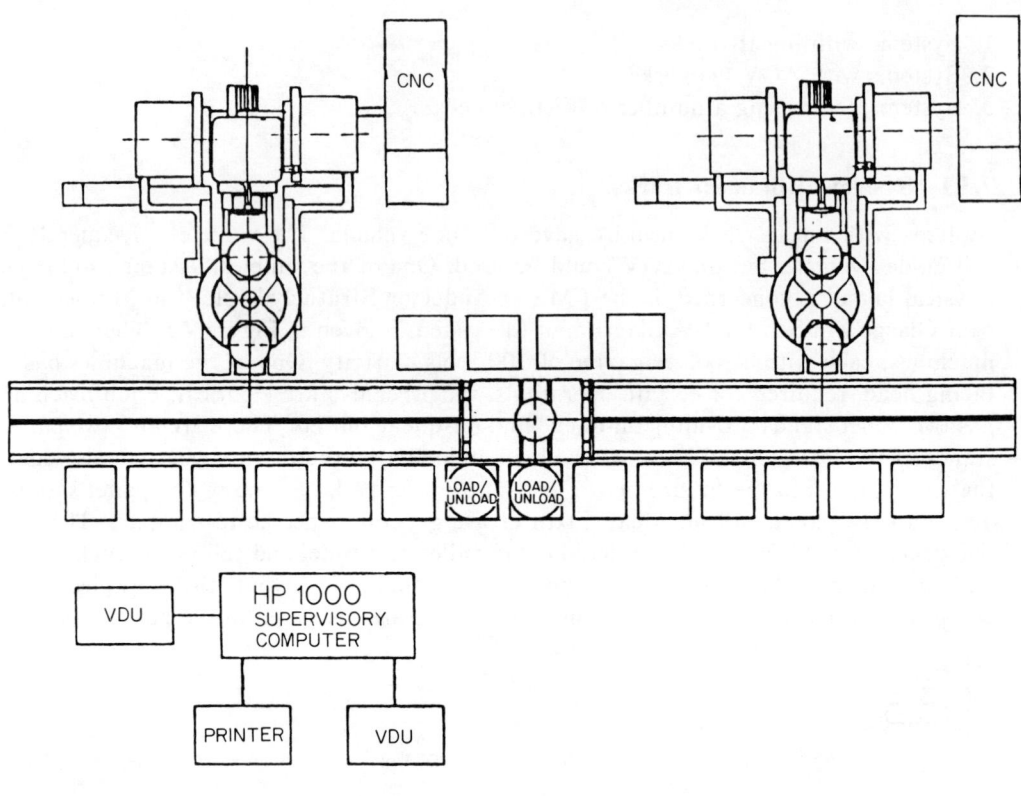

LAYOUT OF VICTOR PRODUCTS FMS

Figure 7.2   FMS at Victor Products (reproduced by permission of Victor Products PLC).

part already at the machine on the other load position, deposit the new part at the machine, and take the finished one away in a single sequence of movements. Fixtures are, in general, specific to one part type and are kept permanently on their pallets. Each load–unload position is dedicated to a particular pallet–fixture combination, and therefore provides some buffer storage capacity.

Cummins Engines Company at Shotts have a system for the manufacture of water pump housings, manifolds and thermostat housings, Figure 7.4. There are three Makino machining centres, with input and output buffers, two load–unload positions, a rail-car and a central storage facility. The storage rack is an interesting feature. It consists of 48 locations on two levels, each dedicated to a particular pallet. The vehicle can raise or lower the pallet to suit the height of the storage rack, machine or load–unload area. The storage rack provides some scope for unmanned operation. Each pallet is dedicated to one part type.

In all of these systems, tools are supplied manually to the machines. Mirrlees Black-stone Ltd in Stockport have an FMS which includes a robot for tool supply, Figure 7.5.[1] A mobile robot runs on a track behind the machines and can place or remove single tools as commanded by the system control. Tools are held in storage racks on the other side of the robot track. Tools enter the system from a rack at the tool preparation

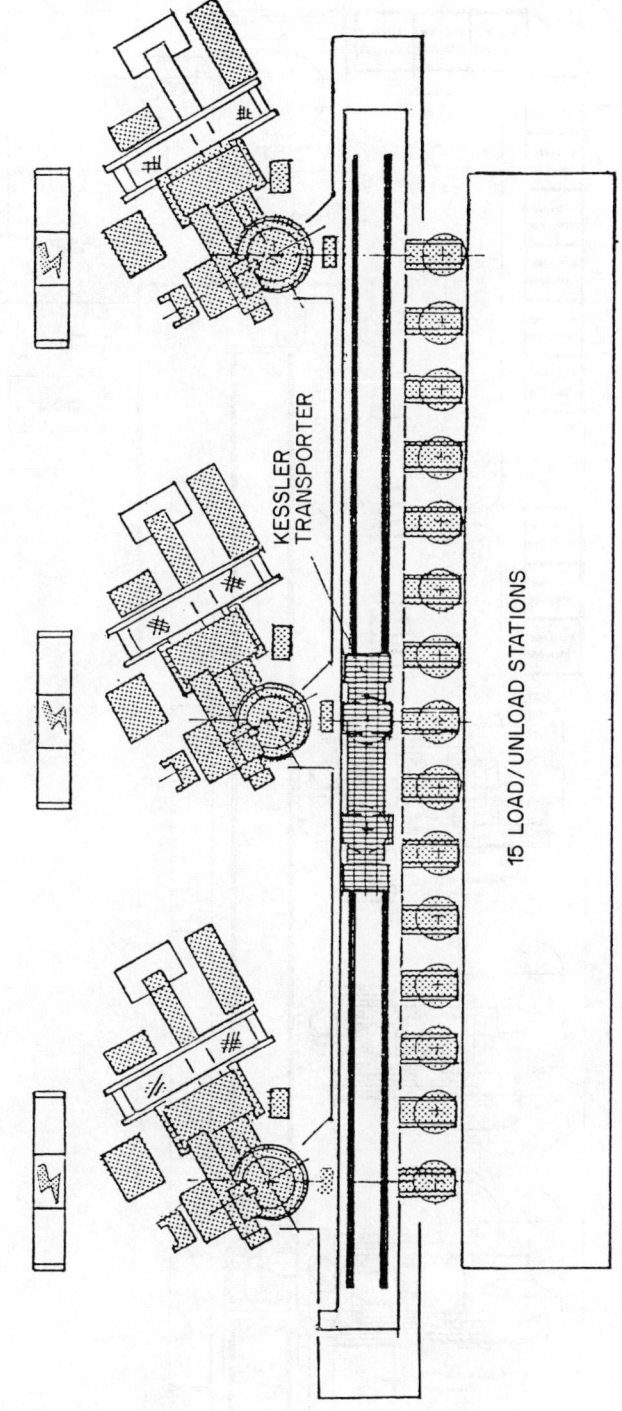

(3) SCHIESS STD. 12DS VERTICAL LATHES

KESSLER TRANSPORTER

15 LOAD/UNLOAD STATIONS

Figure 7.3 Caterpillar vertical turning FMS (reproduced by permission of Caterpillar, Inc.).

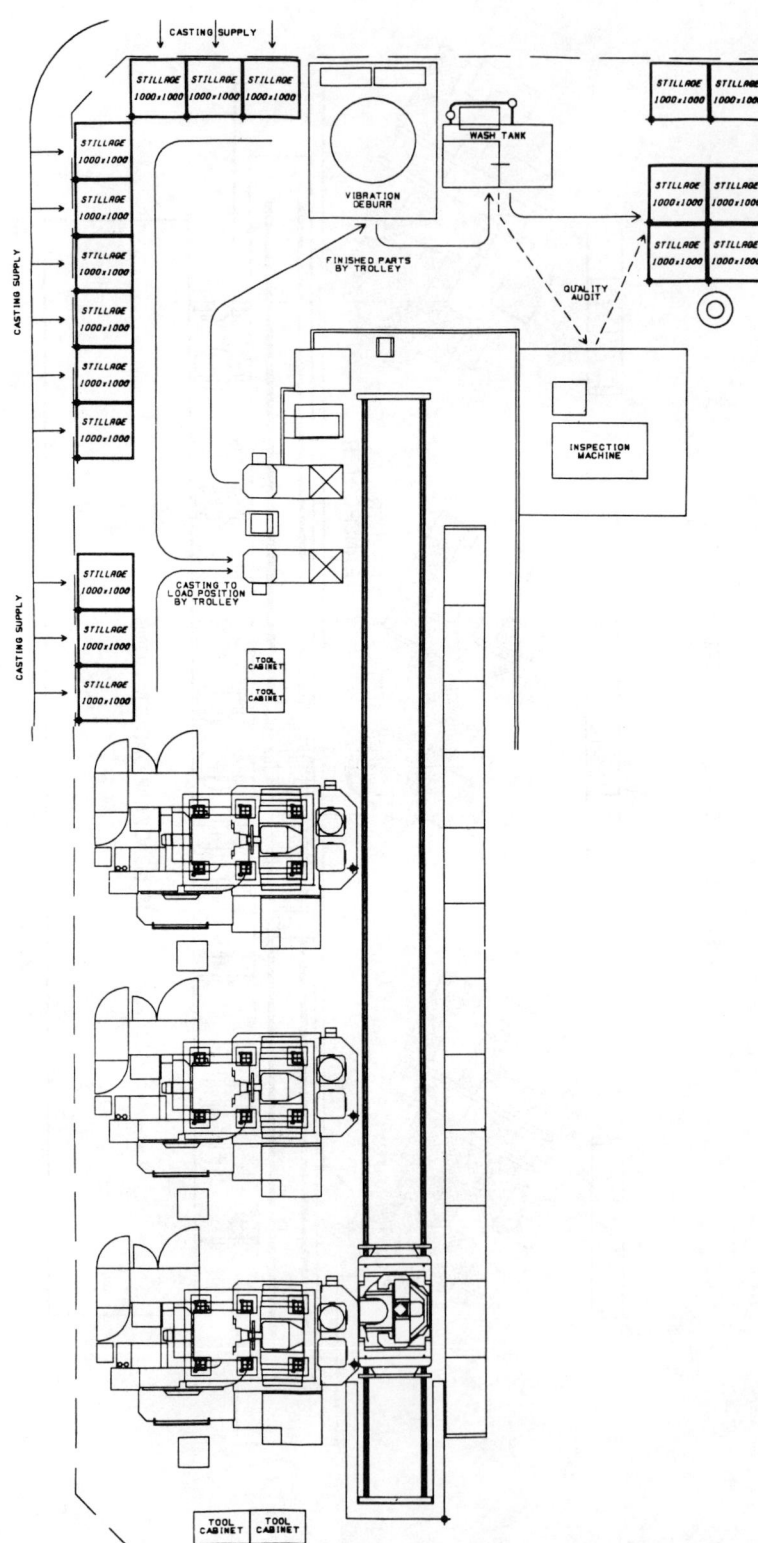

Figure 7.4 Cummins Engines Makino FMS (reproduced by permission of Cummins Engines Ltd).

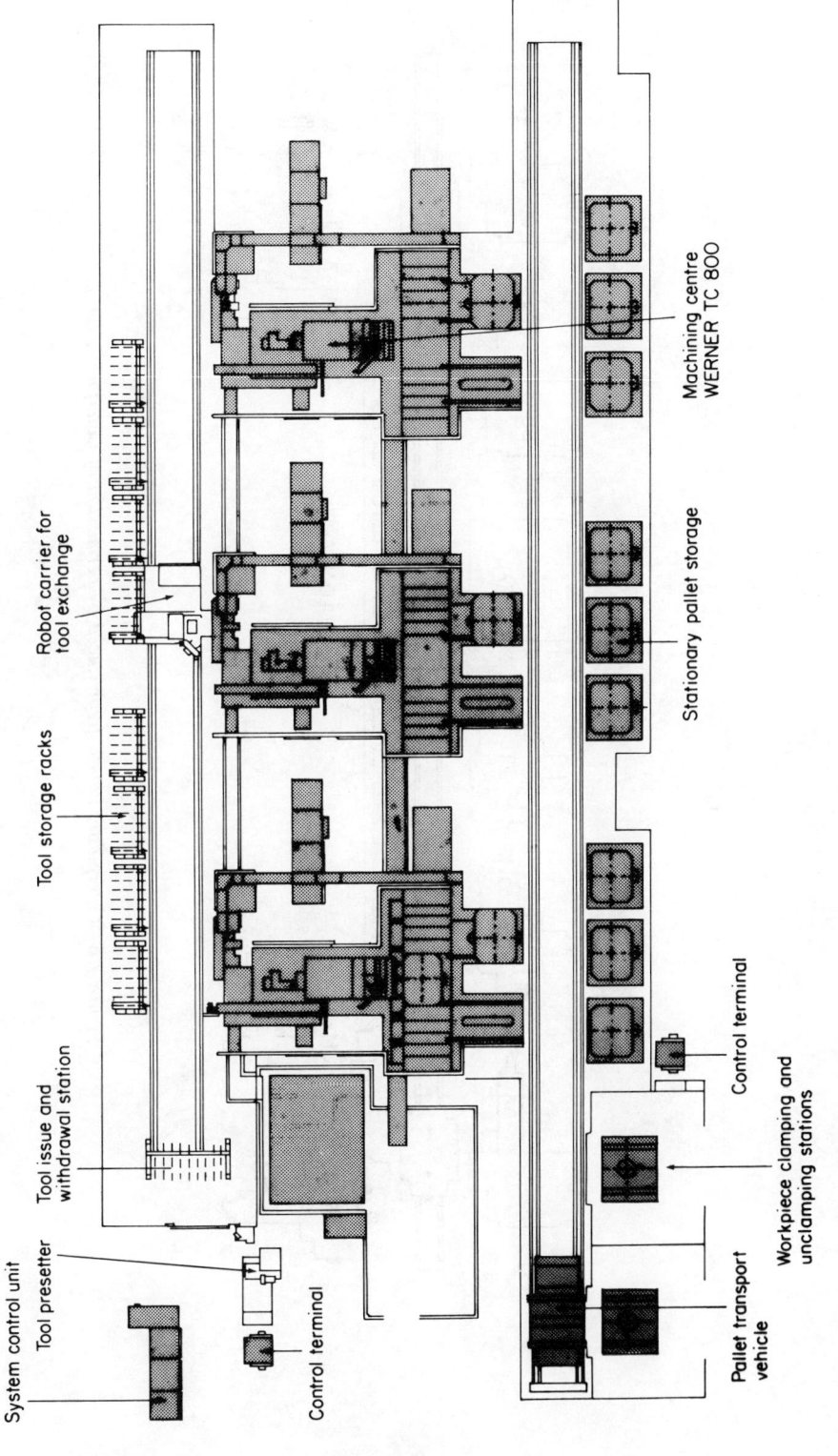

System control unit

Tool presetter

Control terminal

Tool issue and withdrawal station

Tool storage racks

Robot carrier for tool exchange

Machining centre WERNER TC 800

Stationary pallet storage

Control terminal

Workpiece clamping and unclamping stations

Pallet transport vehicle

Control terminal

Flexible manufacturing system FMS 800-3

Figure 7.5 Mirrlees Blackstone FMS (reproduced by permission of Werner and Kolb GmbH).

162

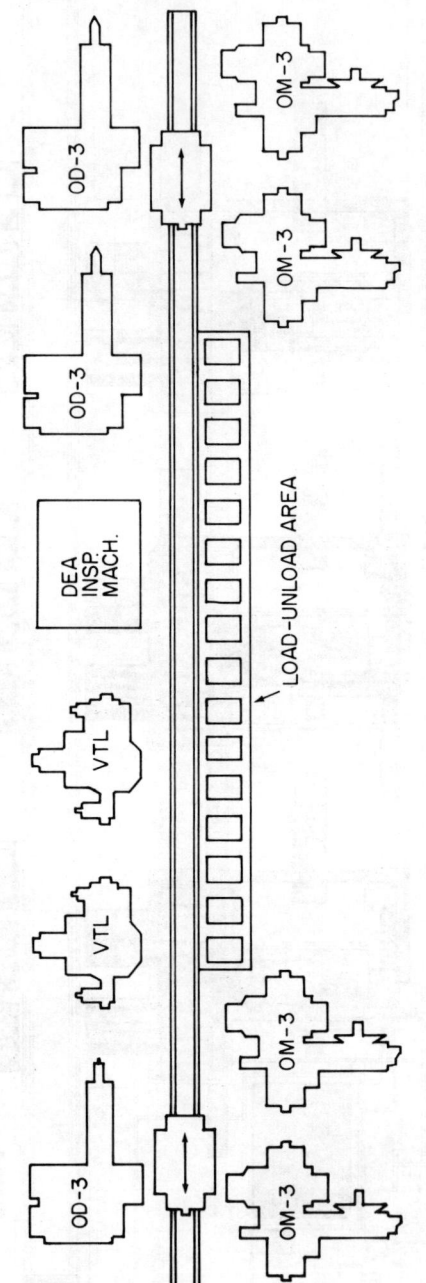

Figure 7.6 Caterpillar, Peoria (reproduced by permission of Caterpillar, Inc.).

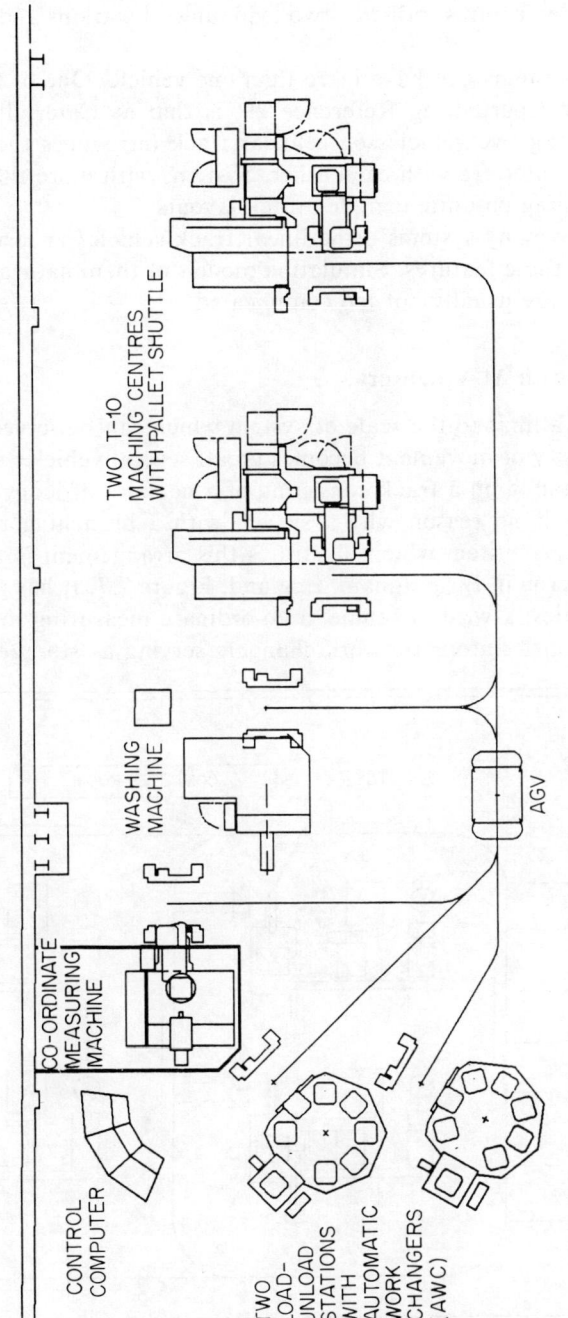

Figure 7.7 Cincinnati Milacron, Birmingham, FMS (reproduced by permission of Cincinnati Milacron Ltd).

area, where they are placed by the tool setter-operator. The robot collects them from that rack and places them in the main rack until required by a machine. When tools need resetting the robot brings them to the rack at the preparation area. The system also has machining centres, rail-car, two load–unload stations and several pallet storage locations.

Linear track systems can have more than one vehicle. One of the first FMSs, which has been widely reported, eg Reference 29, is that at Caterpillar in Peoria, Illinois, Figure 7.6. Having two vehicles on a linear track introduces the problem of ensuring that they do not interfere with each other. Systems with more than one vehicle usually have AGVs running on quite complex track layouts.

There are now many systems using linear track vehicles in service, involving various combinations of these features. Simulation models of them naturally tend to have much in common, and are usually not too complicated.

7.3.2 Systems with AGV networks

There is clearly a limit to the scale of system which can be served by only one vehicle. When the intensity of movement becomes great, several vehicles are needed. AGVs are usually used, running on a track consisting of a network of loops and branches.

Strictly, there is no reason why a system with a branching track should not have only one vehicle. A system which illustrates this arrangement is the small FMS cell at Cincinnati Milacron in Birmingham, England, Figure 7.7. It has two machining centres with pallet shuttles, a wash machine, a co-ordinate measuring machine and two load–unload stations with automatic work changers serving as storage locations for pallets.

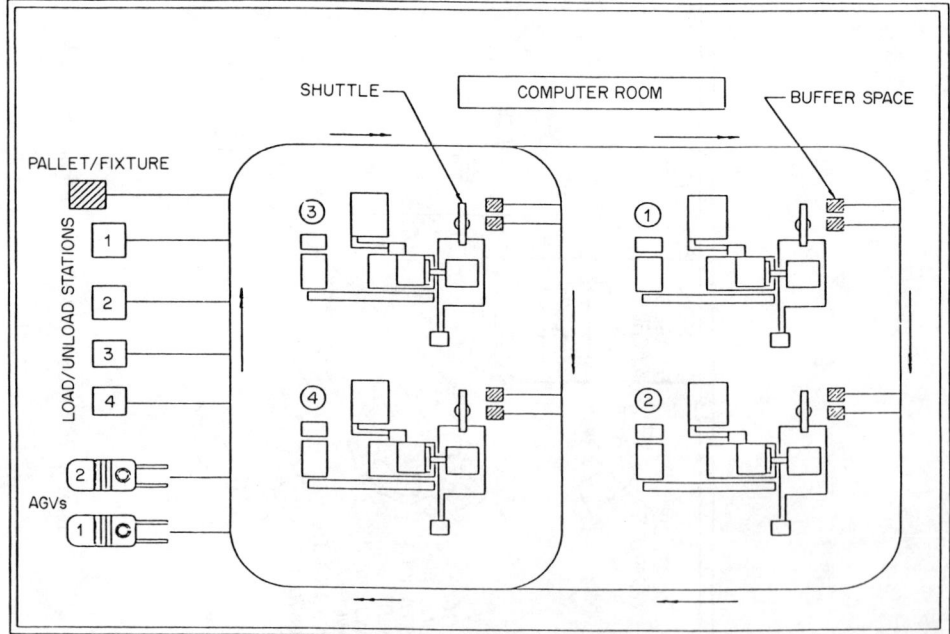

Figure 7.8   Dowty Mining FMS (reproduced from Cresswell and Kempa[14] by permission of the authors and the Institution of Production Engineers).

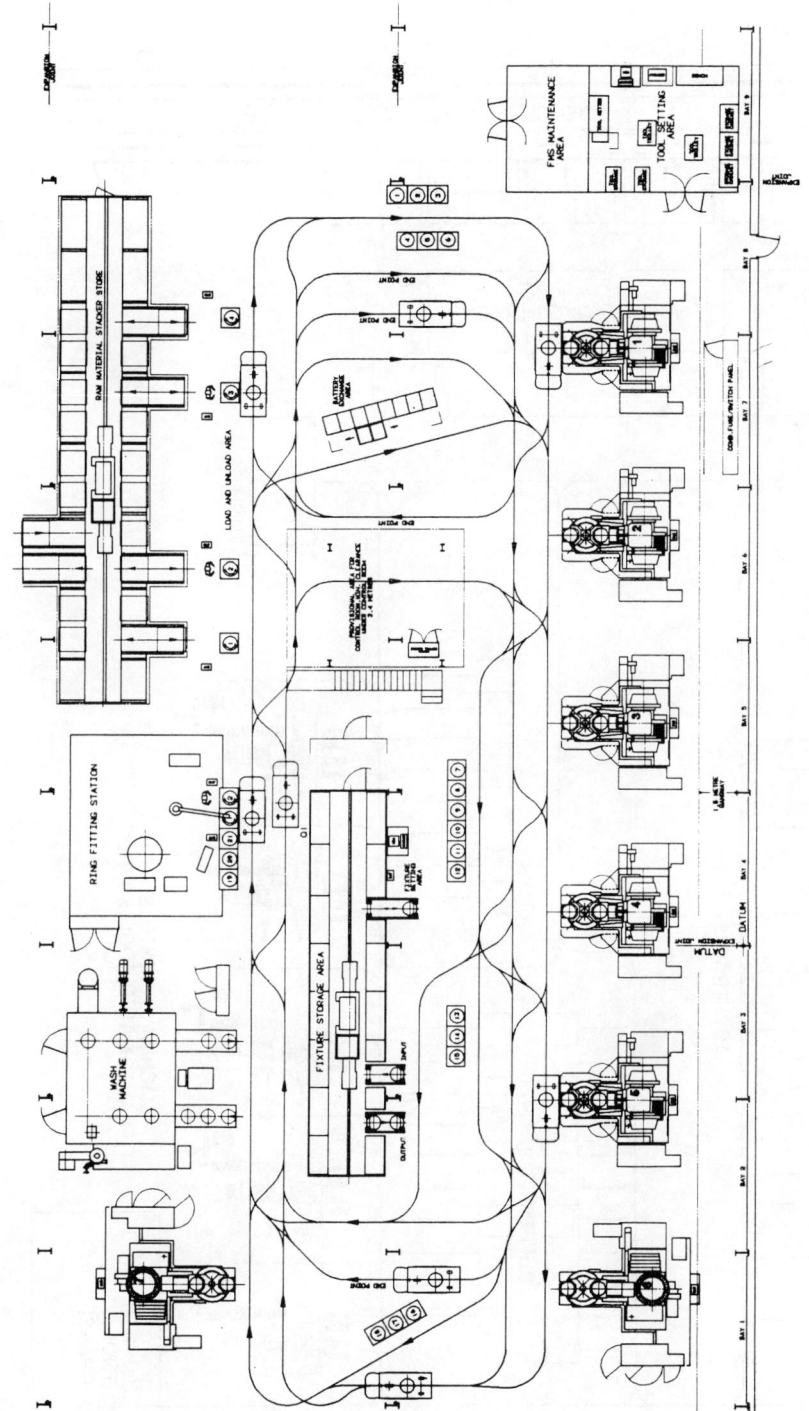

Figure 7.9 Hattersley Newman Hender's FMS (reproduced by permission of HNH and KTM Ltd, Brighton, England).

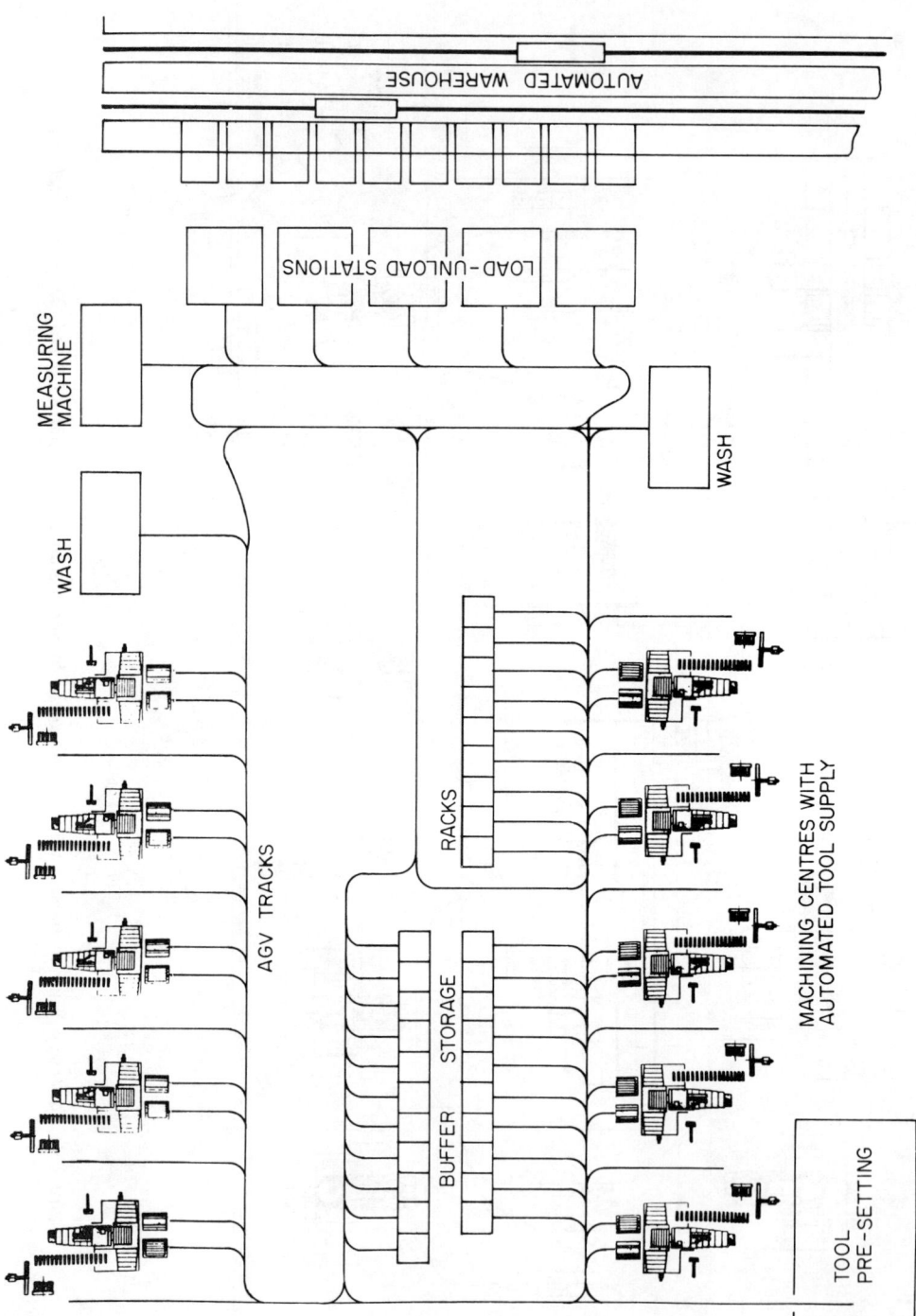

Figure 7.10 JCB Transmissions' FMS (reproduced by permission of JCB Transmissions' Ltd and Scharmann GmbH &m Co).

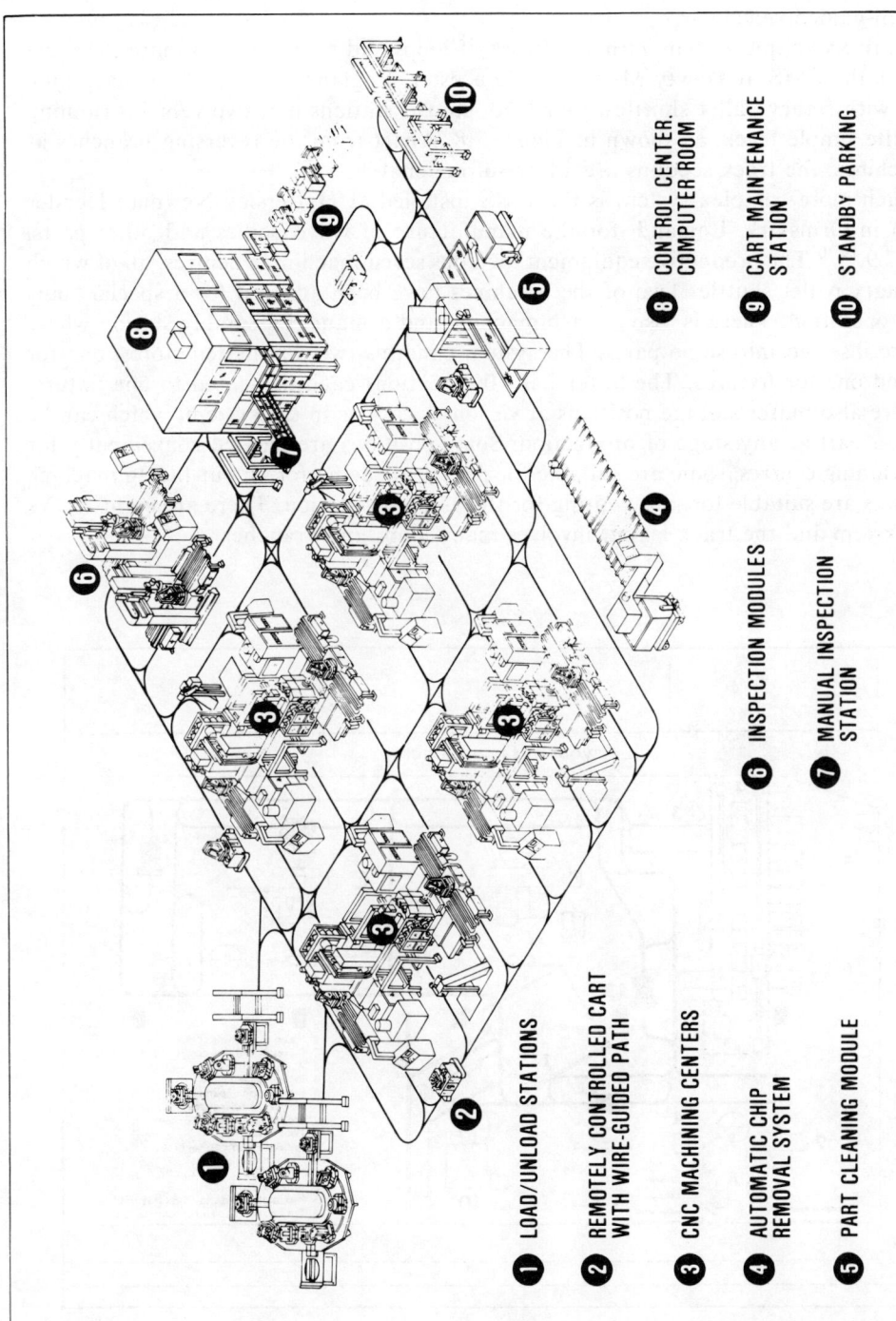

1 LOAD/UNLOAD STATIONS

2 REMOTELY CONTROLLED CART WITH WIRE-GUIDED PATH

3 CNC MACHINING CENTERS

4 AUTOMATIC CHIP REMOVAL SYSTEM

5 PART CLEANING MODULE

6 INSPECTION MODULES

7 MANUAL INSPECTION STATION

8 CONTROL CENTER. COMPUTER ROOM

9 CART MAINTENANCE STATION

10 STANDBY PARKING

Figure 7.11 Vought Aircraft (reproduced by permission of Cincinnati Milacron Ltd).

However, if the work load is so low that only one vehicle can cope with the traffic, it would usually be more economical to arrange the facilities in a line and have a linear track rail-guided vehicle.

A relatively simple system with a network of loops and branches and more than one vehicle is the FMS at Dowty Mining in Gloucester, England.[14] It has four machining centres with rotary pallet shuttles, four load–unload stations and two vehicles running on a quite simple track, as shown in Figure 7.8. Apart from the reversing branches at the machines, the track sections are all uni-directional.

A much more complex system is the FMS installed at Hattersley Newman Hender (HNH) in Ormskirk, England, for the manufacture of valve bodies and other parts, Figure 7.9.[19,30] The productive equipment includes seven machining centres, all of which have rotary pallet shuttle. Two of the machines have been adapted for a special "out-facing" operation. There is also a wash machine and a manual assembly station where rings are inserted into some parts. The system contains two automated stores, one for parts and one for fixtures. The latter has 100 locations each dedicated to one fixture. There are also buffer storage positions at various locations in the system, which can be used by a part at any stage of production. Some positions are located conveniently for the machining centres, some are suitable for pallets queueing for an out-facing machine and others are suitable for jobs waiting for the assembly station. There are eight AGVs in the system and the track layout involves many loops and branches.

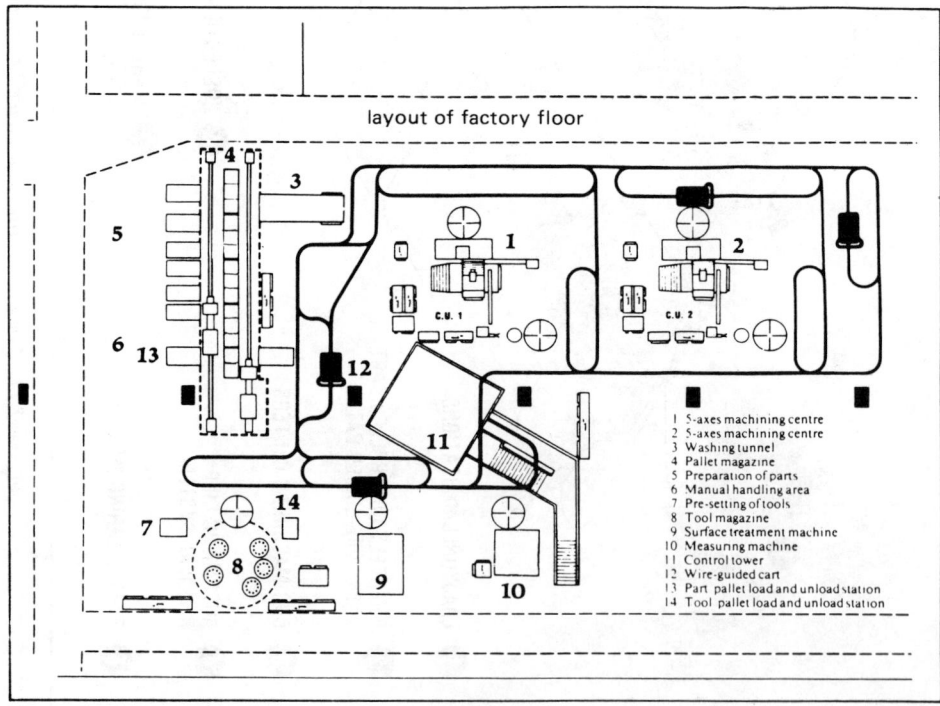

Figure 7.12   Citroen, Meudon (reproduced by permission of Citroen Industrie UK Ltd).

Another example of a system with several vehicles is that at JCB Transmissions in Wrexham, Wales, which is illustrated in Figure 7.10.[22] It has 10 machining centres, with input and output buffers, two washing machines and a co-ordinate measuring machine. There are buffer storage positions for 60 pallets between the two rows of machines. Raw material is kept in an AS/RS, which has ten positions where material to be loaded can be presented to the load–unload operators. There are five load–unload stations, from which components are taken to the machines by one of several AGVs, and to which they are returned for unloading.

Other well-reported systems with AGV networks are those at Vought Aircraft in Texas[4] Figure 7.11 and at Citroen's Meudon plant in France, Figure 7.12.[24]

7.3.3 Systems consisting of several cells

Several systems consist of a number of distinct sub-divisions or cells, each of which has some of the characteristics of an FMS. One example is the Con-rod line being installed by Cummins Engines at Shotts in Scotland.[17] Figure 7.13 presents a layout plan. The system consists of some nine cells mainly linked by conveyor. The first cell is an FMS consisting of three machining centres, a rail-guided vehicle and a robot operated load–unload station. The machines have input and output buffers and there are five buffer storage locations and two stations for manual gauging and refixturing. Gauging is also done automatically at the load–unload robot. The second cell comprises two lines of machining centres and gun-drills, with one return line passing through a wash machine. The machines are linked by conveyors. Loading and unloading is by robot which picks up con-rods from the input conveyor and unloads them after machining to the output conveyor. There then follow three robot cells where various machining, assembly, disassembly and other operations are carried out. These cells are linked by conveyors. The robot in each cell picks the rods up from the input conveyor coming from the previous cell and, after it has been processed, places it on the output conveyor to the next cell. From the third robot cell the rods are carried by conveyor to a group of manually controlled cells for various finishing operations, assembly, balancing and packaging. These manual cells may be automated at a later stage.

Holset Engineering in Huddersfield have a system consisting of seven cells for the manufacture of shaft and wheel assemblies for their turbochargers.[18,32] The manufacturing sequence was broken down into self-contained stages, and a cell created to perform the operations of each stage, involving machining, welding, hardening and tempering, grinding and balancing. The layout of the system is shown in Figure 7.14. Pallets carrying between 14 and 32 parts are moved through the sequence of cells by AGVs, of which there are three. Within each cell there are two pallet stands from which parts are moved among workstations by gantry robot. There is also a store for raw materials, partly machined or finished components. The system has been designed to allow each cell to operate independently of the others, to minimise the effects of any breakdowns.

A system which demonstrates that there is room for ingenuity in designing FMS, and the difficulty facing any would-be classifier of FMSs, is that at Cessna in Glenrothes, Scotland, Figure 7.15.[20] The system has several interesting features. There are three machining centres, each having a pallet shuttle providing one queueing position and the machining position. There is also a wash station and co-ordinate measuring machine and a cell for peripheral operations. Component transport is by a mobile robot. The cell for

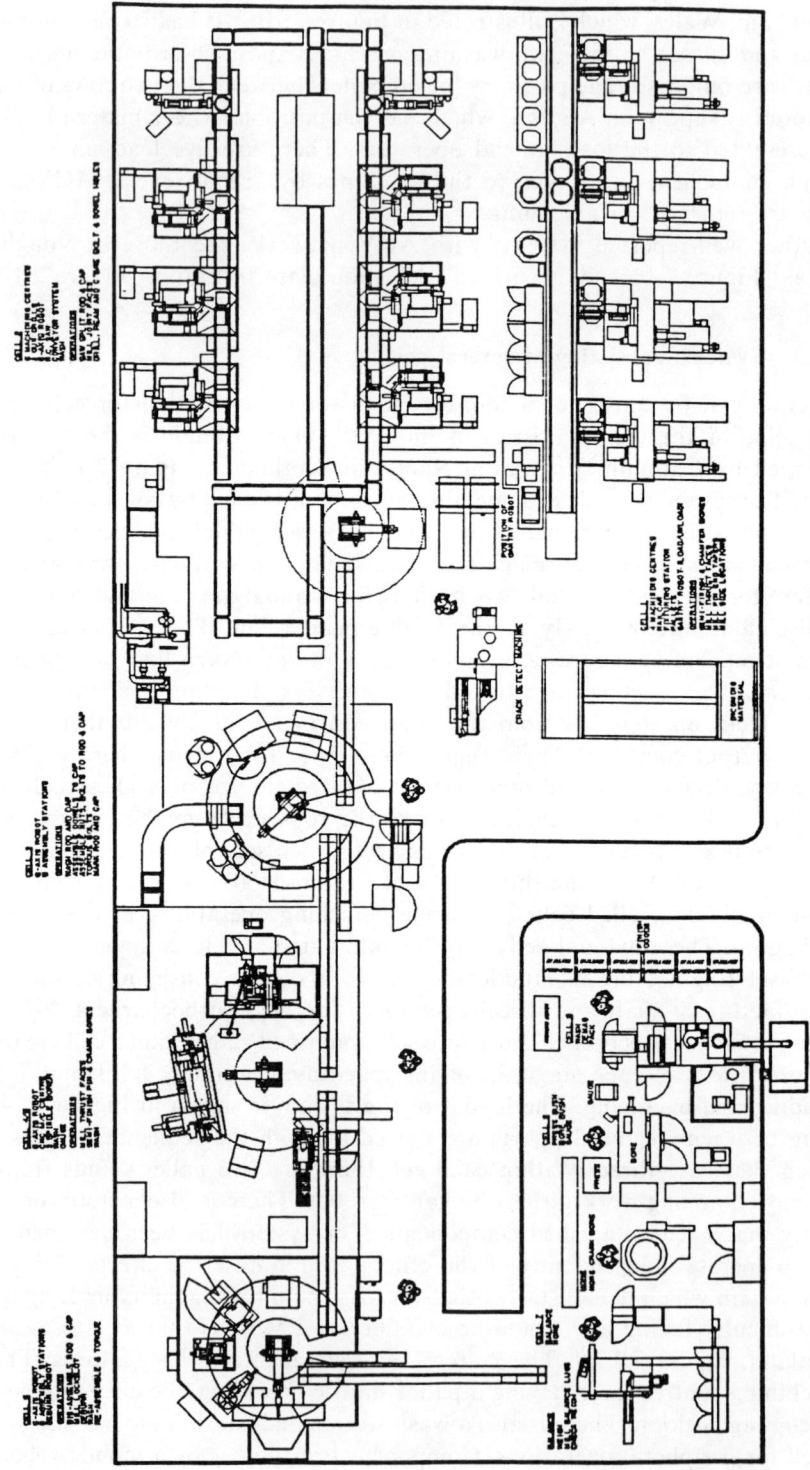

Figure 7.13 Cummins Engines Con-rod Line FMS (reproduced by permission of Cummins Engines Ltd).

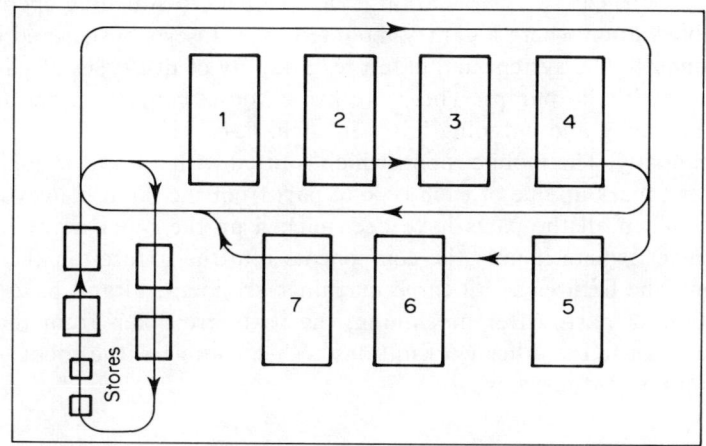

Figure 7.14 *Layout of cells in Holset Engineering FMS (reproduced by permission of Holset Engineering Ltd).*

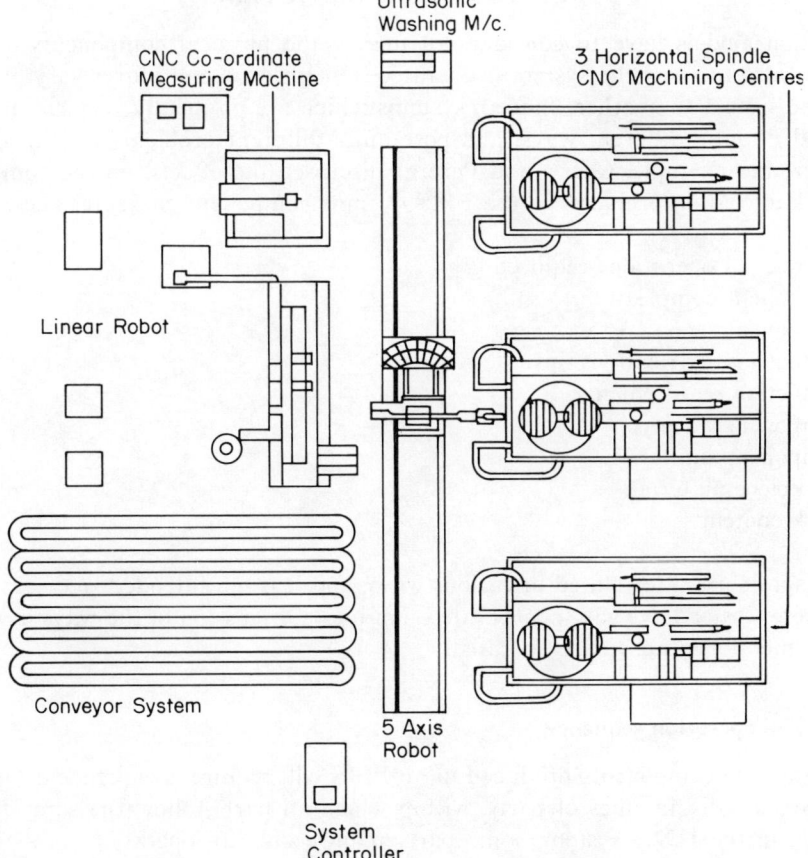

Figure 7.15 *Cessna, Glenrothes (reproduced by permission of Cessna Fluid Power Ltd).*

peripheral operations has two workstations, one where a cross-drilling operation is done, and an assembly station where a gantry-mounted robot inserts bushes and bearings into certain components. The system manufactures a family of five types of part, which are assembled into hydraulic pumps. There are five carousel conveyors, one for each type of part, with incoming and outgoing levels in each. Parts are loaded into, and removed from, these manually. The mobile robot, which is fitted with a tray for holding the parts during transport, picks up one of each type of part from the input conveyor, and places it in the tray. When all the parts have been picked up the robot moves to one of the machining centres, where it puts the components into the fixture, on the pallet in the buffer position. The fixtures at all three machines are identical and have locations for one of each type of part. After machining, the parts are taken from the fixture and carried by the robot to the other workstations. When complete, the robot puts the parts into the appropriate output conveyor.

7.4   OPERATIONAL ASPECTS OF FLEXIBLE MANUFACTURING SYSTEMS

Simulation models have to consider not merely the physical components of the system, but also the way the system is operated. Operational procedures vary enormously from one system to another. Indeed, systems which are physically very similar may be operated in quite different ways, and pose quite different problems for the simulation modeller. For example, Carrie and Perera[9] discussed the effects, on the complexity of the work allocation problems in the FMS, of nine component characteristics, namely:

1. Diversity of operations required.
2. Component complexity.
3. Component size.
4. Product family relationships.
5. Production requirements.
6. Component variety.
7. Component similarity.
8. Workpiece material.
9. Work content.

These factors are interrelated in various ways, and it is not intended to discuss them in detail here. Instead, we will incorporate them into a discussion of the ways in which the systems mentioned above are operated.

7.4.1   The operation sequence

In general, the components produced in an FMS will require a sequence of operations. There are usually families of parts, within which all parts follow the same route. For example, in the HNH system, some parts require only an operation on a machining centre, others require operations on both a machining centre and on one of the special out-facing machines, while a third group are machined on a machining centre, washed, and then have a ring inserted, after which they have an out-facing operation and a second wash.

In some cases, the complexity of the workpieces requires that operations which could be carried out at a single machining centre must be divided among several machines, owing to the work content involved or the limitations of tool capacity on the machines. This situation exists in the Anderson Strathclyde system. The components are large steel castings weighing up to over two tonnes and measuring more than two metres in length. The total work content may be over 20 hours. The processing sequence for a part typically involves three different fixtures, and a total of about 15 operations. In each fixture the part may require roughing, semi-finishing, finishing and facing-head operations, and may also have to return to the load–unload area during processing in one of the fixtures for the clamps to be repositioned, so that the areas which were obscured by the clamps can be machined. Operation times vary from about ten minutes to about eight hours. A further aspect of the complexity of the components is the number of cutting tools needed in their manufacture, possibly over 200. The number of parts being produced on the system has built up from seven at the commissioning stage to about 40 after three years in operation. Even with only a few parts in the system the problem of deciding which operations and tools to allocate to which machine is complex. At an early stage of planning, it was decided, on the basis of an initial set of parts, that two of the five identical machines would perform roughing operations, and the other three finishing operations. The two roughing machines would have nominally similar sets of tools in their magazines so that a roughing operation could be done at either machine. The number of tools needed for finishing operations was too great to permit similar tool sets to be allocated to the machines. Two of the three finishing machines are for semi-finishing operations and one for finishing operations, and their tool sets vary somewhat. Consequently, some operations can be done on only one of them, other operations can be done on two machines. The required quantities of the components average about two or three per month. The number of pallets in the system is much less than the number of fixtures, so that frequent changing of fixtures on pallets is needed. Initially, each part had its own set of fixtures, with one of each. Later, as new parts were introduced into the system, some parts were able to share fixtures. The operational problems therefore depend on determining, for the current schedule, the tools and fixtures needed, the extent to which tools and fixtures are shared by the parts, then allocating them among the machines and parts to be produced and deciding on a rough part launch sequence. This problem can be rather tricky.

Fortunately, few FMSs involve such complex components as at Anderson Strathclyde. In many systems the parts are completed in one operation, as is the case in the Dowty Mining FMS. However, it is more common for parts to require two operations, one from each side of the component. This requires two fixtures, unless the component is sufficiently symmetrical that both stages of machining can be done using one fixture.

7.4.2 Synchronising the processing in two or more fixtures

When the parts require operations in two or more fixtures, the production engineers have to consider whether the succesive stages are to be synchronised in any way. In the case of Anderson's FMS the parts are large, and the pallets can carry only one part. Therefore, when the part returns to the load–unload area when machining in one fixture is complete, it must be unloaded from its fixture. If the fixture into which it requires to be placed is already on a pallet at the other pallet stand, it would be possible to transfer it immediately from one fixture to the other. Synchronising the successive fixtures would demand skilful planning to ensure that the next pallet was ready. Since the part just unloaded from the other fixture would also be looking for its next fixture, and indeed the fixture from which the first mentioned part had been removed will also be looking for a new part a chain reaction of part transfers would build up which could not be achieved with only two pallet stands. Consequently, it is normal in the Anderson's system for the part to be unloaded and laid aside in a holding area, between stages in different fixtures.

If the parts are small then other methods can be adopted. A common arrangement is to mount a four-sided pillar on the pallet with a different fixture on each face. These are often called cubic fixtures. Then, if a part requires four operations in different fixtures, the pallet can carry four parts, one at each stage of machining. In the more usual case, when two operations are required, the pallet can carry two components at each stage. This method has the advantage that whenever the pallet returns to the load–unload area the two parts in the second operation fixture will be unloaded, the two parts in the first operation fixtures will be moved round to the fixtures for the second operation and two fresh components loaded into the operation one fixtures. With this method there is no need to lay aside the partly machined components while waiting for the pallet carrying the next fixture to become available. This arrangement is adopted in Cummins Engines' Makino line, and in Cell 1 of their Con-rod line.

7.4.3 Pallet stands for each pallet

Some of the systems mentioned include a pallet stand dedicated to each pallet, its "home" position. The Caterpillar vertical turning cell and Cummins Makino line are examples. Whether this is a practical idea depends on the variety of parts in the system. In the vertical turning cell the number of parts is quite small and it is sensible to have a pallet for each fixture and a pallet stand for each pallet. Since these also serve as load–unload positions, the problem of fixture synchronisation is somewhat simplified. In the Cummins Makino system, although there are several dozen parts involved, the parts fall into three families and so the total number of fixtures needed is still quite small. Thus, it is again feasible to permanently assign fixtures to pallets and to provide a home position for each pallet. In these two systems the fixtures are in almost constant use due to the production requirements of the parts. The introduction of new parts into the system might be constrained by this arrangement, or it might become untenable, mounting and dismounting of fixtures becoming necessary. In Anderson's system, however, the production requirements for the parts are not large, and therefore each fixture is in use for only a small proportion of time, and many fewer pallets than fixtures are required. This means that fixtures are frequently dismounted from pallets and kept in storage until needed again. This situation arises even although the total number of parts is not great, because each part requires several fixtures. In the HNH system some 700 different parts are produced, but this number will build up to well over two thousand as the complete range of parts are proved out on the system. The provision of a home position for each pallet is out of the question. Instead, fixtures are kept in an AS/RS when not in use.

7.4.4 Part similarity and standardisation of fixtures and tools

The greater the variety of parts the greater will be the tool and fixture variety. To maintain the tool and fixture variety and total number within aceptable limits, it is usually vital to rationalise as much as possible on component design, clamping methods, drill sizes and so on.

In the HNH system, it would have been impracticable to have had a unique fixture for each part type. To reduce the number needed a major rationalisation exercise was carried out. Parts were redesigned to adopt a standard method of location in fixtures, so that as many parts as possible could share a common fixture. This enabled the number of unique fixtures needed to be reduced to around 60. Allowing for the duplication of fixtures for parts having high production requirements, a total of around 100 fixtures exist at present.

Similarly, it is important to standardise on the tools needed to manufacture the parts. The problem of tool variety versus machine magazine capacity has already been referred to in connection with the Anderson Strathclyde system. In the HNH case, rationalising drill and bore sizes reduced the number of tools needed from around 200 to about 140.

7.4.5 Tool allocation

Normally, some tools are used on many parts while others are used on only a few of them. As with fixtures, those tools used on many parts have to be duplicated. Once the total number of tools has been determined the allocation of these tools to machines has to be established. Objectives are to minimise the frequency of tool changing and to maximise the number of machines on which an operation can be performed. In the HNH system tools have been allocated to machines so that the highest volume parts can be machined on any machine, and over 90% of the operations can be performed on more than one machine.

The allocation of tools to machines in Anderson's FMS has already been discussed. Since several hundred tools are used, it is the allocation of tools to machines which determines the routing of parts through the system and what constitutes an operation. In the Dowty Mining system, on the other hand, the entire range of tools can be held in each machine, so that any operation can be done on any machine. If the tool magazine capacity constrains the allocation of operations to machines, many strategies are available to minimise the complications. In the Cummins Makino system there are three basic families of parts. Within each family the tooling is similar, but the total tool variety exceeds the capacity of any machine. The strategy adopted is to equip each machine with the tool sets for two of the three families, so that there is a preferred and alternative machine for every operation.

7.4.6 Grouping parts on pallets

The practice of grouping parts at successive stages on a single pallet has been referred to. This is in fact only one of a number of strategies which can be exploited to simplify operation of the system. Another tactic would be to group together on a pallet several different parts which are used in a common assembly. This enables balanced sets of parts to be produced. This technique is used in the FMS at Cessna. Alternatively if the variety of tools is a constraint then parts could be grouped according to the similarity of tool sets required. This gives a hardware solution to what can otherwise be a complex software problem for work sequencing. Another approach would be to combine, on each of the pallets, groups of parts which give roughly equal work contents. This method is used at Victor Products where several small parts are placed on a single pallet to give about an hour's work content for the machining centres. This can simplify sequencing of operations and the movements of vehicles.

7.4.7   Sequencing the movements of vehicles

The discussion above concerning the complexity of parts in the Anderson Strathclyde system and of trying to co-ordinate the successive stages of movement of parts through the Anderson system, indicates that even although there is only one vehicle in that system, the scheduling problems can be rather difficult. In a system where there are several vehicles, the problem of sequencing the movement of vehicles can be very complex. One of the main questions in the HNH system (which was resolved by simulation) concerned the number of AGVs which the system would require. Simulation showed that eight were sufficient, whereas the supplier's initial estimate, based on previous experience, suggested that 10 or 12 would be necessary. With eight vehicles and a path layout involving many loops and sources and destinations for movements, the number of vehicle sequencing decisions can be very large. As with chess, it is virtually impossible to predict all the situations which could arise even just a few moves into the future. The efficiency of such systems relies heavily on the decision rules which are written into the control software. In this system the paths are all uni-directional. Having bi-directional motion would have greatly complicated the problem.

7.5   PROBLEMS IN THE PLANNING AND CONTROL OF FLEXIBLE MANUFACTURING SYSTEMS

In this discussion various problems in operating FMSs have been identified. On the whole these concern decisions as to how the system should be operated to maximise its efficiency on a day-to-day basis. They all relate to specific systems, whose design was itself the result of detailed analysis of many factors. Carrie[8] gave a list of some of the questions to be answered when planning an FMS:

Which parts are to be produced on the FMS?
What operations are required on them?
What types of machines are to be included?
How many machines of each type are required?
How are the parts to be fixtured, palletised and transported through the system?
What tools are required for each operation?
How should the machines be grouped?
Which operations and tools should be allocated to which machine groups?
How many machines, fixtures, pallets, transporters and tools are required?
How are fixtures and pallets to be assigned to parts?
What buffer storage must be provided?
What rules are to be built into the control software?

Some of these problems concern the design of the system, while others relate to its operation. It is important, especially when building simulation models of systems, to recognise that different types of problem will need different types of model. Consequently we need a framework within which the various questions may be placed so that we can address similar problems with similar types of model. There are many papers in the literature discussing the nature of the planning problems in FMS. One of the most appropriate in the present context is that by Van Looveren et al.,[31] who identify six problems and three levels of planning, as follows:

Strategic planning:
— the screening problem, a preliminary economic evaluation of alternatives to eliminate inefficient designs.
— the selection problem, to identify the alternative with the highest net savings, considering both technical and economic factors.
Tactical planning:
— the batching problem, organising production so that orders are completed on time, taking into account the limited numbers of pallets and fixtures.
— the loading problem, determining which operations will be performed on which machines and with what tools.
Operational planning:
— the release problem, controlling the flow of work into the system taking the overall allocation of resources to part types, and the current status of the system, into account.
— the dispatching problem, concerning the routing of the parts through the system, taking advantage of any alternatives which exist.

These different levels of planning are concerned with different time scales, dealing with long-term prospects, medium-term sales forecasts and current system status respectively, which Van Looveren presents in the form of Figure 7.16.

At the strategic planning level, a simulation model will have to use approximate estimates of production requirements, routings and operation times. As we tackle the lower levels our simulation model will have to become more detailed and our knowledge of operating practices much more specific. It is difficult to draw a clear division between the levels of planning. They are bound to overlap. In building a simulation model of a system we have to be explicit about the decision rules which are used, even although we may not have clear information on which to base them. The rules can be altered and refined as the project proceeds to the more detailed planning levels. Strictly, it is not until the strategic plans have been made that simulation has a major contribution to play, because until a system design has been suggested it cannot be modelled. Analytical methods, such as SPAR,[3] CAN-Q[25] or MVAQ,[28] may be more effective at the initial planning stages. Simulation can, of course, help to evaluate the alternative designs, but the model builder will often have to make assumptions about the decision rules, which are really in the realm of tactical planning.

The main contribution of simulation will be in the area of tactical planning, for it is at this level of planning that the decision rules are decided. Van Looveren et al. illustrate this level of planning with the diagram shown in Figure 7.17. There are many

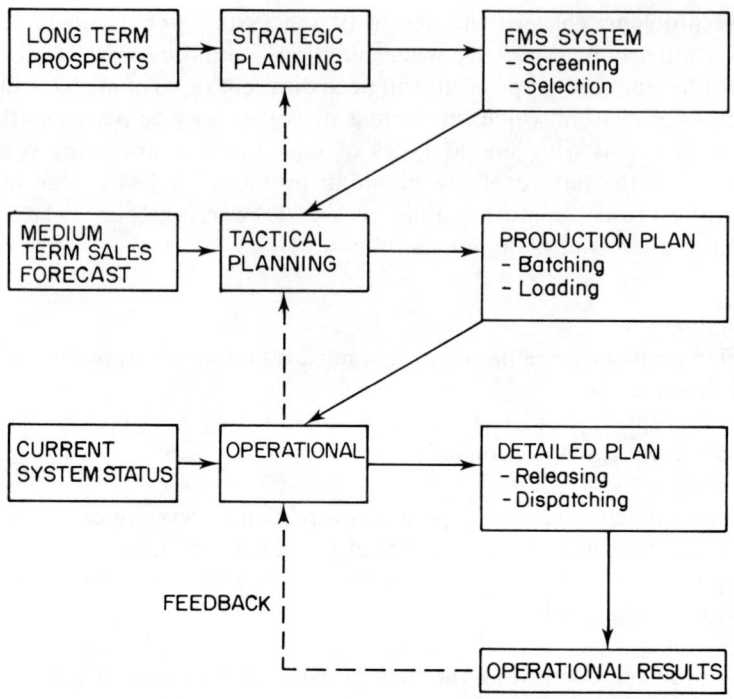

Figure 7.16   Hierarchical planning structure, from Van Looveren et al.[31] (reproduced by permission of Elsevier Science Publishers BV and the authors).

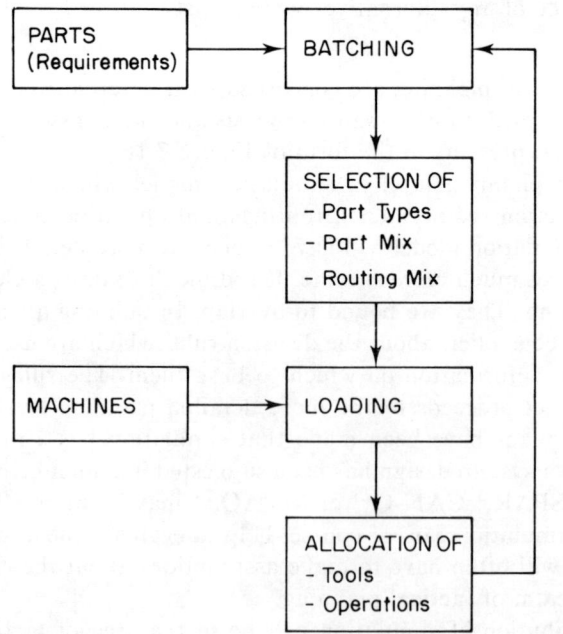

Figure 7.17   Batching and loading, from Van Looveren et al.[31] (reproduced by permission of Elsevier Science Publishers BV and the authors).

contributions to the literature dealing with tactical planning of FMS. For example, Stecke[26] gave an hierarchy of five planning problems:

1: The Part Type Selection Problem: From the set of part types that have production requirements, determine a subset for immediate and simultaneous processing.
2: The Machine Grouping Problem: Partition the machines into machine groups in such a way that each machine in a particular group is able to perform the same set of operations.
3: The Production Ratio Problem: Determine the relative ratios at which the selected part types will be produced.
4: The Resource Allocation Problem: Allocate the limited number of pallets and fixtures of each fixture type among the selected part types.
5: The Loading Problem: Allocate the operations and required tools of the selected part types among the machine groups, subject to technological and capacity constraints of the FMS.

This gives a more detailed breakdown than Van Looveren of his batching and loading problems. In the above discussion on the operational aspects of the systems illustrated we have observed various solutions to the loading problem. For example, in the Cummins Makino line, tools were allocated so that two of the three machines could process any part. The Anderson Strathclyde system was the most complex example of the loading problem among the systems described. As observed, the definition of operations, ie roughing, semi-finishing or finishing was the result of the solution adopted by the system's planners. With most systems there are a number of ways to tackle the problem. Stecke and Solberg[27] state that any solution to the loading problem must comply with certain constraints, namely:

1: Each required operation and all associated tools must be assigned to at least one machine.
2: An operation can be assigned only to those machines capable of performing it.
3: The tools required for the entire set of operations assigned to any machine must not exceed the capacity of the tool magazine of that machine.

and that to improve system performance:

1: The workloads assigned to each machine should be balanced (in some sense) to avoid unnecessary bottlenecks.
2: When feasible, consecutive operations should be performed on the same machine to minimise the number of part movements required.
3: Tool space permitting, operations should be assigned to more than one machine to increase flexibility when routing parts in real time.

As stated by Carrie and Perera,[10] tooling is a major factor in these decisions, in particular the machine grouping and loading problems. As shown in the Anderson Strathclyde system, an operation is a somewhat arbitrary collection of mini-operations by various cutting tools which, for reasons of fixturing or part orientation, are to be done together on one machine. Among the outputs of the tactical planning process is the basic data concerning the organisation of the FMS. This will include a list of the operations needed on each part, the machine or machine group where each operation is to be done, its duration, the list of the tools required and the cutting time of each tool.

At the operational level, the system manager is concerned with the release of parts onto the system, and the relative priorities of parts in buffers, vehicle scheduling and so on. Simulation may have a reduced role to play at this level, since it deals with the current status of the system. Few simulation models are able to access the current status of a system. Therefore, setting up the initial conditions in the model may be too time-consuming for personnel managing the system to make much use of simulation. Instead, operating procedures which these personnel should adopt may be established by off-line simulation modelling, usually as part of the tactical planning process. Indeed, since rules of this type will normally be built into the system control software, it is essential that they are established in advance of building the system. An important question is the extent to which the system software provides facilities for the manager to override the system's normal logic. It is to be expected that, as time progresses, more work will be done in providing simulation tools as part of the real-time software for controlling the FMS. One such approach has been described by Carrie and Perera.[11]

In practice, the role of simulation lies at two levels, which straddle Van Looveren's hierarchy:

System Design: Addressing the system design problems, and ensuring that the system has sufficient capacity to meet its production targets. Simulation modelling at this stage aims to assist in the strategic planning problems, but must consider the decision rules to be used at the tactical level.

System Operation: At this stage the design of the system has been determined, and simulation can assist in determining the best operating procedures for machine grouping, sequencing rules and so on. Thus, simulation is concerned with the interface between tactical decisions and operating procedures.

Some details of a case study, which describes the stages in a simulation project of an FMS, and how the model and the data were revised several times as new information emerged, and how different problems were examined,[7] is described in chapter 13.

SUMMARY

This chapter has given a brief description of a selection of systems, showing the types of element which may be included in an FMS and typical system configurations. We then reviewed some of the operational aspects and problems, and outlined some of the procedures adopted in the systems to ease the day-to-day management problems. It was observed that the problems of planning FMSs fall into distinct categories which can be arranged in an hierarchy. It was suggested that the major role for simulation is at the tactical level. At this level the overall structure of the system has been established and it is desired to establish the best operating procedures, possibly so that they can be written into the control system software. In subsequent chapters, examples of the use of simulation to tackle problems at various planning levels will be presented.

REFERENCES

1. Examples of types of computer-controlled flexible manufacturing systems for boring, drilling and milling machining processes, Werner und Kolb, GmbH, Berlin.

2. *Management Guide to Flexible Manufacturing*, The Institution of Production Engineers, London, 1986.

3. *SPAR User's Manual*, CMS Research Inc., Oshkosh, Wisconsin, USA, 1986.

4. $10 million FMS at Vought to build fuselage components, *The Production Engineer*, January 1984, 39–40.

5. Baxter, R., CIM solves the problem at Victor, *Computerised Manufacturing*, June 1986, 20–23.

6. Browne, J., Dubois, D., Rathmill, K., Sethi, S. and Stecke, K., Classification of flexible manufacturing systems, *The FMS Magazine*, April 1984.

7. Carrie, A.S., FMS simulation case study, Proc. meeting on New software tools for simulation in manufacturing, Belgian Institute of Automatic Control, Working Party on Systems Engineering, Antwerp, May 1985.

8. Carrie, A.S., Some planning problems in FMS, *Proc. Conf. on Planning for Automated Manufacture*, pp113-117, Coventry, September 1986, Mechanical Engineering Publications Ltd, London.

9. Carrie, A.S. and Perera, D.T.S., Work allocation in flexible manufacturing systems, *Proc. Int. Conf. on Computer aided Production Engineering*, pp91–95, Mechanical Engineering Publications Ltd, London, 1986.

10. Carrie, A.S. and Perera, D.T.S., Work scheduling in FMS under tool availablility constraints, *Int. J. Prod. Res.*, 24, 6, 1299–1308, 1986.

11. Carrie, A.S. and Perera, D.T.S., A simulation tool for real time scheduling of FMS, Proc. 3rd National Conf. on Production Research, Nottingham, September 1987, In *Advances in Manufacturing Technology II*, edited by P. F. McGoldrick, pp131–137, Kogan Page, London, 1987.

12. Carrie, A.S., Adhami, E., Stephens, A. and Murdoch, I.C., Introducing a Flexible Manufacturing System, *Int. J. Prod. Res.*, 22, 907–914, 1984.

13. Cresswell, C. and Edghill, J.S., A survey of FMS installations, Technical Note TBS4, Dept. of Mechanical Engineering and Production Engineering, UWIST, 1986.

14. Cresswell, C. and Kempa, R., Simulation case studies for the design and control of manufacturing systems, I.Prod.E. seminar on Manufacturing—Simulating the Future, March 1987, I.Prod.E., London.

15. Dupont-Gatelmand, C., A survey of flexible manufacturing systems, *J. of Manuf. Syst.*, 1, 1, 1982.

16. Findlay, R.C., FMS concepts realised, *Proc. 3rd Int. Conf. on Flexible Manufacturing Systems*, pp397–405, IFS (Publications) Ltd, Bedford, England, 1984.

17. Hunter, A.A., Justifying FMS to provide a competitive edge, *Proc 5th Int. Conf. on Flexible Manufacturing Systems*, pp417–426, IFS (Publications) Ltd, Beford, England, November 1986.

18. Kellock, B., Why this team said: we'll do it our way, *Computers and Manufacturing Technology*, June 1986.

19. Kochan, A., KTM confirms UK leadership, *The FMS Magazine*, April 1985.

20. Kochan, A., Cessna FMS enters production, *The FMS Magazine*, April 1985.

21. Kochan, A., Caterpillar invests heavily in flexible machining cells, *The FMS Magazine*, July 1985.

22. Kochan, A., A few lessons for suppliers, *The FMS Magazine*, October 1985.

23. Murison, A.E., Why we went to flexible manufacturing systems at Victor Products, *Proc. of Conference on Planning for Automated Manufacture*, Coventry, September 1986, Institution of Mechanical Engineers, London, 1986.

24. Purdom, P.B., The Citroen flexible manufacturing cell, *Proc. 2nd Int. Conf. on Flexible Manufacturing Systems*, pp93–104, IFS (Publications) Ltd, Beford, England, 1982.

25. Solberg, J.J., *CAN-Q User's Guide*, School of Industrial Engineering, Purdue University, West Lafayette, 1980.

26. Stecke, K.E., Production planning problems for flexible manufacturing systems, Ph.D. Dissertation, Purdue University, West Lafayette, 1981.

27. Stecke, K.E. and Solberg, J.J., Loading and control policies for a flexible manufacturing system, *Int. J. Prod. Res.*, 19, 481–490, 1981.

28. Suri, R. and Hildebrandt, R.R., Modelling flexible manufacturing systems using mean-value analysis, *Journal of Manufacturing Systems*, 3, 1, 27–38, 1984.

29. Talavage, J., Simulation analysis of an operational computerised manufacturing system, Proc. CAM-78, Int. Conf. on Computer-aided Manufacture, East Kilbride, Scotland, June 1978.

30. Tiernan, T., How to place a very tall order—and win, *Computers and Manufacturing Technology*, June 1986.

31. Van Looveren, A.J., Gelders, L.F. and Van Wassenhove, L.N., A review of FMS planning models, in *Modelling and Design of FMS*, edited by A. Kusiak, pp3–31, Elsevier, Amsterdam, 1986.

32. Webb, S., User project management for Holset FMS, *Proc. 5th Int. Conf. on Flexible Manufacturing Systems*, pp135–144, IFS (Publications) Ltd, Bedford, England, November 1986.

CHAPTER 8

Building FMS models —
1: Load–unload operations, pallets, machines

In chapter 5 a simple model was developed which included machines, components and transport facilities. In chapter 7 the types of feature which may be present in an FMS were discussed, and some examples of FMSs were given. In this chapter we start to consider the logic which will be necessary to incorporate some of these features, beginning with load–unload operations, pallets, machines and some associated aspects. The model of chapter 5 provides a skeleton which we can develop. As we consider new features, we will need to add new types of entities and activities to that basic model and to consider other activities, such as materials movements, in greater detail. We will use activity cycle diagrams to illustrate most of this discussion, since they are largely independent of the software package on which the logic is implemented. To facilitate the use of activity cycle diagrams a consistent notation for the arrows of the different types of entity will be used (so far as possible). Figure 8.1 gives the notation used in this chapter.

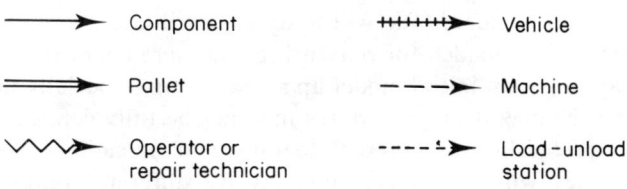

Figure 8.1   Arrow notation for activity cycle diagrams.

8.1   LOAD AND UNLOAD OPERATIONS

In the previous model, when components arrived they were moved directly to the machine for their first operation. In most FMSs the part to be machined has to be loaded on a pallet of some kind before it can be processed, and afterwards has to be taken off the pallet. In other words, loading involves taking a component which has been delivered to the system and preparing it for processing. Once the component has been loaded on a pallet it may be transported to the machine required for its first operation. Similarly, unloading consists of taking a component which has been brought back to the load–unload area after processing, from its pallet and placing it in a rack or on the floor, to await removal to the assembly department, finished parts store or elsewhere. To incorporate load and unload operations in the model we need to consider two aspects:

1. amending the activity cycle diagram for the component,
2. adding the associated new entities and their activity cycles.

8.1.1   Revision of component's activity cycle

The activity cycle developed in the earlier model is reproduced in Figure 8.2. One method of including load and unload operations would be to consider them as just another two instances of the PROCESS activity. However, if we do so, we will need to amend the cycle in the way it describes the arrival into and departure from the system. In the original cycle the component waited for handling on arrival in the system. Now it must be loaded on a pallet before it can be moved to a machine, and therefore will wait for PROCESS instead of MOVE. Similarly, after the final operation, completed components will not require a movement after their final operation, since they will already be at the load–unload station, and can LEAVE immediately. Remember that in figure 8.2 LEAVE served two purposes, firstly to move the component out of the system after its final operation, and secondly to record any statistics about components as they left the system. Now LEAVE will only involve the gathering of statistics. If we refer to components which have just been delivered to the FMS as NEW components, and those which are ready to leave as OLD, the cycle would become as shown in Figure 8.3.

Notice that the test for the queue into which the component is placed after PROCESS has been expressed in terms of the nature of the operation just performed. This is exactly equivalent to the previous test, since if the loading and unloading activities are included in the count of operations required, after unloading has been done there will obviously be no more operations to be done. (We will postpone until later, consideration of the case when the component is unloaded for refixturing and more operations.) This approach is often an easy way to get an initial model up and running, especially in the initial stages of a project when the operating procedures may not be fully defined.

However it is preferable to represent loading and unloading as separate types of activity because they will almost certainly involve different combinations of entities than do other types of PROCESS. Adding separate activities called LOADUP and UNLOAD, and a queue of FINISHED components, will give the component activity cycle shown in Figure 8.4. LOADUP will be used rather than LOAD, since LOAD is an ECSL keyword, and models will be presented using ECSL[1] since it is quite easy to read programs written in that language.

The LEAVE activity now serves only to collect statistics about jobs as they are completed, and that can be done just as well in the UNLOAD activity, since once a component is unloaded and its statistics collected we are probably no longer interested in it. Thus the LEAVE activity (and consequentially the OLD queue) can be omitted, giving the activity cycle shown in Figure 8.5.

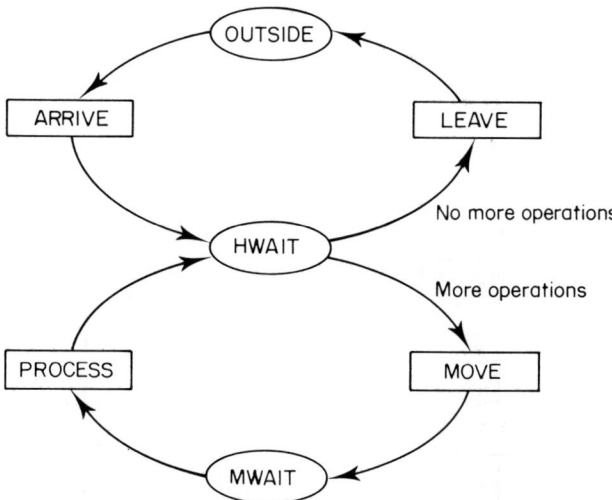

Figure 8.2   Component's activity cycle diagram, from figure 5.2.

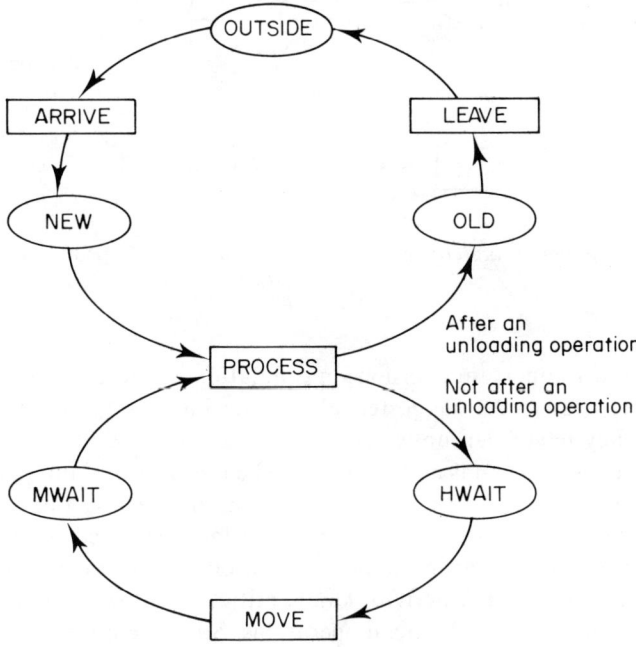

Figure 8.3   Component's activity cycle diagram in which PROCESS includes loading and unloading activities.

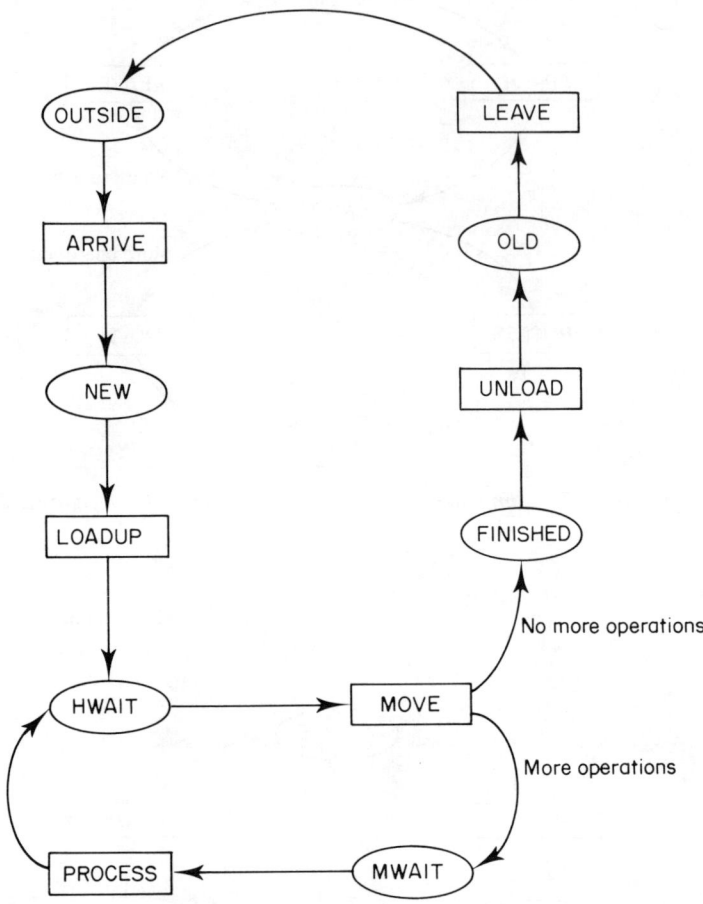

Figure 8.4   Component's activity cycle diagram when LOADUP and UNLOAD are separate activities.

One of the most important decisions in simulation is where the boundary of the system under study is drawn. As suggested above our interest in the components probably ceases when they have been unloaded and can be taken to inspection, assembly, stores or elsewhere. These diagrams assumed this to be the case, and implied that the boundary is drawn immediately after the unloading operation, and so the LEAVE activity was deleted. If, however, we were interested in the facilities used to transport components to and from the system, and perhaps some automated storage system, then we would widen the model and further activities and entities would be introduced, just as we have now done by adding load and unload operations. Since we intend to introduce additional features to the model one at a time, we will retain our original, albeit restricted, boundary of the system and proceed to consider the other entities which may be involved in LOADUP and UNLOAD.

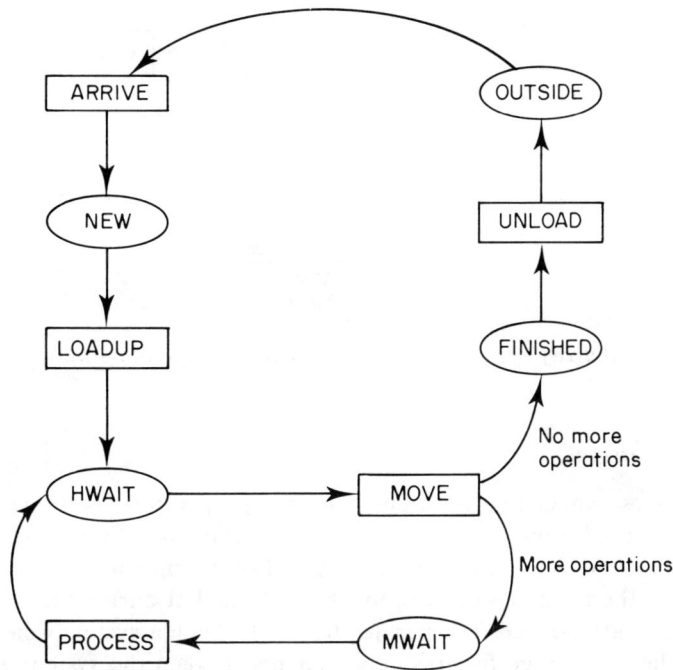

Figure 8.5   Component activity cycle diagram with LEAVE omitted.

8.2   OPERATORS

Usually loading and unloading is done by human operators. They will be either idle or working (loading or unloading a component). Since there is no restriction on the sequence in which they perform these operations their activity cycle will be as in Figure 8.6.

However, the operators may have other tasks in the system. These may involve patrolling around the machines, loading tools, inspecting components, administrative tasks and so on. Frequently, it is difficult to get a clear statement of these other tasks and especially of their duration. If data is available, they can be added to figure 8.6 as additional independent operations. If they are likely to constitute only a small proportion of the operators' work load we may ignore them, or we might add an activity which summarises all the other tasks, as in Figure 8.7.

Figure 8.6   Activity cycle diagram for load–unload operators.

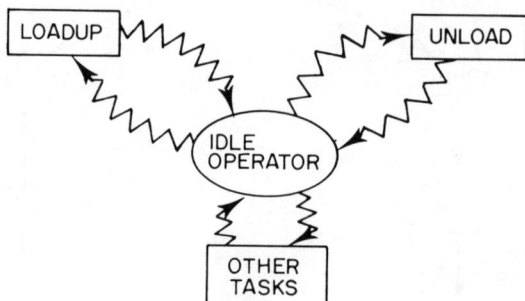

Figure 8.7   Activity cycle diagram for load–unload operators who have other duties.

8.2.1   Shift patterns

One of the most important questions concerning operators, is do we need them to be in attendance at all times? If they have other duties then by implication they will not be available all the time, although each period of absence will be of short duration. In many systems the machines operate for three shifts, but during the night shift there are no operators in attendance. To operate effectively, such a system would have to provide a sufficient buffer storage for fresh components to keep the system going through the night. There are technical and economic questions involved, but for simulation modelling we need logic to switch off the operators during the night while the rest of the entities carry on. In fact, we need to consider the more general case of a varying number of operators in each shift. For example, there might be a full crew on duty during the day shift, a partial crew during the second shift, and none at all at night.

Generally, packages do not provide routines for handling shift patterns automatically, and the modeller has to write his own code. An exception to this is SIMAN,[2] which provides a function block called ALTER, to adjust the capacity of resources and is ideally suited to handling shift patterns. If the package does not provide such a function, logic to handle shift patterns can be defined in various ways.

A simple method is to introduce a new activity, called a SHIFTCHANGE, and a dummy entity, CHANGE. We also need a variable, men, to indicate the number of operators available on each shift. The function of the shiftchange activity is to keep track of which shift is in progress and set the number of operators available. The dummy entity, change, is used to indicate when the time of the next change of shift will occur. We amend the logic of load and unload activities to depend, not only on an operator and a component, but also on the number of operators busy being less than the number of men available. The ECSL coding for this, if 2 men are available for the first shift, 1 for the second and none on the third shift, might be:

```
BEGIN SHIFTCHANGE
TIME OF CHANGE LE 0
N = 0
CHAIN
  SHIFT EQ 1
  N = 2
```

```
    MEN = 1
    OR SHIFT EQ 2
    N = 3
    MEN = 0
    OR SHIFT EQ 3
    N = 1
    MEN = 2
 SHIFT = N
 TIME OF CHANGE = 480

 BEGIN LOADUP
 BUSY LT MEN
 FIND FIRST OPERATOR A IN IDLEOPS
 DURATION = LOADTIME
 BUSY + 1 AND BUSY − 1 AFTER DURATION
 OPERATOR A FROM IDLEOPS INTO IDLEOPS AFTER DURATION
```

An alternative would be to place the operators into a queue called "off-shift" or "off-work". However, to do so raises other details. For example, when a shift changes and operators leave the factory, would we have to check that they were not in the middle of an activity? Would we send them home when they had finished that activity? Or, would we not start an activity if it could not be completed before the end of the shift? As with so many of these questions, when we examine the fine detail of procedures we may become involved an excess of detail which does not justify the gain in accuracy obtained. It is best to use the simplest method consistent with sufficient accuracy.

8.3 AUXILIARY EQUIPMENT

Sometimes auxiliary equipment such as lifting gear, may be involved. If so, then we must consider another of the fundamental questions in simulation: Do we need to include the equipment in the model? Would its absence affect our results? There are no simple answers to these questions, and every case has to be considered carefully. In general however, if loading or unloading is never likely to be delayed for want of the auxiliary equipment then it is probably safe to omit it. If the equipment is in short supply, or may be moved to another location, or for any other reason is likely to hold up the operation, then it must be included. Perhaps one of the objectives of the simulation model is to answer these questions.

  Even if auxiliary equipment might affect the operations, it may not be necessary to include all the details in the model. In the job shop model developed in chapters 4 and 5 we omitted the machine operators because we decided that if the machine was available then the operator must also be available. It was only necessary to include one of the two types of entity. We can exploit this approach to deal with auxiliary equipment by including only the most restricting item. For example, in the Anderson Strathclyde system the castings are rather heavy and an overhead crane is available at the load–unload area for the operators. Because there is only one crane but two operators the crane was included in the model, but the operators omitted, on the reasoning that the availability of the crane would limit loading and unloading more than the operators.[3]

8.4   PALLETS

We have already referred to pallets, which are almost always used to carry prismatic components for location on machines and transporters. Let us consider how they can be modelled, and their activity cycle. In its simplest form the activity cycle diagram will be as in Figure 8.8.

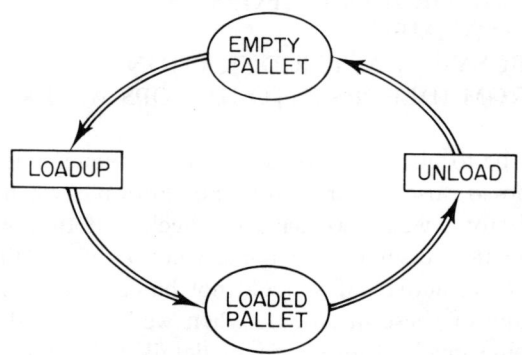

Figure 8.8   Simple form of activity cycle for pallets.

For entities with a simple cycle of this type it is often possible to model them by a single variable, representing the number of empty pallets. This variable would be decremented by one every time a LOADUP is performed and incremented by one every time an UNLOAD is done. However if there are any other factors involved, such as the type of pallet or the number of components currently on a pallet, then it would normally be necessary to treat the pallets as entities proper, with their cycle and whatever attributes are needed to hold information about them.

Although this cycle may be adequate, we may seek to describe what happens to pallets as they are moved around the system. It will be evident that once a component has been loaded on a pallet, the component and pallet will always travel together as they are moved away from the load station, around the machining stations and eventually back to the load station. Perhaps the pallet's cycle should be similar to the component's between the LOADUP and UNLOAD activities. However, since the component cannot move independently of the pallet, it seems wasteful to specify the cycles of both types of entity in detail. This is a rather common consideration in model building, and we have three choices:

1. model both entities in detail,
2. model one in detail and the other in reduced detail,
3. omit one or other entity from the model altogether.

The best decision will depend on the characteristics of the system to be modelled and the aspects of it which are of particular importance in modelling it. It will also depend on the software package in use and how easily it permits these options to be realised.

Modelling both entities in detail presents no particular problem and merely involves duplicating the tests and actions in the logic. Some packages provide a facility for automatically duplicating the logic for one type of entity once the logic for another type has been described. For example, CAPS[1] allows the cycle of one entity to be linked to another for a series of two or more activities, between a pair of named queues.

If we desire to model one entity in reduced detail, then we must decide which one to model in full detail and which to model in reduced detail. Either the part or the pallet can be treated in this way. If we model the pallet by the simple cycle above, the combined diagram would be as in Figure 8.9.

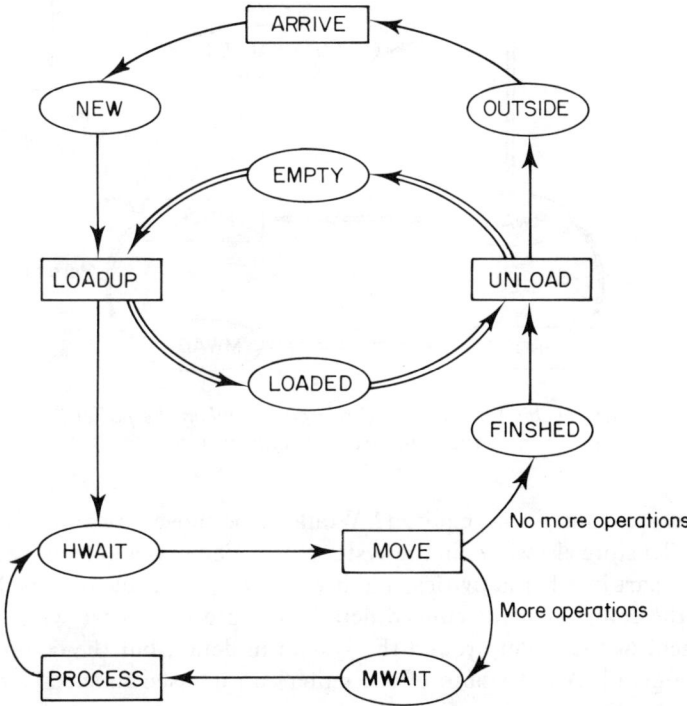

Figure 8.9   Activity cycles for pallet and component when the component is modelled in detail but the pallet in summary only.

If we adopted the alternative approach and modelled the pallet's movements in detail, the component could be placed in a queue "ON A PALLET" while it is on a pallet. This would yield Figure 8.10.

It is largely a matter of convenience which of these alternatives is adopted. If the pallet has no activities independent of the component then the first method will be more straightforward. But if the pallet has independent actions then they must be modelled. For example, if there were no more components ready for loading on the pallet after an UNLOAD operation, what would be done with the pallet? Would it be stored at the

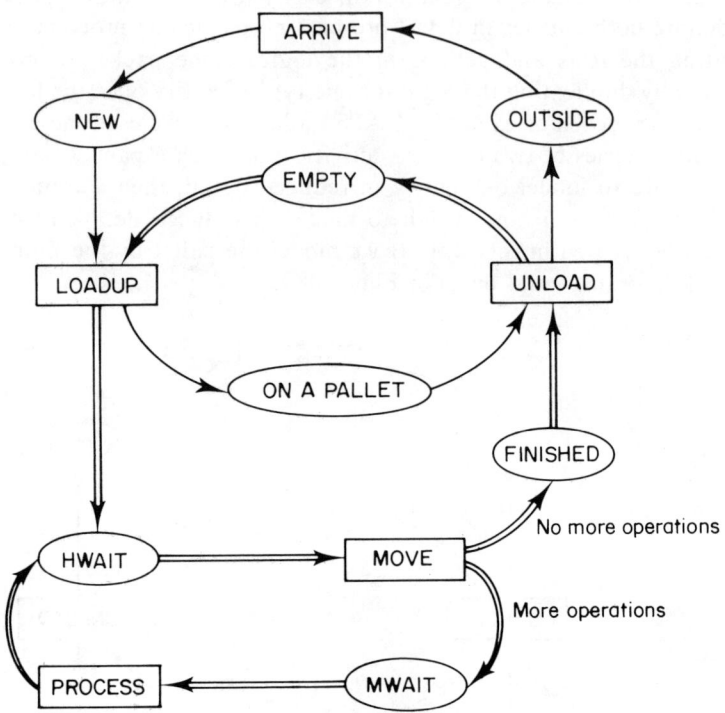

Figure 8.10   Activity cycles for pallet and component when the pallet is modelled in detail but the component in summary only.

load–unload area until it is required? Would it be moved away from the load–unload area to a pallet store elsewhere in the system? If the empty pallet will be moved elsewhere then it takes part in other activities, and it would be preferable to model the pallet in full detail but the component in reduced detail. Because of this we will, in general, model the movement of the pallet around the system in detail, but the component will not be modelled in detail. An attribute of the pallet can be used to specify which component is on the pallet.

The third choice identified above was to omit one or other entity from the model altogether. This will only be permissible if the omitted entity would never restrict the performance of the system. For example, if each type of component had a special type of pallet and there was room for storing all the pallets at the load–unload area then a pallet would always be available for loading a new component whenever a finished component had been unloaded. Then the pallet might be omitted. Similarly, there are circumstances when the component entity could be omitted. For example, if it were assumed that there would always be a supply of components to be machined on the system, then it would be unnecessary to test for the presence of a NEW component, and hence the component entity might be omitted. These possibilities will be discussed at greater length shortly.

8.5 LOAD–UNLOAD STATIONS

We have said that loading and unloading operations will be done at the load–unload area, without indicating what this area consists of or what facilities are located there. When pallets are brought to the load–unload area they require to be located on some base plate or table. These locations are referred to as load–unload stations. It is conceivable, although rare, that loading operations are done at some locations and unloading at different locations. We will omit this possibility and assume that both loading and unloading can be done at any station. We can call them load stations for short.

Before a load or unload operation can be done there must be a pallet at the load station. Before a pallet can be brought to a load–unload area there must be a vacant load station. Bringing a pallet to the load–unload area makes a vacant load station occupied and taking a pallet away makes an occupied one vacant. Because the availability of load stations may constrain the operation of the FMS they usually will have to be included as entities in the model. From the foregoing comments it is evident that it might be possible to describe their activity cycle as in Figure 8.11. Perhaps surprisingly, the load and unload operations do not appear in the activity cycle of the load stations. These activities affect the pallet and the component, if any, which occupies the load station. The states of these entities determine whether a loading or unloading operation can be performed.

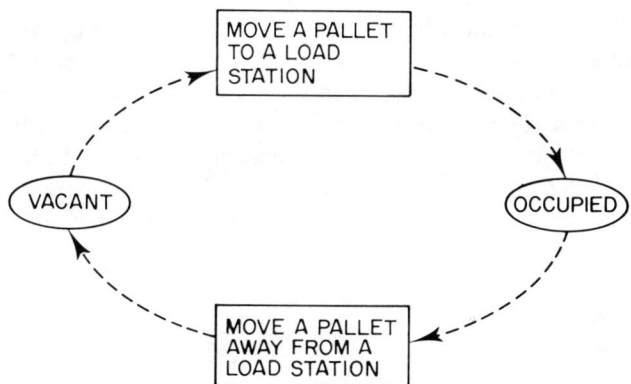

Figure 8.11   Activity cycle diagram for load–unload stations.

As observed with pallets, an entity having such a simple cycle can often be modelled by a single variable. This would define the number of vacant load stations and would be incremented whenever a pallet is moved away and decremented when a pallet is brought to the load–unload area. Whether this can be done in a particular case depends on the nature of the system concerned.

In chapter 7 we observed that there were two kinds of load station:

1. General load stations,
2. Dedicated load stations.

Chapter 7 included diagrams of two Cincinnati Milacron systems which had automatic work changers (a chain of eight pallets) at the load–unload stations. At first sight these might be considered as a third type, but they can be treated as general load–unload stations with a buffer. Buffer storage will be considered in chapter 9.

8.5.1   General load stations

General load stations can be used by any pallet, or type of pallet. There may be ten or more pallets in a system but only two load stations. If there are more pallets than load stations, it is necessary to test for the presence of a vacant load station before a pallet can be moved to the load–unload area for unloading. This may often be achieved by the use of a single variable as described above. Indeed, treating the load stations as entities will normally only be necessary if they must be distinguished from one another for some reason. One possible reason might be that they are in different locations and the time taken by a vehicle to move a pallet to them will depend on which load station is involved. Another reason might be that we wish to record statistics about each of them separately, such as the proportion of time when they are occupied.

8.5.2   Dedicated load stations

Dedicated load stations are ones which can only be used by one pallet, type of component, fixture or similar restriction. If each load station is dedicated to one pallet then it is no longer necessary to test for the availability of a vacant load station before a pallet can be brought to it for unloading. Since no other pallet may use the station then whenever the pallet requires to be moved to the station it cannot be there already and therefore the station must be available. In these circumstances it should be possible to omit dedicated load stations from the model. The way pallets, parts, and fixtures are assigned to load–unload stations varies considerably from one FMS to another. Each system to be modelled must be studied carefully. The test for availability of a load station can be omitted only if there can never be more than one entity requiring to use any load station at any time.

8.6   MOVEMENT TO AND FROM LOAD STATIONS

It follows from the cycle diagram in figure 8.11 that movements to and from load stations affect these entities, whereas movements between other workstations do not. This means that these two different types of movement involve different sets of entities, and therefore it is no longer correct to represent them both by a single MOVE activity. In fact we need to have three movement activities. Let MOVE be used for movements between workstations, INTRODUCE for movements from load stations to workstations and RETURN for movements from workstations back to load stations. These new activities, together with a new queue to separate LOADUP and INTRODUCE, give the pallet cycle shown in Figure 8.12. When the cycles for the other entities, that is the load stations, operators, components, vehicle and machines are added to figure 8.12 we obtain the diagram in Figure 8.13.

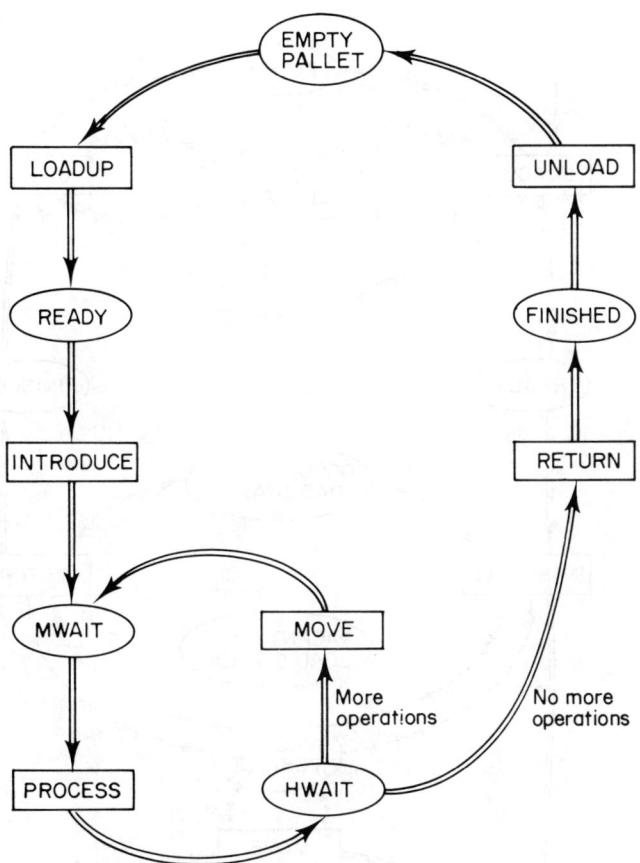

Figure 8.12 *Activity cycle diagram for pallet when movements to and from load and unload stations are separately defined.*

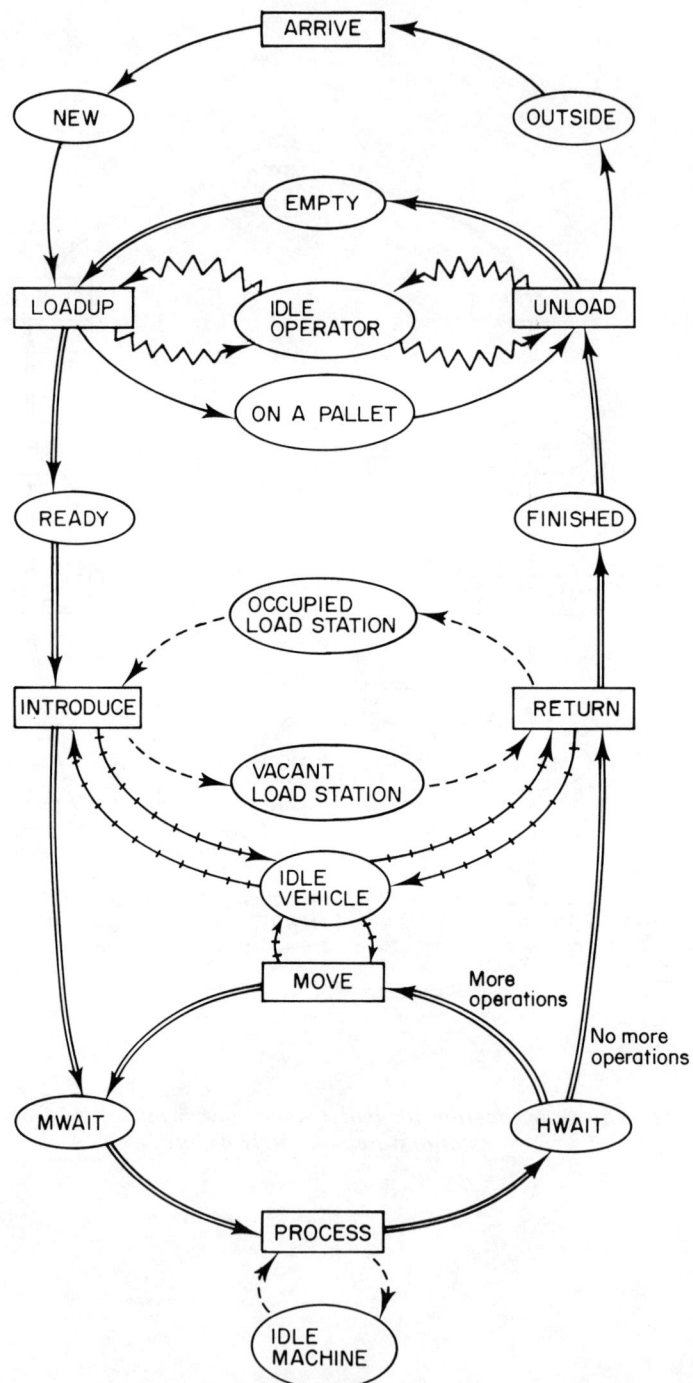

Figure 8.13   Consolidated activity cycle diagram including pallet, component, load stations, load–unload operators, machine and vehicle.

8.7   COMPONENT ARRIVAL MECHANISMS

In controlling the arrival of "customers" into a simulation model, the traditional approach, stemming from queueing theory, is to assume that customers arrive at random intervals. The duration of the time between arrivals is frequently assumed to be distributed according to the negative exponential distribution, whose shape is as shown in Figure 8.14. The interval between arrivals will quite often be short but may from time to time be a much longer. The distribution has been found useful in a wide range of applications, and is defined by a single parameter, the mean time between arrivals. You will remember that this distribution was used in the model in chapter 5.

For modelling manufacturing systems we need to consider whether this approach is appropriate. Depending on our objectives we could adopt one of two basic strategies:

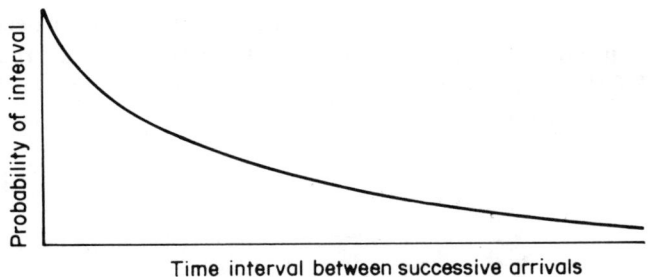

Figure 8.14   Approximate shape of negative exponential distribution.

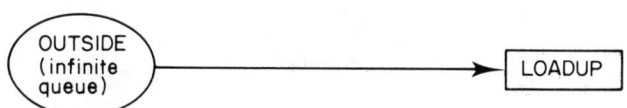

(a) Strategy 1: assuming there are always components waiting

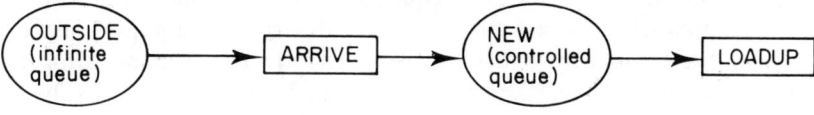

(b) Strategy 2: controlled rate of arrival of components

Figure 8.15   Comparison of component arrival strategies. (a) Strategy 1: assuming there are always components waiting. (b) Strategy 2: controlled rate of arrival of components.

1. assume that there are always components waiting, or
2. control the rate of arrivals by some means.

The first strategy is more appropriate if we seek to know the maximum capacity of the system. The second is suitable if we wish to find out how the system will respond to different work schedules. The fundamental difference between the strategies is that the LOAD activity will never be constrained by a shortage of work in the first method. Figure 8.15 illustrates this point. Now let us consider the implications for our model of these two approaches.

8.7.1   Assuming that there are always components waiting for loading

If we assume that components are always waiting we will no longer need to test for the availablility of a suitable component. As indicated in figure 8.15(a), this eliminates the need for the ARRIVE activity and the NEW queue. The component's activity cycle would simplify to that shown in Figure 8.16.

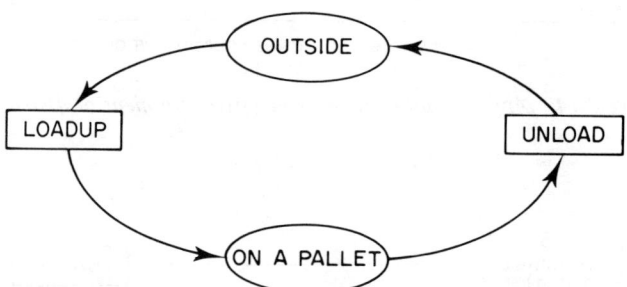

Figure 8.16   Simplified component cycle if it is assumed that components are always waiting.

In fact, the component entity now serves no function in the model which cannot be obtained from attributes of the pallet. The time to process a component can be deduced by recording, in an attribute of the pallet, the time when loading and unloading are done. Similarly, the sequence of operations required on the component can also be stored in attributes of the pallet, or in a table, as will be described in the next section. In these circumstances, the component entity can be omitted from the model altogether.

A further simplification of the logic of the model would normally be possible. If the component entity is omitted from figure 8.13 it will be seen that the conditions necessary for a LOADUP activity to commence will exist as soon as UNLOAD has been completed. Since these activities use the same set of entities, ie a pallet and an operator, they can be combined in a single activity, in which machined components are removed from the pallet and fresh ones put in their place. This would enable figure 8.13 to be redrawn as in Figure 8.17.

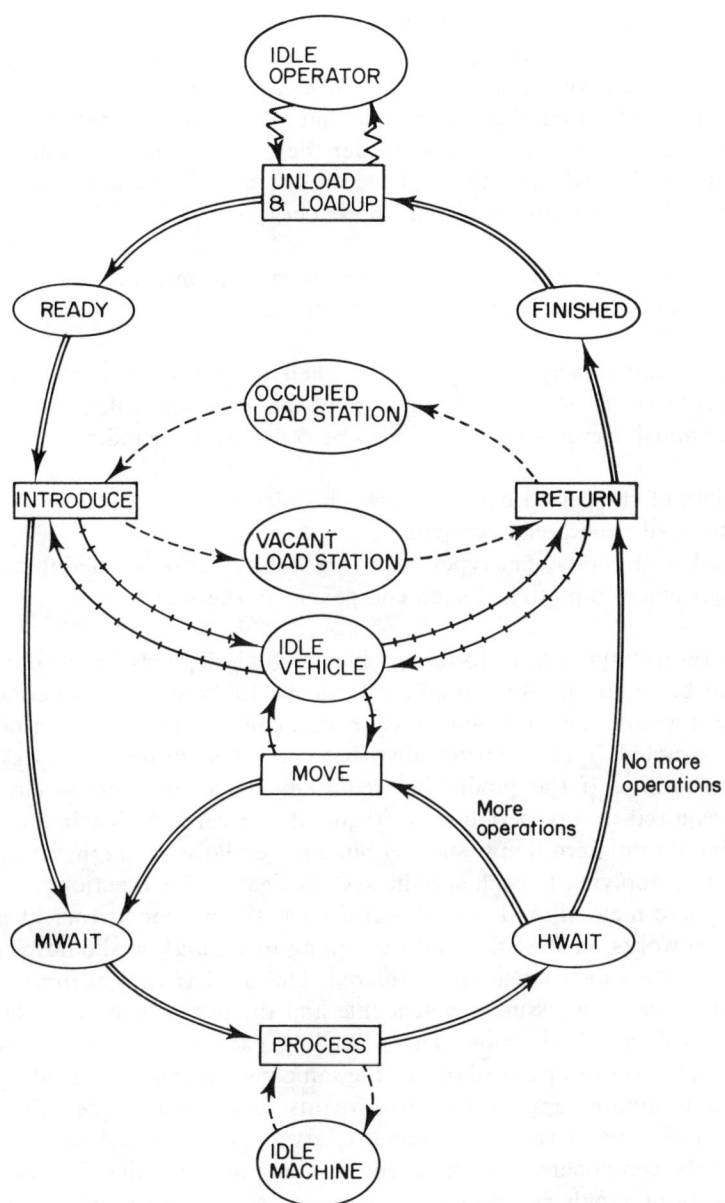

Figure 8.17   Revised activity cycle diagram omitting the component entity.

8.7.2   Controlling the rate of arrival of components

Now consider the case where the rate of arrivals is to be controlled, so that the performance of the system under defined load conditions may be investigated. From the discussion above, one approach would be to introduce new components for processing according to a negative exponential distribution. However, in a manufacturing environment work is usually done to a schedule, so that many orders would arrive simultaneously at regular intervals, such as a week. Under these conditions, the time between arrivals is deterministic, but the quantity and types of components scheduled each week would vary. Once again there are two basic approaches:

1. Create arrivals of individuals or batches at random intervals,
2. Schedule quantities of work at regular intervals.

The second method may be appropriate when the response of the system to typical schedules is to be investigated. The nature of the work scheduled each period will have to be determined. Several factors have to be considered, including:

1. the length of the period between new schedules,
2. the total load generated each period,
3. the number of component types occurring in any period's schedule, and
4. the requirement quantity of each component in the schedule.

There are several approaches to this problem. Much depends on the data available and how it can be organised. One method would be to sample, each period and for each component type, the quantity scheduled of that component in that period. The quantity might be sampled from a normal distribution, a histogram or whatever distribution seems appropriate. If the production requirements of any component are small, the quantity required in any period may frequently be zero. A distribution which gives a finite probability of zero items, such as binomial or Poisson, might be appropriate. The circumstances applying to each specific system clearly play a major part in determining the appropriate method, and we will not discuss the matter further at present. One of the standard works on statistics or on sampling in simulation should be consulted.

Several subtle points must be considered. One is that in real life there is clearly a distinction between the issue of a schedule and the actual supply of the raw material, fixture, control tape and so on. The preceding paragraph seems to assume that the materials and so on will be available at the same instant that the schedule is issued. This may be an acceptable approximation to reality, but a more acceptable method might be to have individual arrivals at random intervals and use a schedule to determine the priority of the components on the schedule. This might require data to be available on the reliability of suppliers, process planners and so on, and probably widens the scope of the simulation project beyond what is practicable. Clearly, how this matter should be handled will depend on the factors relating to the specific system and the objectives of the simulation project.

Another point to be considered is that schedules normally call for components to be completed in the stated period, whereas we are concerned with methods of creating new components. Some method must be evolved to determine when work should start on a component so that it can be completed by the specified period. This seems to suggest that we need to know how well the system will perform before writing the model, which is obviously not practicable. Indeed, one of the objectives of the simulation model may be to discover what the lead times for the system will be.

This raises the subject of "Just-in-time" methods of controlling production, presently a topic of much interest. JIT methods are often termed "pull-systems", because they cause production to be initiated to meet a requirement at a successive workstation. Creating arrivals in the form of raw components at the start of the system relates closely to the traditional "push-systems". Simulation has an important role to play in comparative studies.

This discussion shows that handling component arrivals can be a rather complex subject. We have reviewed the principal aspects of the topic and will give examples, but a fully comprehensive treatment is beyond the scope of this text.

8.8   MACHINES AND WORKSTATIONS

In chapter 5 an elementary activity cycle for the machines was used, as reproduced in Figure 8.18. This cycle applies to workstations in general, and states that the machine is either working, activity PROCESS, or doing nothing, queue IDLE. Let us consider whether this is an adequate representation of the situation. Suppose the workstation is a machining centre, although the approach will apply to any type of workstation with suitable amendments. There are two basic questions which we must ask, namely:

1. Are there any other activities in which the machine engages?
2. Should the PROCESS activity be broken down into more specific activities?

Figure 8.18   Elementary activity cycle diagram for machine.

If we consider the operations of the machine in more detail we might develop a list of activities such as:

1. Place workpiece on worktable.
2. Transfer control program from executive computer.
3. Change tools in the magazine.
4. Execute operations defined in the control program.
5. Gauge the workpiece.
6. Move workpiece away from the machine.

We could go into greater detail. Executing the control program could be broken down into many different activities, such as load tool into spindle, move spindle to start of cut, make cut, etc. We have to decide how far we need to break things down for the purpose of our model. We have already identified some of the principles involved in earlier discussions. The main points are:

1. Whether one of these small actions will inevitably follow on from another.
2. Whether circumstances could arise which would prevent this happening.

For example, once the machining method has been proved, it might be reasonable to assume that the execution of the control program would run successfully once it has been initiated, and therefore there would be no need to break it down into actions dealing with individual tools. Similarly, any gauging will probably be done automatically. Let us assume that the tool magazine of the machine is large enough to hold any tool that may be called up, so that there will be no need to change the tools in the magazine. We might still need to replace worn tools, but let us overlook that possibility for the present. In the same way, we might be justified in assuming that once the control tape has been down-loaded from the executive computer the execution of the program will follow. This would let us define the operations as follows:

1. Place workpiece on worktable.
2. Transfer control program from executive computer and execute it.
3. Move workpiece from the machine worktable.

The main cause for one activity not following directly on another is that it involves a different set of entities. As already suggested, the entities which will be included in the model depend very much on the purpose of building the model. If we are concerned with the operation of the system in a "macro" fashion we would probably not include cutting tools as entities in the model. In that case the simplification just presented would be a reasonable one. On the other hand, we may be particularly interested in the consumption of tools and tool failure. In that case it would be necessary to include tools and probably to record each action of each tool. Since there may be hundreds, or even thousands, of tools in a complex FMS this would add greatly to the model. We will come back to modelling of tools at a later stage, but for the present let us assume that we can treat the execution of the control program as a single activity.

However, we must look more closely at the first and third segments of the process as given above, namely, moving the workpiece on and off the machine worktable. In the model in chapters 1, 4 and 5, we assumed, but perhaps did not state the assumption, that once a job was delivered to the machine, processing could be done. If the operator of the machine placed the workpiece in the machine, machined it and then unloaded it, then the assumption was reasonable, since we had already assumed that each man and his machine were inseparable. However, in an FMS, some mechanism, such as a pallet shuttle, would be used to place a workpiece on a machine and remove it from the machine. Also, the workpieces are brought to, and taken away from, the machine by some vehicle. These pallet shuttles and vehicles are also part of the model and would normally be treated as entities. The machine's activity cycle will therefore become as shown in Figure 8.19. The pallet is clearly involved in all three of these actvitiies, so we should show the relevant part of that entity's cycle. If there is no buffer position at the machine the result will be as in Figure 8.20. We have to show separate queues for each entity between TRANSFER PALLET TO MACHINE and PROCESS, and between PROCESS and TRANSFER PALLET FROM MACHINE, even although both entities are about to enter the same activity. This is necessary to conform to the rule of ACDs that a queue contains one type of entity only. The diagram shows that

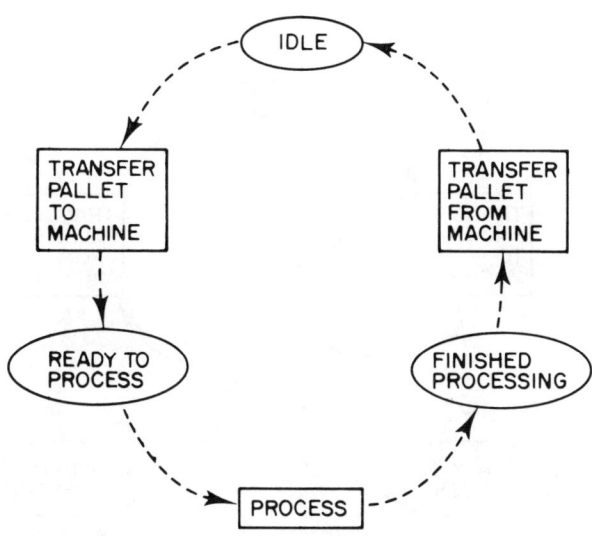

Figure 8.19   Expanded activity cycle diagram for machine.

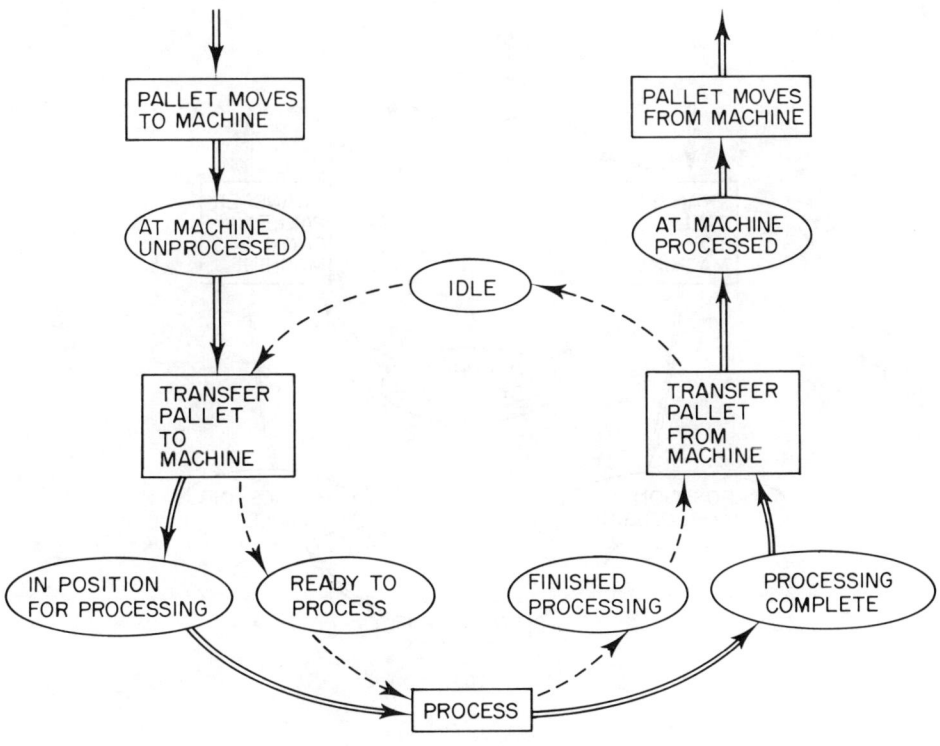

Figure 8.20   Activity cycle diagram for machine and pallet.

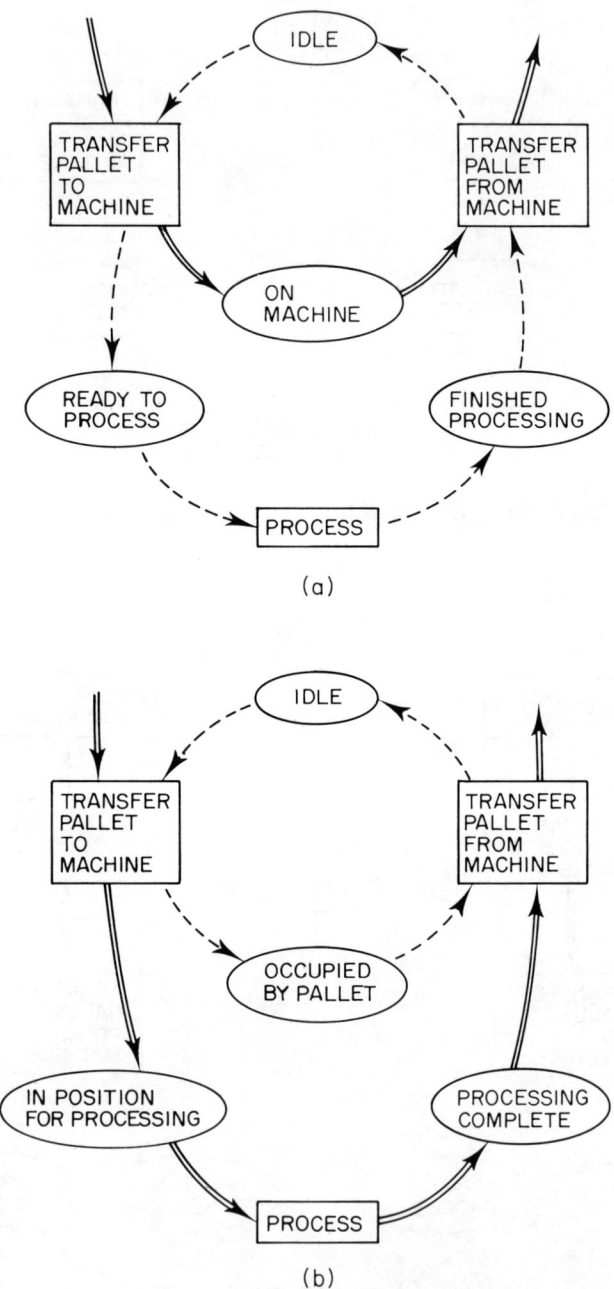

Figure 8.21   Activity cycle diagrams for machine and pallet with (a) simplified cycle for the pallet and (b) simplified cycle for the machine.

the pallet and machine engage in the same sequence of activities from TRANSFER PALLET TO MACHINE until TRANSFER PALLET FROM MACHINE. When entities have parallel cycles some simplification may be possible. Both cycles contain a queue prior to processing, the activity PROCESS and a queue after processing has been completed. If we do not wish to model both cycles in detail then we could substitute a single queue for that part of one or other entity's cycle. This queue would represent either the time when the pallet is on the machine or the time when the machine is occupied by the pallet. Figures 8.21(a) and 8.21(b) show these two alternatives.

As with several such alternatives, which to choose is mainly a matter of convenience. The main factor is probably the ease with which the cycles can be defined with the software package in use. Another factor is the statistics which are to be collected. For example, if we wish to know what proportion of the time a machine is occupied by a pallet, but are less interested in whether the machine is active, or is waiting for the pallet to be moved away from it, then the second approach might be better. This is because the statistic which we need is given directly by observing the time during which machines remain in the queue OCCUPIED BY PALLET. On the other hand, if we need to have a breakdown of the time spent actually working on the components, then we would need to record separately the durations of the PROCESS and TRANSFER activities, and the time spent in the various queues.

When activity cycle diagrams were first described, it was explained that the condition for an activity to begin is that the required entities are in the respective queues leading into the activity. Therefore, when the activity TRANSFER PALLET TO MACHINE has been completed, the conditions for PROCESS to begin must exist. Therefore PROCESS will immediately follow TRANSFER PALLET TO MACHINE. However, the queues IN PLACE FOR PROCESSING and READY TO PROCESS are not necessarily "dummy" queues. The two activities cannot be joined together as one composite activity, because they will involve different sets of entities. For example, the vehicle which brought the pallet to the machine might also be involved in transferring the pallet to and from the machine, but would not be involved in processing the part. Alternatively, the pallet might be transferred to the machine from a buffer station. This case will be discussed in chapter 9.

8.9 BREAKDOWNS

While considering machines and workstations it will be timely to consider a topic which is one of the least deterministic aspects of the performance of manufacturing systems, namely breakdowns of equipment. One of the most frequently asked questions about proposed systems is how seriously will the output of the system be affected by a breakdown of a machine, or any other workstation or piece of equipment? Would the system stop working altogether? Can work be re-assigned to other machines so that production can proceed at a reduced level? If a system is flexible, then it ought be capable of adjusting its operations to minimise the effects of breakdowns.

In incorporating breakdowns in the model, we must consider two aspects:

1. What logic must be added to the model to handle the breakdowns?
2. What logic must be added to model the decision rules for taking remedial action when a breakdown occurs?

8.9.1   Breakdown logic

Breakdowns are unpredictable, and therefore are assumed to occur at random intervals, distributed according to some probability distribution. The average time between breakdowns is called the mean time between failures, or MTBF. The shape of the distribution depends on the characteristics of the equipment concerned, and can often be modelled by one of the standard distributions, such as normal or negative exponential. Alternatively, a histogram could be built up from past records. Samples can be drawn from a probability distribution or from a histogram by the method described in chapter 3. These samples give the time from the completion of the repair, after one breakdown, until the occurrence of the next. The repair time can also be expressed as a probability distribution or histogram, and samples can be drawn, when a breakdown occurs, to determine the time at which normal operation will begin again.

The sequence of actions would be as follows:

1. assume that at time zero the machine is working.
2. draw a sample from the time between breakdowns distribution, and add it to the current clock time.
3. schedule a start of breakdown event at this time.
4. when the clock reaches this time take the machine out of service.
5. draw a sample from the repair time distribution, and add it to the current clock time.
6. schedule an end of breakdown event at this time.
7. when the clock reaches this time return the machine to normal service.
8. go to step 2 and continue cycling through these steps until the end of the simulation run.

In implementing this method, two situations should be recognised:

1. when the availability of maintenance technicians is ignored,
2. when the model must include the availability of the technicians.

The first situation is obviously simpler to model. All we are concerned with is ensuring that sufficient time out of service is allowed. We can do this by adding the repair time to the time of the PROCESS activity. We need to define an attribute for each machine which is subject to breakdown, to record the time of its next failure. If that time will occur during the current PROCESS then the duration of the activity is adjusted by a sample from the repair time distribution, and a sample drawn to determine the time for the subsequent failure. ECSL coding to achieve this might be:

```
DURATION = PROCESSTIME
CHAIN
  CLOCK + DURATION GT BREAKTIME
  DOWNTIME = SAMPLE(REPAIRTIME,SA)
  BREAKTIME = DURATION + DOWNTIME + SAMPLE(RUNTIME,SB)
  DURATION = DURATION + DOWNTIME
  OR CONTINUE
MACHINE FROM READYTOPROCESS INTO FINISHEDPROCESSING
                                      AFTER DURATION
PALLET FROM INPLACE INTO PROCESSINGCOMPLETE AFTER
                                      DURATION
```

In this method, DOWNTIME is the total time for which the machine is out of action, including any time waiting for the technician to arrive.

In the second approach we take a closer look at the repair process, and include the repair technician (or groups of maintenance personnel in some form) in the model. We have to check for the availability of the technician before starting the repair operation. Using the same extra attribute and overall logic, if the PROCESS activity is due to be completed before the breakdown then the machine moves to FINISHED PROCESSING in the normal way, but if the breakdown will interfere with operation the machine will be placed in a BROKEN DOWN queue. This could be represented by expanding the activity cycle diagram of the machine, figure 8.19, to include the contents of Figure 8.22. The duration of REPAIR is the actual repair time and excludes the time waiting for the technician, which is represented by the queue BROKEN DOWN. Note that the queue BACK IN SERVICE is included here to represent a queue such as IDLE or PROCESSING COMPLETE, depending on the effects of the breakdown.

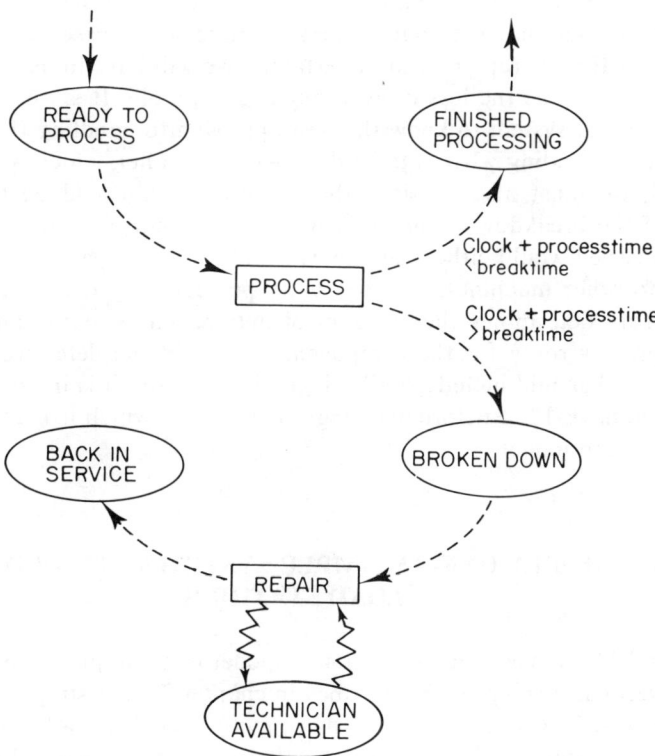

Figure 8.22   Addition to machine's activity cycle to handle breakdowns.

With either approach, we must adopt the full machine activity cycle. The simplification shown in figure 8.21(b) cannot be used.

8.9.2   Actions for recovery after breakdowns

One of the trickier aspects of modelling breakdowns is to decide what to do with the component, pallet, fixture, etc, which are involved in the PROCESS activity during which the machine broke down. This can add considerably to the complexity of the model. In the first method we merely delayed the completion of the operation, and there was no problem in expressing the logic.

However, with the second method we must take some explicit action. The pallet was omitted from figure 8.22, so that this question could be considered separately. For one thing we do not know when the machine will be back in service and so cannot directly place the pallet into PROCESSING COMPLETE at the correct time. However, if we assume that the pallet and its load can proceed according to plan, but merely be delayed in time, then the FINISHED PROCESSING queue can be the back in service condition, and if the pallet's cycle had been simplified as in figure 8.21(a), then the pallet would be withdrawn from ON MACHINE later as though nothing had happened.

In reality, things are likely to be more complicated. For example, the component may have been scrapped or the pallet or fixture damaged. If the component is scrapped, is it removed from the pallet at the machine or will it travel back to the load–unload area to be removed? Will it travel to an inspection station for a decision to be made? The pallet as a whole may be taken out of the system, in which case the machine should be returned to IDLE after repair. Almost certainly, we will have to make some simplifying assumptions, so long as the loss of accuracy is acceptable. It is important to study the practice in the real system to know the best approach to handling these problems.

In addition to deciding what is to be done with the pallet, which was on the machine, and its load, we must also consider the actions which would be taken to minimise the effects of the breakdown. The work which would normally have been done by the machine concerned would either have to wait until it was repaired, or would have to be re-assigned to other machines. The flexibility possessed by the system and its control software largely determines the freedom of action. The system can probably use an alternative process route for the components, to avoid the defective workstation. The simulation model should include similar logic. Here again, it is important to know what would be done in real life before deciding the extent to which it is necessary to include these effects in the model.

8.10   MODEL ONE—A SIMPLE SYSTEM WITH DEDICATED
LOAD STATIONS

For our first FMS model, let us develop a model of a simple system, rather like the Caterpillar vertical turning FMS described in chapter 7, consisting of three machines, a vehicle and twelve load stations, as shown in Figure 8.23. The load stations are dedicated to particular part types and also serve as pallet storage locations. We will make a few simplifying assumptions. Whereas in the real system most of the parts required two operations, and some fixtures were used by two component types, we will assume that each part requires only one operation and every pallet is used by only one part type. The vehicle in the real system had two load-carrying positions, but we will assume that it can carry only one pallet at a time. So the main assumptions will be:

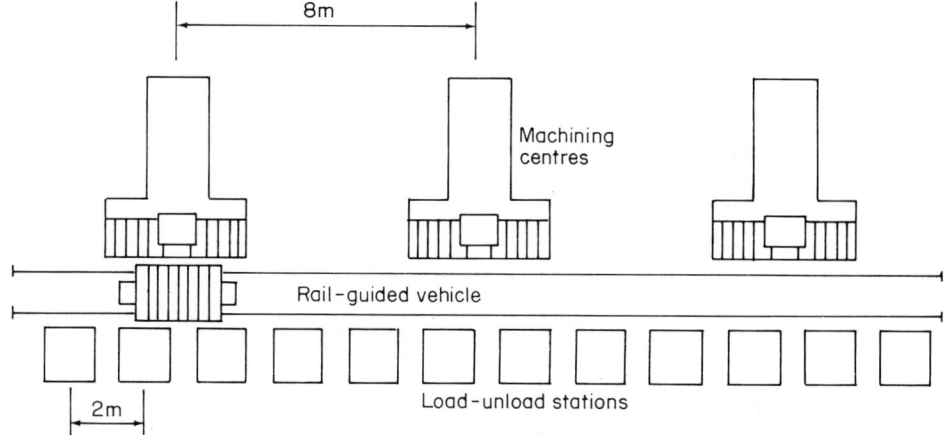

Figure 8.23 Layout of system for Model One, a simple system with dedicated load stations.

1. the vehicle has only one load-carrying position
2. each part requires only one operation
3. operations can be performed on any machine
4. machines do not break down
5. each load station serves only one part type and pallet
6. fixtures are permanently mounted on the pallets.

We could add to this list, if we consider the fine details of how the system operates, probably *ad infinitum*. For example, we could add that cutting tools last for ever, or at least long enough for us not to have to consider them. However, we have to make a start, and the key is to have a reasonably sound basis, which the above list probably provides.

8.10.1 Activity cycle diagram

The entities in the model will be:

machines
components
pallets
load-unload positions
operators
vehicle.

Since there an equal number of load stations and pallets, and each load station is dedicated to a particular pallet, the availability of a load station can never prevent any activity from occurring. Therefore we can omit the load stations from the model.

We can form the activity cycle diagram for the system by referring to figures 8.13 and 8.21(b) and cutting out the load stations and the possibility of more operations. This gives the diagram of Figure 8.24. The use of figure 8.13 implies that the rate of arrival of the components will be controlled in some way.

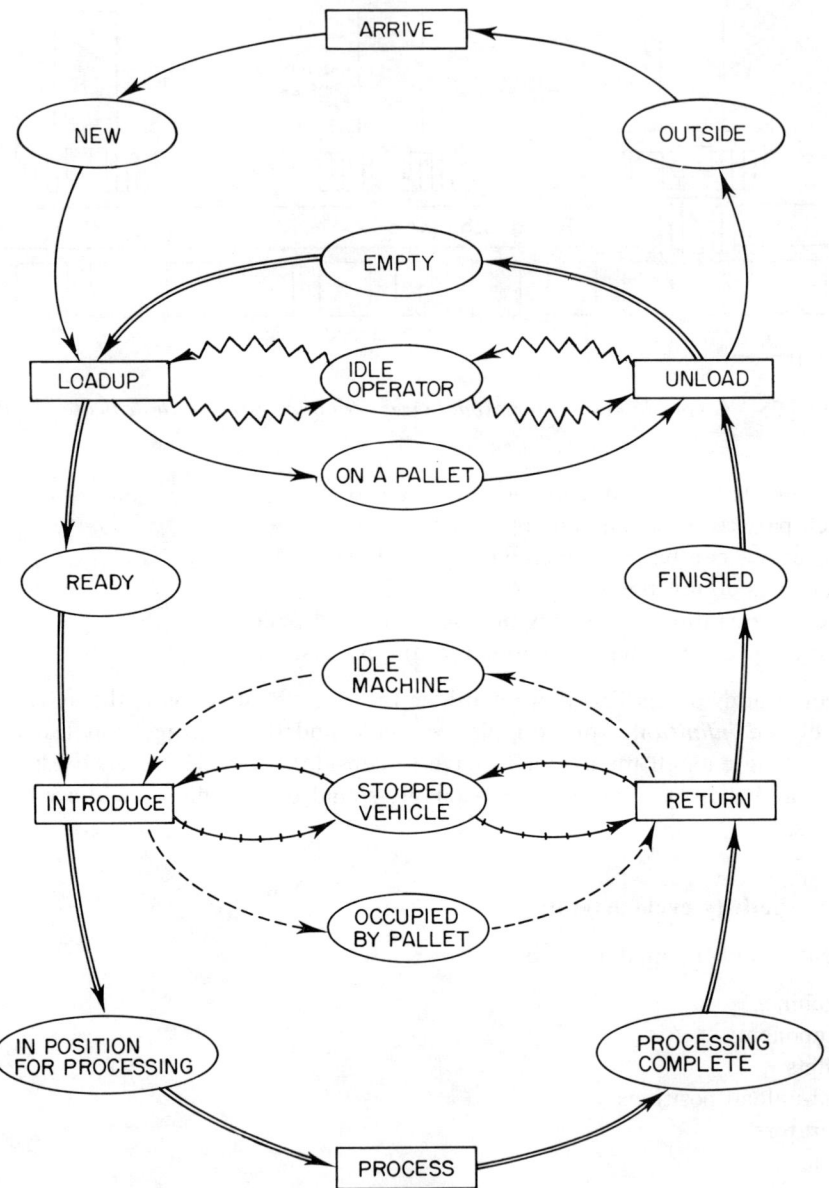

Figure 8.24   Activity cycle diagram for the system for Model One, a simple system with dedicated load stations.

8.10.2 Queue disciplines

If we were to use CAPS to build the model we would need to specify the various queue disciplines. In the model we must ensure that when we load a component on a pallet we use the correct type, and when we unload the pallet we put the correct component aside. In real life this is obviously the case, but because the pallet and component are separate entities and are held in different queues, we have to logically link them together in the model by some explicit method. Similarly, we must keep track of which pallet is at which machine so that, when the operation is complete, we set the correct machine back to idle and return the correct pallet to its correct load station. The rules will be very similar to those used in chapter 5. The queues which need to have a priority rule defined will therefore be NEW, OCCUPIED and ON A PALLET. As in chapter 5 we have to define attributes of the entities concerned so that they can be matched.

NEW: We must select a component from NEW of the same type as the pallet in EMPTY. We define a type attribute for each component, say CTYPE, and a type for each pallet, say PTYPE. Then, when we are testing to see whether a load can be done, we search for a component whose value of CTYPE matches the value of PTYPE of a pallet in EMPTY. PTYPE will be a permanent attribute of the pallet. CTYPE will be assigned as each component arrives.

OCCUPIED: Similarly, we will define an attribute of the pallet to record which machine it is at, say MCAT, and an attribute of the machine, say MNO, so that we can set MCAT of the pallet to MNO of the machine. When an operation on a pallet is complete we then know that machine MNO is the one being set to IDLE. MNO will be a permanent attribute. MCAT will be set in INTRODUCE to MNO of the machine selected, and will be set back, to say zero, when the pallet returns to the load station.

ON A PALLET: Once again we need to record the identity of the component which is on the pallet, so that we unload the correct one. Since there can only be one component of each type on a pallet, and only one pallet of each type, there will only be one component with the correct value of CTYPE in ON A PALLET so we can use the same attributes and test as for NEW.

8.10.3 Data to be used

We shall use fictional data for part requirements and the time for the operations and movements. Suppose the forecast monthly requirements of the components and the machining times per piece are:

Component number	Quantity per month	Machining time
1	320	18
2	270	60
3	260	36
4	180	40
5	140	36
6	120	24
7	100	32
8	80	28
9	72	20
10	64	20
11	48	30
12	40	32

We can do a simple calculation to assess the load on the machines, by multiplying the quantities by the times and dividing by the time available in the working month. Assuming 480 minutes per shift, up to 3 shifts per day, 20 days per month and 100% efficiency, the machine utilisation (expressed as percentages) would be:

Utilisation of machines (%)

		Number of shifts		
		1	2	3
Number	1	597	298	199
of	2	298	149	99
machines	3	199	99	66

This shows that three machines will have to work three shifts, since 2 machines on 3 shifts or vice versa leaves no margin for losses.

We can do similar calculations for the load–unload operator. Suppose that the total load and unload time is 6 minutes per component, then the utilisation of the operator will be:

Number of shifts per day:	1	2	3
Utilisation of operator (%):	105	52	35

The operator will be somewhat under-utilised if he works three shifts. Of course, he may have other duties to perform, such as tool preparation. Unless these duties are demanding we can be confident that the operator will not be a constraint on the output of the system. We might investigate whether having the operator present on two shifts only would be advisable.

Similarly, if moving a pallet from load station to a machine or vice versa takes 2 minutes, then we obtain:

Number of shifts per day:	1	2	3
Utilisation of vehicle (%):	70	35	23

Clearly there will be no capacity problem with the vehicle, if the system works three shifts. Strictly, it is questionable whether we would be justified in using a single time estimate for all movements irrespective of the location of the source and destination. The distances travelled vary up to about 20 metres, which at a speed of say 40 metres per minute, which is quite modest for a rail-guided vehicle, the maximum travelling time will be less than one minute. If transferring the pallet between vehicle and machine or load station takes around 30 seconds, at both source and destination, the times for the shortest move and the longest move will lie between one and two minutes, even allowing for acceleration and deceleration. Hence a single estimate of two minutes seems not unreasonable, for the present at least. We will examine materials movements in more detail in chapter 9.

Another essential piece of data is the rate of arrival of the components. We can divide the time available by the number of components to be produced, and determine the average time between successive arrivals for each type of component. For example, for component 1, 320 parts are required per month, 80 per week, or 16 per day, which is equivalent to one every $480/16 = 30$ minutes if one shift is worked per day, or $480 \times 3/16 = 90$ minutes if three shifts are worked. The complete set of inter-arrival times are:

Average times between arrivals, in minutes

Component number	1 shift per day	2 shifts per day	3 shifts per day
1	30	60	90
2	35	71	106
3	36	73	110
4	53	106	160
5	68	137	205
6	80	160	240
7	96	192	288
8	120	240	360
9	133	266	400
10	150	300	450
11	200	400	600
12	240	480	720

As observed in a preceding section, we can either generate arrivals of each component type using separate arrival generators, or we can generate arrivals at the overall rate, and sample for the component type. It will be simpler in the model to do it in the latter way. The overall rate of arrival can be computed, based on the overall total monthly production, as follows:

Number of shifts per day:	1	2	3
Mean time between arrivals:	5	11	17

The results which we will wish to collect include facility utilisations, the time for a component to pass through the system, number produced, and queue lengths. Some other statistics may be collected to check on the correct working of the model, rather than to assess the productivity of the system. We could use a run-in period of a day, and a run length of a week. Strictly, the results obtained using these times should be assessed for stability and the periods revised if necessary.

Finally, we must define the initial conditions. Again for the sake of simplicity, we can assume that at the start of the run there are no components in the system, all pallets are empty, and the machines, operator and vehicle are idle.

8.10.4   Program listing

The listing of an ECSL model for this system is given in Figure 8.25. The model was developed using CAPS and refined by manual editing of the code. A substantial proportion of the program is concerned with collecting the statistics and printing out the values obtained. Most of the names for the statistical variables have been assigned by CAPS. User-assigned names include TIHIST, TIMHIST and MADE. TIHIST is the histogram for the time in the system of all the parts, while TIMHIST are the time in the system histograms for each type of component. TIHIST and TIMHIST are similar histograms, but have to be given different names, because TIMHIST is an attribute of pallet, while TIHIST is not an attribute. The time in the system of components is measured from the start of loading to the end of unloading. MADE is a counter for the number of each type of component produced.

```
THERE ARE 50 COMPON SET NEW ONPALL OUTSID WITH
+TIMIN CTYPE
THERE ARE 1 OPERAT
THERE ARE 12 PALLET SET READY COMPLE FINSHD EMTY WITH
+PTYPE MCAT MACHTIME MADE HIST TIMHIST ( 11 10 20)
THERE ARE 3 MACHIN SET OCCUPI IDLE WITH MNO
THERE ARE 1 VEHICL
THERE ARE 1 ZZARRI
HIST TIHIST( 11 10 20 ) TYPEDIST( 12 1 1)
FUNCTI PICTUR SAMPLE NEGEXP MEAN TOTAL
RECYCL
RUNINZ = ONEDAY AND PREVCLOCK = ONEDAY
ACTIVITIES ONEWEEK + ONEDAY

DURATION= CLOCK - PREVCLOCK
ADD NEW TO HIST ZANEW DURATION
ADD ONPALL TO HIST ZBONPALL DURATION
ADD READY TO HIST ZCREADY DURATION
ADD COMPLE TO HIST ZDCOMPLE DURATION
ADD FINSHD TO HIST ZEFINSHD DURATION
ADD EMTY TO HIST ZFEMTY DURATION
ADD TIME IN EMTY TO HIST ZGEMTY ( 11 10 20 )
ADD OCCUPI TO HIST ZHOCCUPI DURATION
ADD TIME IN OCCUPI TO HIST ZIOCCUPI( 11 10 20 )
FOR OPERAT WITH TIME OF OPERAT LT 0
   ADD DURATION TO AZOPERAT
FOR IDLE
   ADD DURATION TO BZMACHIN
FOR VEHICL WITH TIME OF VEHICL LT 0
   ADD DURATION TO CZVEHICL
FOR ZZARRI WITH TIME OF ZZARRI LT 0
   ADD DURATION TO DZZZARRI
PREVCLOCK = CLOCK
```

Figure 8.25   ECSL program listing for Model One. (a) Definitions and Recording sections. (b) Activities section. (c) Finalisation and Data sections.

```
BEGIN RETURN
TIME OF VEHICL LE 0
FIND FIRST PALLET A IN COMPLE
   FIND FIRST MACHIN B FROM OCCUPI
      MNO EQ MCAT
MCAT OF PALLET A= 0
DURATION = MOVETME
PALLET A FROM COMPLE INTO FINSHD AFTER DURATION
MACHIN B FROM OCCUPI INTO IDLE AFTER DURATION
TIME OF VEHICL = DURATION
ADD 1 TO RETURN

BEGIN INTROD
TIME OF VEHICL LE 0
FIND FIRST PALLET A IN READY
FIND FIRST MACHIN B IN IDLE
MCAT OF PALLET A= MNO OF MACHIN B
DURATION = MOVETIME
AADURATION= DURATION + MACHTIME OF PALLET A
PALLET A FROM READY INTO COMPLE AFTER AADURATION
MACHIN B FROM IDLE INTO OCCUPI AFTER DURATION
TIME OF VEHICL = DURATION
ADD MACHTIME OF PALLET A TO TMACHTIME
ADD 1 TO MADE OF PALLET A
ADD 1 TO INTROD

BEGIN UNLOAD
TIME OF OPERAT LE 0
FIND FIRST PALLET B IN FINSHD
   FIND FIRST COMPON A FROM ONPALL
      CTYPE EQ PTYPE
TIMIN OF COMPON A= CLOCK - TIMIN OF COMPON A
ADD TIMIN OF COMPON A TO TIMHIST OF PALLET B
ADD TIMIN OF COMPON A TO TIHIST
DURATION = ULOADTIME
COMPON A FROM ONPALL INTO OUTSID AFTER ULOADTIME
TIME OF OPERAT = ULOADTIME
PALLET B FROM FINSHD INTO EMTY AFTER ULOADTIME
ADD 1 TO UNLOAD

BEGIN LOADUP
TIME OF OPERAT LE 0
FIND FIRST PALLET B IN EMTY
   FIND FIRST COMPON A FROM NEW
      CTYPE EQ PTYPE
TIMIN OF COMPON A= CLOCK
DURATION = LOADTIME
COMPON A FROM NEW INTO ONPALL AFTER LOADTIME
TIME OF OPERAT = LOADTIME
PALLET B FROM EMTY INTO READY AFTER LOADTIME
ADD 1 TO LOADUP

BEGIN ARRIVE
TIME OF ZZARRI LE 0
FIND FIRST COMPON A IN OUTSID
CTYPE OF COMPON A= SAMPLE ( TYPEDIST, SB )
DURATION= 0+NEGEXP( OVRLART, SA )
COMPON A FROM OUTSID INTO NEW
TIME OF ZZARRI = DURATION
ADD 1 TO ARRIVE
REPEAT
```

Figure 8.25(b).

```
FINALISATION
TYPE **"Final report from simulation MODELONE "/
TYPE 'RETURN was started' RETURN ' times'
TYPE 'INTROD was started' INTROD ' times'
TYPE 'UNLOAD was started' UNLOAD ' times'
TYPE 'LOADUP was started' LOADUP ' times'
TYPE 'ARRIVE was started' ARRIVE ' times'
TYPE 'Utilization of OPERAT'+4,(1-AZOPERAT/( 1. *(CLOCK -RUNINZ)))
TYPE 'Occupancy of MACHIN'+4,(1-BZMACHIN/( 3. *(CLOCK -RUNINZ)))
TYPE 'Utilization of MACHIN'+4,(TMACHTIME/( 3. *(CLOCK -RUNINZ)))
TYPE 'Utilization of VEHICL'+4,(1-CZVEHICL/( 1. *(CLOCK -RUNINZ)))
TYPE 'Utilization of ARRIVE'+4,(1-DZZZARRI/( 1. *(CLOCK -RUNINZ)))
TYPE 'Average number in queue NEW     '+4,MEAN (ZANEW)
TYPE 'Average number in queue ONPAL    '+4,MEAN (ZBONPAL)
TYPE 'Average number in queue READY    '+4,MEAN (ZCREADY)
TYPE 'Average number in queue COMPL    '+4,MEAN (ZDCOMPL)
TYPE 'Average number in queue FINSHD   '+4,MEAN (ZEFINSHD)
TYPE 'Average number in queue EMTY     '+4,MEAN (ZFEMTY)
TYPE 'Average number in queue OCCUPI   '+4,MEAN (ZHOCCUP)
TYPE 'Average time in system - overall      '+4,MEAN (TIHIST)
FOR PALLET I
  TYPE 'Average time in system - type 'I   ,+4,MEAN (TIMHIST)
FOR PALLET I
  TYPE 'Type ' I ' Number made ' MADE OF PALLET I
FOR PALLET I SUM MADE
TYPE 'TOTAL MADE 'I
TYPE 'Histogram of length of queue NEW    '/PICTURE(ZANEW )
TYPE 'Histogram of length of queue ONPALL'/PICTURE(ZBONPAL)
TYPE 'Histogram of length of queue READY '/PICTURE(ZCREADY)
TYPE 'Histogram of length of queue COMPLE'/PICTURE(ZDCOMPL)
TYPE 'Histogram of length of queue FINSHD'/PICTURE(ZEFINSHD)
TYPE 'Histogram of length of queue EMTY   '/PICTURE(ZFEMTY )
TYPE 'Histogram of delays at EMTY   '/ PICTURE(ZGEMTY )
TYPE 'Histogram of length of queue OCCUPI'/PICTURE(ZHOCCUP)
TYPE 'Histogram of delays at OCCUPI'/ PICTURE(ZIOCCUPI)
TYPE 'Histogram TIHIST of TIMIN   '/ PICTURE(TIHIST)

DATA
TIME OF ZZARRI 0
OUTSID1 TO *
EMTY 1 TO *
IDLE 1 TO *
ONEWEEK 7200
ONEDAY 1440
OVRLART 17
MOVETIME 2
LOADTIME 3
ULOADTIME 3
TYPEDIST 1694 320 270 260 180 140 120 100 80 72 64 48 40
MACHTIME       18  60  36  40  36  24  32 28 20 20 30 32
SB 333
SA 111
MNO 1 2 3
PTYPE 1 2 3 4 5 6 7 8 9 10 11 12
 END
```

Figure 8.25(c).

8.10.5 Initial results

The results obtained on running the model, with the exception of the histograms which have been omitted to save space, are quoted in Table 8.1 and broadly match those predicted. The slight differences between the number of times the various activities were started is not an error. For example, that 440 components were introduced to machines and only 437 returned from them does not mean that 3 components vanished from the machines. During the sampling period 440 components were taken to machines, and most of them would be among the 437 which returned. The first few returned would have been introduced before the sampling period started. Similarly, the last few components to be introduced would not be returned within the sampling period. The two statistics are independent, though correlated, observations, and over a long period should become equal. The facility utilisations are very close to the predicted values. The difference between the occupancy and the utilisation of the machines is that, whereas the utilisation is calculated on the processing time involved, the occupancy includes the time when the machines are in the FINISHED PROCESSING queue and in the INTRODUCE and RETURN activities, ie all the time not in IDLE. There is a slight difference in the number of components produced in total and of some types, from that expected. The expected quantities per week, a quarter of the monthly requirement, compared with those actually produced and the number which would have been produced if the total actually produced were in strict proportion to the requirements, are:

Component:	1	2	3	4	5	6	7	8	9	10	11	12	Total
Expected:	80	67	65	45	35	30	25	20	18	16	12	10	423
Actual:	80	62	61	53	33	40	35	25	13	14	14	10	440
Pro-rata:	83	70	68	47	36	31	26	21	19	17	12	10	440

The main deviations are that proportionally fewer components of types 2, 3 and 9 and more of 4, 7 and 11 have been produced than required. An important question arises here: Are these differences statistically significant, or are they within the limits of statistical sampling? Standard statistical tests may be applied. A Chi-squared test can be applied to the mix of components, comparing the actual numbers produced and the quantities which would have been expected in a total of 440. Chi-squared has the value $(83 - 80)^2/83 + (70 - 62)^2/70 + \ldots + 0/10 = 11.56$. There are eleven degrees of freedom, since twelve values are involved. With eleven degrees of freedom this value of Chi-squared is far from significant, and we can conclude that there is no evidence that the mix is not acceptable. The difference between the expected number of arrivals is only 4%, but a deviation of this magnitude, on such a large number of samples from the time between arrivals distribution, is a cause for some concern. In chapter 3, courses of action such as performing additional runs using different seeds for the random number streams, or longer run lengths should be performed to ensure the stability of the results. In the extreme, the validity of the random number generators in the package may be questioned. Unfortunately, the simulation engineer generally has no means of amending the generators, and can either accept their validity or refer his doubts to the software supplier for comment, and perhaps improvement in a later release of the package. For our present purposes we will accept the values obtained.

Table 8.1   Initial results from Model One

RETURN was started	437 times
INTROD was started	440 times
UNLOAD was started	437 times
LOADUP was started	442 times
ARRIVE was started	441 times

Utilization of OPERAT	.3662
Occupancy of MACHIN	.7643
Utilization of MACHIN	.6816
Utilization of VEHICL	.2436
Utilization of ARRIVE	1.0000

Average number in queue NEW	1.5463
Average number in queue ONPAL	2.8568
Average number in queue READY	.5381
Average number in queue COMPL	.0091
Average number in queue FINSHD	.0254
Average number in queue EMTY	8.7769
Average number in queue OCCUPI	2.1718

Component Type	Number made	Average time in system
1	80	37.0000
2	62	73.9344
3	61	56.8852
4	53	56.4151
5	33	59.0909
6	40	42.3077
7	35	49.4285
8	25	42.5000
9	13	34.6153
10	14	37.1428
11	14	48.5714
12	10	44.0000
Overall	440	51.1899

8.10.6   Experimentation

This is a very simple model, and there is little scope for unpredictable interaction among the entities, and consequently the results conform to those predictable by simple calculations. At the strategic planning level, we have established the capacity of the system. However, there is some scope for experimentation at the tactical and operational levels.

1.   Operator and vehicle priority rules

There are two implicit priority rules in the model, which may affect the results. These refer to the operator and vehicle. Both of these entities do two activities. The ordering of these activities in the model will cause the vehicle to perform RETURN in preference to INTRODUCE, and the operator to UNLOAD completed components in preference to LOADUP fresh components. It would be wise to discover whether the results are affected by this ordering. Simulation runs were performed with the order of LOADUP and UNLOAD, and of INTRODUCE and RETURN reversed. Table 8.2 gives selected results. Although the differences are small, they confirm what we would intuitively expect. The length of queue NEW is reduced, and FINISHED increased, by giving priority to loading rather than unloading. COMPLETE is much reduced when RETURN has priority over INTRODUCE, with a slight reverse effect on READY. Similarly, the time in the system is reduced by giving priority to UNLOAD and RETURN. The output of the system increases by one component when LOADUP has priority over UNLOAD. This may be a marginal effect in that one more occurrence of the INTRODUCE activity may just have happened to fall within the sampling period, rather than due to any real difference.

Table 8.2   Results with alternative operator and vehicle priority rules

Vehicle Priority:	Operator Priority:	UNLOAD > LOADUP	LOADUP > UNLOAD
RETURN	Average number in queue NEW	1.5463	1.4433
>	Average number in queue READY	.5381	.5350
INTRODUCE	Average number in queue COMPL	.0091	.0073
	Average number in queue FINSHD	.0254	.0629
	Average time in system—overall	51.1899	51.5981
	TOTAL MADE	440	441
INTRODUCE	Average number in queue NEW	1.4229	1.3943
>	Average number in queue READY	.5194	.5100
RETURN	Average number in queue COMPL	.0184	.0166
	Average number in queue FINSHD	.0261	.0547
	Average time in system—overall	50.7780	51.0502
	TOTAL MADE	440	441

The results show generally only marginal effects by changing the priority rules. This can be attributed to the fact that the utilisation of the vehicle and operator are sufficiently low for these entities to almost always waiting for a pallet or component to become available for an activity. It is likely that, with higher utilisation of vehicle or operator, a greater effect would have been observed. The remaining runs are based on LOADUP having greater priority than UNLOAD and INTRODUCE having greater priority than RETURN, which seemed to give a good compromise between output and time in system.

2.   Assuming components are always available

The results indicate that the output of the system is not limited by the capacity of the machines, vehicle or operator, but by the rate of arrival of components. We could increase this rate in stages until the system becomes saturated. However, we can go directly to the same result by using the technique described in section 8.7, that is, assuming that there would always be components waiting. In principle this means ensuring that there are always components in the queue NEW, so that the existence of NEW components never constrains the system. To do so, we can eliminate the component entity altogether and join the UNLOAD and LOADUP activities. Queue EMTY will not exist. TIMIN, which was an attribute of the component will have to become an attribute of the pallet. The revised logic for the new combined unload and load activity is as follows:

```
BEGIN LOADUP
TIME OF OPERAT LE 0
FIND FIRST PALLET B IN FINSHD
TIMIN OF PALLET B= CLOCK + ULOADTIME - TIMIN OF PALLET B
ADD TIMIN OF PALLET B TO TIMHIST OF PALLET B
ADD TIMIN OF PALLET B TO TIHIST
TIMIN OF PALLET B= CLOCK + ULOADTIME
DURATION = LOADTIME + ULOADTIME
TIME OF OPERAT = DURATION
PALLET B FROM FINSHD INTO READY AFTER DURATION
ADD 1 TO LOADUP
```

The results are given in Table 8.3. They show that output has increased from 441 to 603, or 37%, and that the occupancy of the machines is 100%, indicating that the system

Table 8.3   Results when it is assumed that there are always components available for loading, and the component entity is deleted from Model One

RETURN was started	604 times
INTROD was started	603 times
LOADUP was started	603 times
Utilization of OPERAT	.5027
Occupancy of MACHIN	1.0000
Utilization of MACHIN	.8656
Utilization of VEHICL	.3350
Average number in queue READY	8.4306
Average number in queue COMPL	.0627
Average number in queue FINSHD	.0666
Average number in queue OCCUPI	2.8325

Component Type	Number made	Average time in system
1	51	144.5098
2	45	162.7273
3	50	144.1176
4	50	145.6863
5	51	142.1568
6	51	145.6863
7	51	144.9019
8	51	142.1568
9	51	144.9019
10	52	144.5098
11	50	144.8000
12	50	144.8000
Overall	603	145.7214

is now working at maximum potential. However, the most striking result is that the numbers of components of each type produced are roughly equal, as are the time in system values, which have almost trebled. We have achieved maximum output but lost control over the product mix. In fact this is to be expected, since, whereas in the earlier runs the mix of components arriving into the queue NEW was controlled by sampling from the TYPEDIST histogram, in the new run this was not done. What we need is a decision rule to ensure that components are produced in the correct proportions to meet the scheduled requirements.

3. A schedule-based priority rule

In the runs whose results are quoted in tables 8.1 and 8.2, the operator selected the next pallet to load using the first-in-first-out rule, and not in relation to the production requirements. He might deal with a component which has been waiting a long time even although it may not be very urgent, in the sense that it was not needed to meet the scheduled requirement, at the expense of other more urgent parts. In the run which gave the figures in table 8.3, the operator works according to the order in which pallets arrive from the machines, and he is generally waiting for the pallet to arrive. Consequently the important decision is which of the waiting pallets the vehicle takes to the machines. In the initial runs this too was determined using the FIFO rule. Thus the vehicle selected the pallet which had been waiting longest to be taken to a machine, rather than the most urgent.

We need a rule which will ensure that the operator or vehicle selects the next pallet to load according to the schedule of requirements of the components. A simple technique for keeping the production of each component type in line with its requirements is to compare the number produced at any time with the number required and concentrating on those components of which fewer in proportion have been made. To do this we need additional attributes of the pallet.

Let REQD be the number required of each type,
let LOADED be the number so far loaded up, and
let RATIO be the ratio of LOADED to REQD.

We can make the vehicle select, to INTRODUCE to a machine, the pallet with the minimum value of RATIO. REQD, the monthly requirements, will be defined as data, with similar figures to TYPEDIST. LOADED and RATIO will be calculated in INTRODUCE every time a pallet is selected. The amended version of the INTRODUCE activity is as follows:

```
BEGIN INTROD
TIME OF VEHICL LE 0
FIND PALLET A IN READY WITH MIN RATIO
FIND FIRST MACHIN B IN IDLE
LOADED OF PALLET A + 1
X = 1000 / REQD OF PALLET A
RATIO OF PALLET B = X * LOADED OF PALLET A
MCAT OF PALLET A = MNO OF MACHIN B
```

AADURATION = MOVETIME + MACHTIME OF PALLET A
PALLET A FROM READY INTO COMPLE AFTER DURATION
MACHIN B FROM IDLE INTO OCCUPI
TIME OF VEHICLE = MOVETIME
ADD MACHTIME OF PALLET A TO TMACHTIME
ADD 1 TO MADE OF PALLET A
ADD 1 TO INTROD

Note that because the keyword ADD is used to increment it, MADE will only be incremented after the run-in period, whereas LOADED is incremented by one every time the activity occurs from the start of the run. RATIO must be an integer attribute, so the constant 1000 has been introduced to ensure that RATIO will have meaningful values, and the variable X has been introduced to avoid integer overflow in the arithmetic.

The results obtained with this revision are given in Table 8.4. We see that the proportions of the components produced are once more in line with the schedule requirements. The time in system values are higher for component types with smaller requirements.

Table 8.4   Results when it is assumed that there are always components available for loading, and the component entity is deleted from Model One—using a Schedule-based priority rule

RETURN was started	558 times
INTROD was started	558 times
LOADUP was started	557 times
Utilization of OPERAT	.4638
Occupancy of MACHIN	1.0000
Utilization of MACHIN	.8841
Utilization of VEHICL	.3097
Average number in queue READY	8.4906
Average number in queue COMPL	.0416
Average number in queue DONE	.0455
Average number in queue OCCUPI	2.8450

Component Type	Number made	Average time in system
1	100	71.8000
2	91	79.0110
3	100	73.8000
4	60	122.0000
5	43	166.7442
6	37	191.0811
7	30	209.3333
8	26	210.0000
9	23	210.0000
10	21	210.0000
11	15	210.0000
12	12	210.0000
Overall	558	125.2244

This follows from the fact that pallets of low demand components will remain in READY at the load station longer than those for high demand components. This can be verified by comparing the lengths of the queue READY in tables 8.3 and 8.4.

These comments apply to the case in which components are assumed to be always available. We should also add the rule to the more normal case where the rate of arrivals is controlled. Some restructuring is appropriate. Instead of placing the rule in the INTRODUCE activity, it would be more sensible to place it in LOADUP, which is the activity in which the operator selects the next pallet to load, from among those in EMTY. This causes INTROD to revert to its previous form and LOADUP to include the statements for the RATIO calculation. The results thus obtained are presented in Table 8.5. They are very similar to those obtained initially, tables 8.1 and 8.2. This

Table 8.5   Results from Model One—using a schedule-based priority rule, LOADUP higher priority than UNLOAD and INTRODUCE higher priority than RETURN

RETURN was started	438 times
INTROD was started	441 times
UNLOAD was started	438 times
LOADUP was started	442 times
ARRIVE was started	441 times
Utilization of OPERAT	.3662
Occupancy of MACHIN	.7679
Utilization of MACHIN	.6825
Utilization of VEHICL	.2441
Utilization of ARRIVE	1.0000
Average number in queue NEW	1.3770
Average number in queue ONPAL	2.8755
Average number in queue READY	.5173
Average number in queue COMPL	.0186
Average number in queue FINSHD	.0543
Average number in queue EMTY	8.7581
Average number in queue OCCUPI	2.1822

Component Type	Number made	Average time in system
1	81	36.7500
2	62	74.1935
3	61	56.8852
4	53	56.4151
5	33	58.4848
6	40	42.3077
7	35	48.8571
8	25	44.1666
9	13	33.0769
10	14	35.7142
11	14	52.8571
12	10	50.0000
Overall	441	51.4155

we can probably attribute to the simplicity of the system and the low utilisation of the operator. This allowed him to process NEW components more or less as they arrived, and so the sampling of product mix in ARRIVE was itself an effective control. In most systems it would be essential to have a rule of this type to ensure that priorities were kept consistent.

In this particular case, the requirement was defined as the expected average monthly production, and the simulation was run for one week. Under normal circumstances the schedule would vary from week to week or month to month, so that new values of REQD would have to be defined each period. In addition, other scheduling rules can be evaluated. For example, we could use the difference between REQD and MADE. This rule would cause very few of the small volume components to be made until most of the requirement of the high volume ones had been made.

4. *Unmanned shifts*

Another question at the tactical level is whether we could operate the system unmanned for a shift, or part of a shift. If the operator can leave the system with as many pallets as possible loaded with fresh components, then the system can continue to operate until all the components are complete. Using the method described earlier, the following additions to the model were made:

```
THERE ARE 1 CHANGE
....

BEGIN SCHANGE
TIME OF CHANGE LE 0
TIME OF CHANGE = TSHIFT
SHIFT + 1
CHAIN
   SHIFT EQ 4
   SHIFT = 1
   OR CONTINUE
CHAIN
   SHIFT EQ 3
   MEN = 0
   OR MEN = 1

BEGIN LOADUP
MEN GT 0
TIME OF OPERAT LE 0
....

BEGIN UNLOAD
MEN GT 0
TIME OF OPERAT LE 0
....
```

The results are in Table 8.6. They show that the output of the system has fallen by 10% compared to table 8.5. Despite the operator's utilisation not being high, 33% over three shifts or 50% on the two shifts actually worked, he has not been able to supply the system with enough work for an unmanned shift. We can check this figure by a rough calculation. When the third shift begins, all three machines could be working, with a maximum of 9 pallets waiting at the load stations with fresh components. Referring to the operation times of the components above, the maximum amount of work which could be stored would be the sum of the nine longest operations, namely 318 minutes or just under two hours for three machines. The average amount of work waiting would be rather less than this. We would expect to get no output on about six hours of the third shift, or 6/24ths of the whole day, which is 25% of capacity. In fact, the loss is less than half that amount, indicating that more work must have been done on the two manned shifts than in the runs when all three shifts are manned. The extra output

Table 8.6   Results from Model One when operator works two shifts only

RETURN was started	397 times
INTROD was started	397 times
UNLOAD was started	400 times
LOADUP was started	398 times
ARRIVE was started	441 times
Utilization of OPERAT	.3320
Occupancy of MACHIN	.6870
Utilization of MACHIN	.6025
Utilization of VEHICL	.2205
Utilization of ARRIVE	1.0000
Average number in queue NEW	35.5329
Average number in queue ONPAL	5.9512
Average number in queue READY	2.3362
Average number in queue COMPL	.0331
Average number in queue FINSHD	1.5537
Average number in queue EMTY	5.7166
Average number in queue OCCUPI	1.9509

Component Type	Number made	Average time in system
1	67	76.5671
2	44	120.0000
3	55	95.8182
4	49	100.6122
5	35	98.8889
6	37	81.3513
7	35	101.1111
8	25	89.1666
9	13	70.0000
10	13	68.4615
11	14	90.0000
12	10	95.4545
Overall	397	92.9000

on shifts one and two is due to a queue of NEW components accumulating at the load–unload area during the night shift, ready for the operator to load during the day. The system can handle the extra daytime load because the output is not constrained by the capacity of the system, but by the demand for the components. There are some significant changes in specific parameters. For example, the average length of the queue of NEW components has risen from an average 1.5 to 35.5. The maximum length of the queue (obtained from the histogram which has not been printed) was 73. This change arises partially from the accumulation of NEW components during the unmanned shift. However, an important factor is that, although the rate of arrivals is as before, the production is reduced, thereby indicating that the system cannot cope. The queue NEW would eventually become very large. In other words, the system is not in a steady state and the results are not indicative of the long-term situation.

Table 8.7   Results from Model One when operator works and components arrive on two shifts only

RETURN was started	409 times
INTROD was started	409 times
UNLOAD was started	410 times
LOADUP was started	409 times
ARRIVE was started	473 times
Utilization of OPERAT	.3408
Occupancy of MACHIN	.7058
Utilization of MACHIN	.6193
Utilization of VEHICL	.2272
Utilization of ARRIVE	1.0000
Average number in queue NEW	47.2710
Average number in queue ONPAL	6.9858
Average number in queue READY	2.5347
Average number in queue COMPL	.0323
Average number in queue FINSHD	2.3334
Average number in queue EMTY	4.6733
Average number in queue OCCUPI	2.0040

Component Type	Number made	Average time in system
1	65	85.1515
2	44	118.6363
3	54	98.1481
4	53	99.0566
5	35	95.2941
6	40	89.0000
7	35	100.0000
8	28	100.7143
9	13	71.5384
10	16	88.7500
11	16	100.0000
12	10	100.0000
Overall	409	96.4878

It is perhaps unrealistic to have the operator working only two shifts, but components being delivered to the system continuously. If the system is automated to the extent that deliveries are made automatically on all three shifts, then one would expect that the loading task would be done automatically by a robot. On the other hand, if transport relies on human personnel, then presumably they will work the same shifts as the operator, and components would not arrive during the unmanned shift. The results obtained when the rate of arrivals is stepped up to one component every 11 minutes, to give the same total daily quantities over only two shifts, are given in Table 8.7. They show a slight improvement in output, but still well below the figure when all shifts are manned. The number of arrivals during the run is well above the output, so the system is even further from the steady state than in the previous run. The number of arrivals, 473, is also somewhat higher than the expected level of 436 (ie $2/3 * 7200/11$). This greater than expected number of arrivals has already been commented on above.

8.10.7   Comments on the system design

This model is a very simple one, with little scope for unpredictable results. However, two conclusions concerning the system design can be drawn from the results.

We can observe that the utilisation of the machines is some 10% less than the occupancy. The difference is due to the fact that the machine must be idle while the vehicle moves the pallet to and from it. This effectively adds 4 minutes to all the cycle times, which is roughly 10% on the average cycle time. Therefore, to achieve the maximum output from a system, the bottleneck activity, in this case the machines, should be provided with buffer locations so that it will not be held up while a fresh component is supplied to it. We will examine the modelling of buffers in the next chapter.

Secondly, we can see that the demand for some components is much greater than for some others. Yet each has its own load station. The utilisation of the load stations is very low. The expected time for loading and unloading at each station is:

Load–unload capacity required in minutes per day

	Number of shifts per day		
Component type	1	2	3
1	96	48	32
2	81	40	27
3	78	39	26
4	54	27	18
5	42	21	14
6	36	18	12
7	30	15	10
8	24	12	8
9	21	10	7
10	19	9	6
11	14	7	4
12	12	6	4

These are clearly small amounts of time in a work shift. The load stations appear to be more important as pallet storage locations than as load stations. Indeed, one might wonder on the wisdom of installing a load station for each component type. In the next chapter a model will be presented of a system which includes non-dedicated load stations. In practice there may be other reasons for having dedicated load stations. If most of the components require two operations in different fixtures, the use of dedicated load stations allows the pair of pallets holding the successive fixtures for each component to be located at adjacent load stations, thus simplifying the transferring of components from one fixture to the next. Various aspects of fixturing methods will be discussed in chapter 10.

SUMMARY

In this chapter we have presented the logic associated with modelling load-unload operations, pallets, machines, breakdowns and some auxiliary items.

Various aspects were combined in a model of a simple hypothetical system (though inspired by a real system) with dedicated load stations. Several versions of the model were run to examine the effects of vehicle and operator priority rules, a schedule-based priority rule, unmanned shifts and assuming that there was an inexhaustible supply of components. As a result, certain conclusions about this design of system were drawn.

EXERCISES

1. Estimate the number of machines which would be needed to cause either the vehicle or the operator to become the bottleneck.

2. Estimate the number of load stations which would be needed if they are not dedicated to particular pallets.

3. Estimate the maximum number of components of each type which can be produced, if only one pallet of each type is available. Compare these values with the quantities actually produced in these experiments.

REFERENCES

1. Clementson, A.T., *ECSL User's Manual*, Cle-Com Ltd, Birmingham, 1985.

2. Pegden, C.D., *Introduction to SIMAN*, Systems Modeling Corp., State College, Pennsylvania.

3. Carrie, A.S., Adhami, E., Stephens, A. and Murdoch I.C., Introducing a Flexible Manufacturing System, *Int. J. Prod. Res.*, 22, 907–914, 1984.

CHAPTER 9

Building FMS models—
2: Machine buffers and central pallet storage

In this chapter we shall examine in more detail further features of flexible manufacturing systems, in particular buffer storage positions, both at work stations and centrally. As in chapter 8, the simple activity cycles which were used in the introductory chapters will be refined. Once again, we will add new entities and activities to the activity cycle diagrams, and it will be helpful to lay out in advance the arrow notation for the cycles of the various entities, as in Figure 9.1.

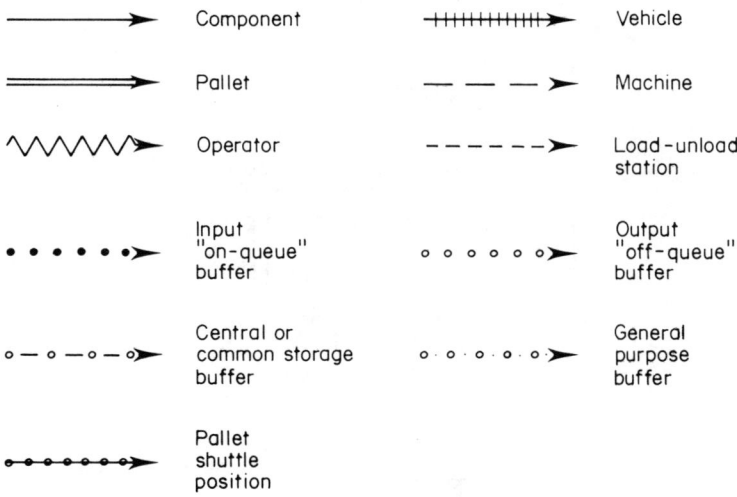

Figure 9.1   Arrow notation for activity cycle diagrams.

9.1   BUFFER STATIONS AT WORKSTATIONS

In chapter 8, we assumed that the pallet would be transferred directly between the vehicle and the machine. While experimenting with Model One, we observed that to

229

maximise the output of the system there should be buffer positions between vehicle and machine. Buffers allow the vehicle to bring a pallet with a fresh component to the machine while the previous one is being processed, and allows the machine to start work on the next job after completing an operation, without having to wait until a vehicle comes and takes the completed job away. In fact most of the systems described in chapter 7 had buffers at the machining centres.

Common types of buffer are:

1. Pallet stands.
2. Pallet shuttles.
3. Carousel-type pallet magazines.
4. Conveyor buffers.

We will now proceed to discuss how these arrangements may be included in the model.

9.1.1   Pallet stands

In this arrangement there are a pair of pallet stands at each machine, as in Figure 9.2(a). One serves as a buffer store for pallets waiting to go on the machine, and the other for pallets which have been processed by the machine. They are often referred to as the "on-queue" and "off-queue" buffers. The twin pallet stand arrangement is the simplest of buffer systems to model. Each pallet stand can hold only one pallet and clearly the buffer is either occupied or vacant. The conversion from occupied to vacant

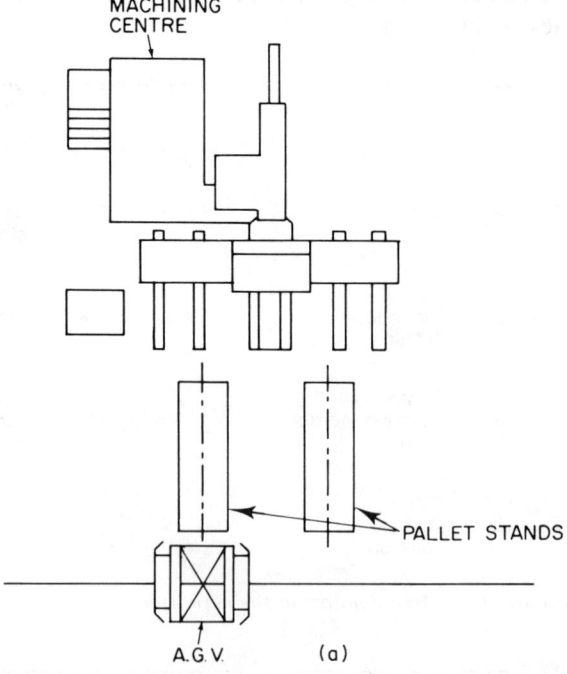

Figure 9.2   *Pallet stands as machine buffers, may be either free-standing units as in the Ander-son Strathclyde FMS (a), or may be an integral part of the machine as in (b) (reproduced by permission of Scharmann GmbH & Co.).*

(b)

Figure 9.2   (continued).

and vice versa depends on the movement of a pallet to or from the pallet stand. Thus their activity cycle will be as in Figure 9.3. Now, when we combine these cycles with those for the machine and pallets, presented in figure 8.20, we obtain the diagram of Figure 9.4. When figure 9.4 is compared with the earlier versions it will be seen that there is an additional queue and activity in the pallet's cycle both prior to and after processing. The machine no longer has to be vacant for the pallet to be moved to the machine, and of course this is exactly the purpose of the buffer. Note that either of the simplifications shown in figure 8.21 could have been used.

With many machining centres the buffer positions are an integral part of the machine, Figure 9.2(b), such as in the Cummins Makino line, and the Mirrlees Blackstone system. With these machines there is space only for one pallet in the on-queue and one in the

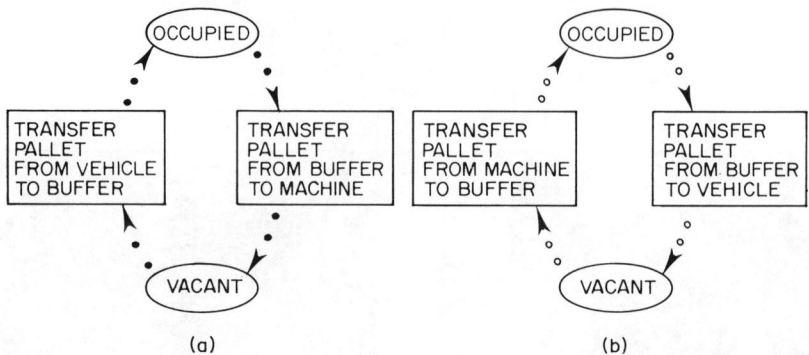

Figure 9.3   ACD for pallet stands at machines, (a) for "on-queue" pallet stand, (b) for "off-queue" pallet stand.

off-queue. However the logic is quite general, and applies no matter how many spaces are available. There would of course be as many on-queue and off-queue buffer entities as there are spaces in them. For example, in the Vought Aerospace system the machining centres have in-line pallet stands with room for two pallets in both input and output sides.

Although the pallet stands are normally assigned to act as on-queue and off-queue buffers, they are usually mechanically identical and could serve either purpose equally well. In fact, it is usually the control software which makes the distinction, to simplify the control logic, and we can take advantage of this in the model. However, we should consider the case when a pallet stand can perform either function, which might be invoked if the mechanism failed on one of the stands. Figure 9.5 gives the cycle for such general purpose buffers.

In this case it is not a sufficient condition for moving a pallet to a machine that there is a vacant buffer position. If the machine is processing a component, then the remaining buffer must be kept empty so that the machine can discharge its pallet. A technique sometimes used is to keep a count of the number of pallets at each machine and its buffers, and ensure that it never exceeds the number of buffer positions. Thus if the machine has two pallet stands then we would never allow more than two pallets at the machine and its buffers. Similarly, if there is only one buffer position then not more than one pallet would be allowed at a time at the machine. Rather than try to express the logic by amending the activity cycle diagram, it may be more conveniently handled by means of attributes of the machines. These could be binary attributes, one for each pallet position, having a value of zero to indicate that no pallet is present or one to indicate that a pallet is present. Alternatively, a single attribute could hold the count of pallets present as discussed above.

This logic would ensure that the machine would be able to discharge its pallet. However, it does impose a limitation because, when the machine has just discharged a pallet to the off-queue buffer, and taken in a fresh one from the on-queue buffer, the vehicle will not be able to bring a new pallet to the now vacant on-queue buffer until it has

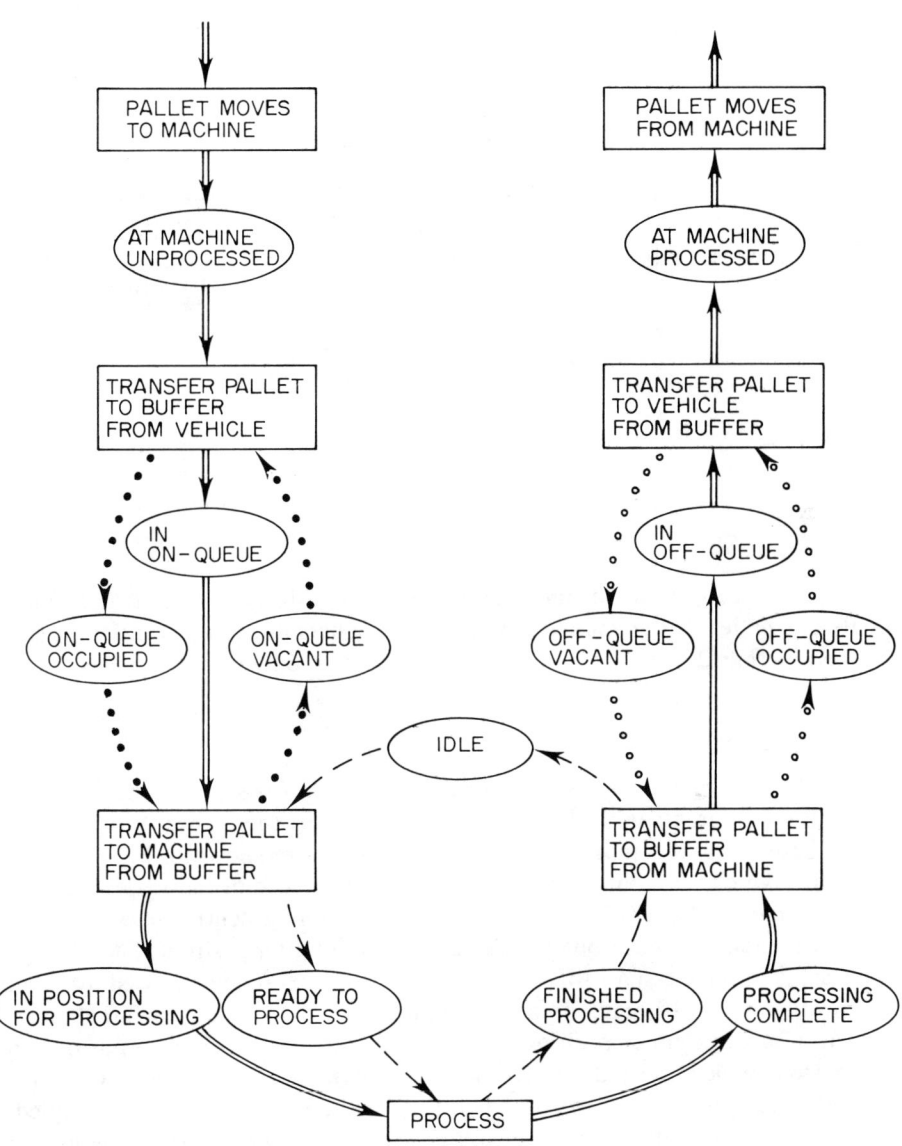

Figure 9.4   Activity cycle diagram combining machine buffers with machine and pallet cycles.

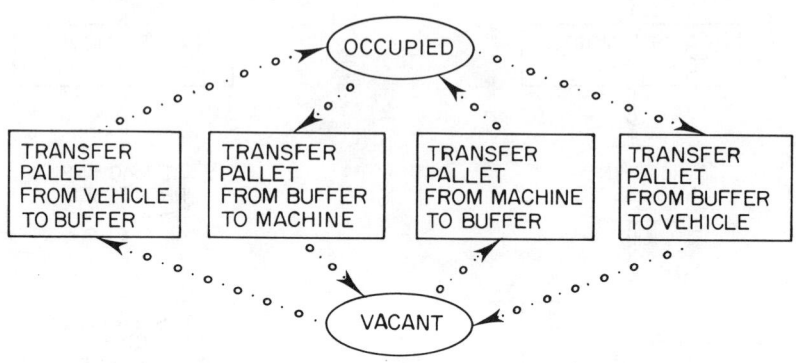

Figure 9.5   Activity cycle diagram for general purpose machine buffers.

removed the pallet from the off-queue buffer. Strictly, this assumes that the vehicle can carry only one pallet. If the vehicle can carry two pallets at a time, a different situation arises. We will consider this case when discussing the modelling of vehicles.

9.1.2   Pallet shuttles

These are rotary mechanisms, which normally have two positions, one of which, the "outer" position, interacts with the vehicle, and the other, the "inner" position, with the machine. There are two common arrangements, which are shown in Figure 9.6(a) and 9.6(b), both of which were illustrated in the systems described in chapter 7. Figure 9.6(a) is the arrangement adopted for KTM's FM100 machining centre, in which the inner position is the machining position and the outer one is a buffer position. The same general layout occurred in the Cessna FMS. Figure 9.6(b) is the arrangement used by Cincinnati Milacron with their T-10 machining centres, for example in their Birmingham plant's FMS. In this case the inner position does not serve as the machining position, but as a position from which the pallet is transferred to the machine table. As with pallet stands, both arrangements permit the operation of the machine to be de-coupled from the transport system. We will now discuss how to model these arrangements, treating the KTM FM100 system, figure 9.6(a), first.

When a pallet is brought to the machine by a vehicle it is placed on the outer position. When the machine finishes processing a workpiece, the shuttle rotates bringing the completed pallet to the outer position and the fresh pallet to the inner, working, position. The machine can then work on the new component while the previous component waits for a vehicle to take it away. Unlike the pallet stand arrangement, the new component and the old one are interchanged in a single action. We might attempt to show this in an activity cycle diagram as in Figure 9.7, in which one pallet moves from the outer position to the inner, and another pallet does the reverse.

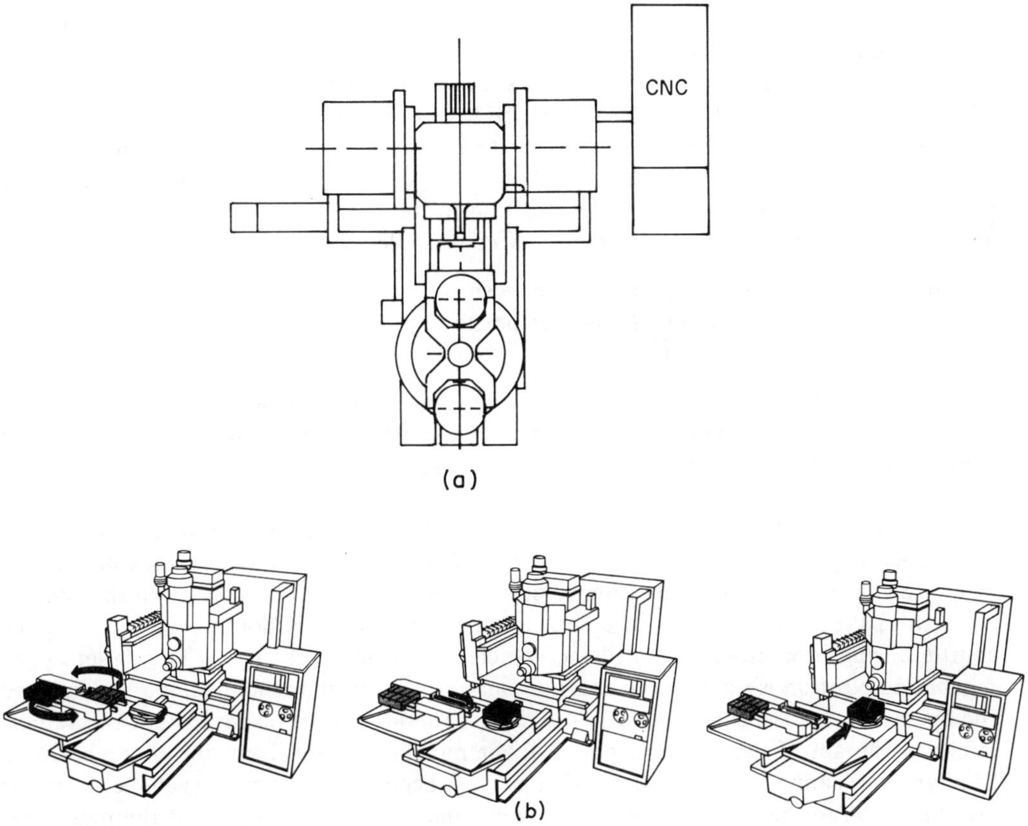

CNC

(a)

(b)

Figure 9.6   Rotary pallet shuttles (a) as on KTM FM100 (reproduced by permission of KTM Ltd, Brighton, England) and (b) as on Cincinnati Milacron T-10 (reproduced by permission of Cincinnati Milacron Ltd).

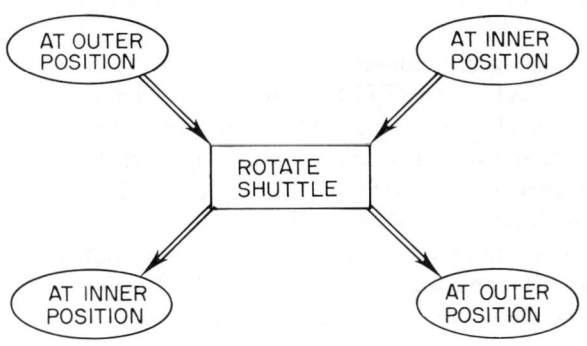

AT OUTER
POSITION

AT INNER
POSITION

ROTATE
SHUTTLE

AT INNER
POSITION

AT OUTER
POSITION

Figure 9.7   Effect of rotating the shuttle on pallets.

However, while this seems at first sight to be an accurate description of what happens, it is ambiguous and inadequate. The logic of activity cycle diagrams interprets this as implying that the shuttling activity requires there to be a pallet in the queue AT OUTER POSITION and also one AT INNER POSITION. However, the shuttle action can be done when a pallet at the outer position is to be moved to the inner position, and does not necessarily require a pallet to be at the inner position. Similarly, shuttling can take place to move a pallet from the inner position to the outer one, whether or not there is a pallet at the outer position. There are three possible situations:

1. moving one pallet from outer to inner position,
2. moving one pallet from inner to outer position,
3. switching two pallets round.

Strictly, there is a fourth possibility, namely rotating the shuttle while neither position is occupied by a pallet. This is not a "normal" activity, probably only performed during commissioning or maintenance. It will not alter the state of the system, since it does not affect any other entities, and will be ignored.

In addition to identifying these three situations we need to observe the conditions when they may occur. The tests to determine whether we can deposit or pick up a pallet cannot be based simply on the presence of an occupied or vacant position on the shuttle, since the pallet does not leave the shuttle when processing is done. It remains on the shuttle during processing and until the vehicle comes and picks it up. The pallet at the machining position is ready to be shuttled outwards as soon as the machine has finished the operation, but we cannot shuttle the pallets round without ensuring that the vehicle is not in the act of depositing a pallet at, or picking one up from, the outer position. Similarly, we cannot move a pallet which has just been deposited by the vehicle, from the outer position to the inner one, while the machine is processing a part at the inner one. Thus the tests for the shuttle rotation activity become interrelated with the activities of transferring a pallet to or from the vehicle, and processing the component.

These considerations are not easily depicted in an ACD if we build them into the pallet's activity cycle. Instead we can introduce a pair of entities representing the positions on the shuttle, and observe the effects of the three types of rotate shuttle activity. Figure 9.8 shows the logic. Notice that each shuttling activity uses precisely two entities, one at the outer position and one at the inner position, so that the ambiguity in the earlier diagram, figure 9.7, has been removed.

Now that the ROTATE SHUTTLE activity has been broken down into its three variants, the pallet's cycle could be added to figure 9.8. This could be done in one of two ways. It could be shown in full, in which case its cycle parallels that of the shuttle position, involving queues (similar to AT OUTER LOADED, AT INNER READY, AT INNER PROCESSING COMPLETE and AT OUTER PROCESSING COMPLETE), and the shuttling and machining activities. Alternatively, it could be shown in condensed form, with a single queue "ON SHUTTLE" being used to refer to pallets on the shuttle at any position and in any condition, using a method similar to that shown in figure 8.21(a).

The machine's activity cycle is also affected by the presence of a pallet shuttle. The earlier diagrams included an activity TRANSFER PALLET TO MACHINE, either directly from the vehicle (figures 8.19 to 8.21) or from a pallet stand (figure 9.3). When a pallet shuttle is present there would be no separate activity of this sort, because in

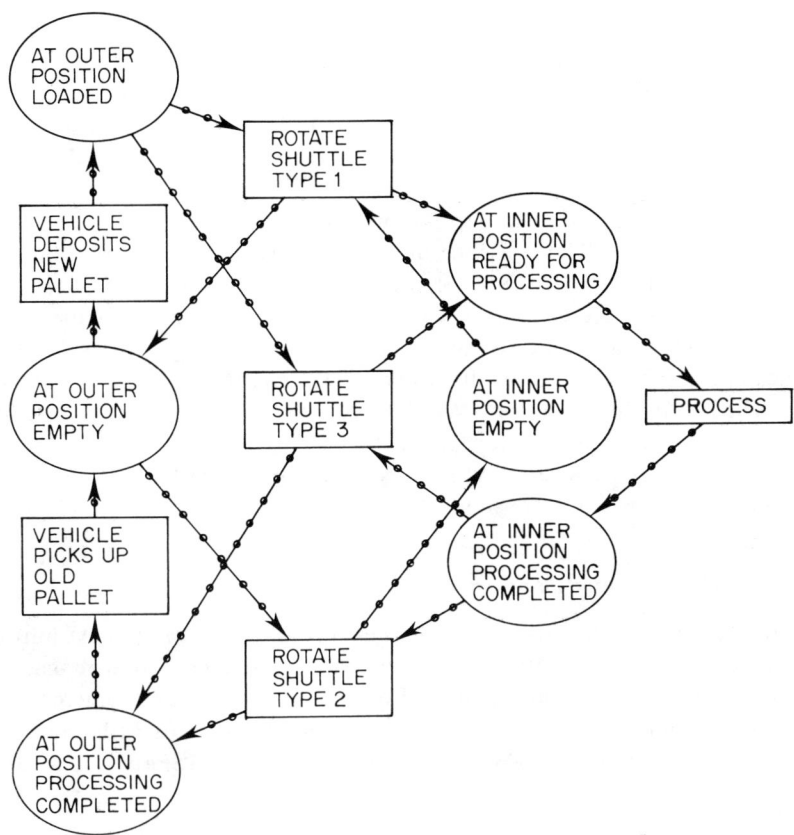

Figure 9.8   Activity cycle diagram for pallet shuttle positions.

the FM100 arrangement the inner position is the machining position. The machine's
activity cycle would reduce to the simple version in figure 8.18.

It may seem strange to show three different ROTATE SHUTTLE activities, since
they are all the same basic activity, but occurring under different circumstances. The
method presented here has been used to maintain consistency of approach and to present
as much information as possible in a single diagram. It would be possible to combine the
three variants into one activity, by adding attributes or variables to the model, but then
the conditions for the activity to occur and the effect on the entities cannot be shown
in an ACD. The advantage of the ACD is that the conditions for an activity occurring
are defined by the logic of the cycles. Also, different software packages provide different
facilities for implementing the logic of a model, so it would be difficult to give a piece
of coding which would be generally applicable. In fact, it is probably easier to write
the code for this logic in any given simulation language than it has been to draw the
diagram.

In most packages the concept of a set is quite general and not restricted to queues.
Also, most packages provide a procedure for rotating the contents of a set, ie taking
the entity from the head of the set, moving the remaining entities up one place and

adding the entity, which was taken from the head, back in at the bottom of the set. This procedure can be very helpful in handling pallets in an indexing mechanism. A dummy entity, which could be called a gap, can be used to keep track of vacant positions. A pallet shuttle can be represented in this way, involving a set with two members. One member represents the pallet or gap at the inner position and the other member the outer position. Rotating the set then has exactly the effect of rotating the pallet shuttle. When a pallet is taken off the shuttle, a gap entity has to be put in its place, and when a pallet is put on the shuttle the gap is removed, thus ensuring that there are always two members in the set. The presence of a pallet for machining or picking up can be detected by examining the attributes of the entity at the head or tail of the set. We will not pursue this idea at present but will refer to it again in the following sections.

Most of the foregoing applies to the second arrangement, figure 9.6(b). The main difference lies in the fact that there will still be a TRANSFER PALLET TO MACHINE and TRANSFER PALLET FROM MACHINE, as in figure 9.4. The test for whether the shuttle can be rotated will depend on the occurrence of either of these activities rather than the PROCESS activity, as was shown in figure 9.7. The machine's pallet cycle would be similar to that in figure 8.19.

9.1.3   Carousel-type pallet magazines

The carousel type of pallet magazine is a generalisation of the pallet shuttle arrangement. In fact, some pallet shuttles have four pallet positions. Pallet magazines usually have six, eight, ten or even more pallets. Figure 9.9 shows the arrangement. Terminology varies from company to company. For example, Cincinnati Milacron use the term Automatic Work Changer, or AWC. Their Birmingham FMS, and the Vought system,

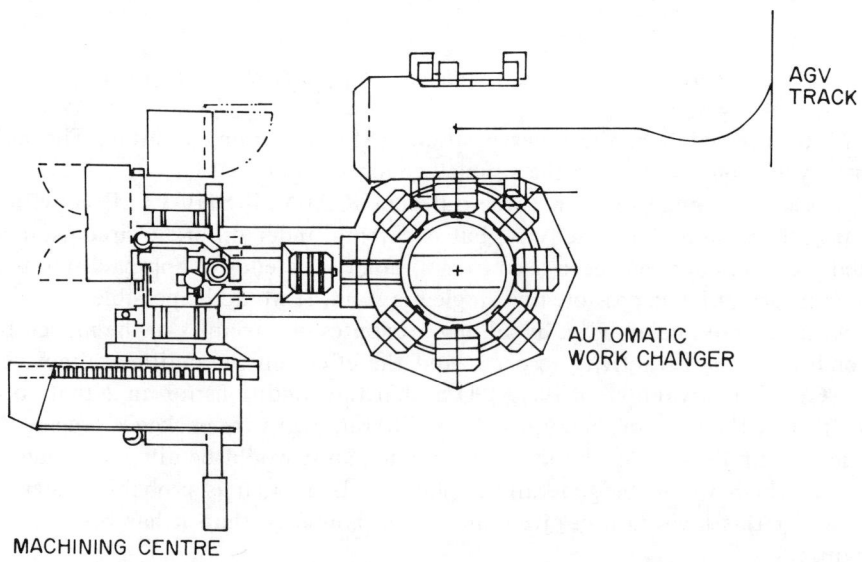

Figure 9.9   *Carousel-type pallet magazine (or Automatic Work Changer) (reproduced by permission from material supplied by Cincinnati Milacron Ltd).*

are examples of systems in which AWCs were positioned at load–unload stations, rather than at machining centres. As with pallet shuttles, one position interacts with the vehicle, and one position serves either as the work position or interacts with it. There may be an equal number of spaces between the inner and outer positions on the incoming side as on the outgoing side, or the numbers may be unequal. AWCs are used when it is desired to give a substantial supply of work to a machine, for example for unmanned operation, perhaps in systems without automated handling. Another function which they serve is to provide a "home" position for each pallet, as with dedicated load stations. Unlike pallet shuttles they usually have a random positioning capability rather than a fixed sequence queue arrangement.

To model pallet magazines we have to generalise the logic developed for the pallet shuttle. The logic will be very similar to that for pallet shuttles, except that there are now intermediate positions on the incoming and outgoing arms of the chain, as well as the inner and outer positions. It would not be convenient (though possible) to extend the approach used in figure 9.8, because there would now be many possible combinations of movement of occupied or vacant positions to and from the inner, outer and intermediate positions. Instead we can adopt the concept of rotating the sets of pallets or gaps, which was introduced in the section on pallet shuttles, together with a statement of the rules for rotating the magazine. In general, we would want to rotate the magazine if there are any loaded pallets on it, until a loaded pallet is brought to the machine for machining, or to the outer position for picking up, unless the machine is working on a component, or a vehicle is picking up or depositing a pallet. In other words provided the outer and inner positions are both inactive, and there is at least one pallet in the magazine, we should rotate the magazine. We can depict this in an activity cycle diagram by using dummy entities, one to indicate that the outer position is inactive, ie not involved in a pick up or deposit, and one to indicate that the inner position is inactive, ie not involved in a machining activity. We can also use a queue PALLET IN MAGAZINE to enable the presence of a pallet to be tested. Figure 9.10 shows this method and the relevant portion of figure 9.8.

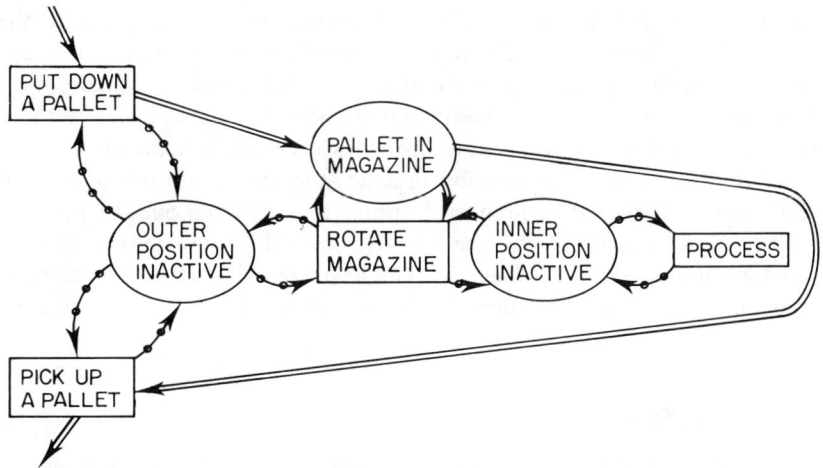

Figure 9.10   Dummy entities to prevent rotation of chain when a pallet is being picked up, put down or machined.

In addition to depicting when the magazine should be rotated we should state what the rotation activity consists of. This cannot be shown directly in the activity cycle diagram, so instead a logical statement can be used:

1. WHILE   there is at least one loaded pallet in magazine
   AND   the inner position is inactive
   AND   the outer position is inactive
   REPEAT
              BEGIN the activity ROTATE THE MAGAZINE
   UNTIL
              a loaded pallet occupies the inner position
              OR
              a loaded pallet occupies the outer position.

2. BEGIN   ROTATE THE MAGAZINE:
   REMOVE   the item at head of the incoming queue,
   ADVANCE   the contents of the incoming queue,
   REMOVE   the item in the outer position,
   ADD   that item to the tail of the incoming queue,
   REMOVE   the item at the head of the outgoing queue,
   ADD   that item to the outer position,
   ADVANCE   the contents of the outgoing queue,
   REMOVE   the item at the inner position,
   ADD   that item to the tail of the outgoing queue,
   ADD   the item originally at the head of the incoming queue to the inner position.
   END

This statement is in two parts, the conditions necessary for the action, and the action itself. Perhaps it seems merely to state the obvious, and, if the preceding discussion has been clear, so it should. Nevertheless our model must explicitly perform all these steps. It is a good habit to state the logic in plain English in this way, or an equivalent, to help us to set down the logical requirements clearly. Notice that, expressed in this way, we actually have four sets, namely the incoming and outgoing queues and the inner and outer positions. If our software package provides a convenient means of checking the attributes of items at any given position in a set, then we could use one set for all the pallets, and then there would be no need to remove items from one set and add them to another. They would remain in the one set but rotate their positions until an item with the required attribute value was found at whichever position of the set corresponded to the inner or outer position. This permits us to model the random access capability of AWCs.

9.1.4   Conveyor buffers

In systems in which pallets are moved on conveyors there will usually be spurs from the main conveyor at each workstation to permit storage of several pallets waiting to be processed, and possibly for pallets waiting to rejoin the main conveyor after processing.

Conveyor buffers can be thought of as a generalisation of pallet stands, in which there may be several pallets of material waiting processing at the work centre, rather than only one or two, as discussed above in connection with the Vought system.

Frequently, the material will be transferred from the pallet at the head of the queue to the machine's chuck or worktable for processing by a robot. If this is the case then some of the characteristics of the pallet shuttle arrangement are introduced. The machine's work position is analogous to the inner position of the shuttle, and the pallet position at the head of the queue is in the position to the outer position of the shuttle, with the robot being analogous to the shuttle mechanism itself. New logic would be required to model the movements of pallets as they rejoin the main conveyor line.

We will not discuss this further at present. In practice, the approach is often determined by particular features of the system being modelled. Sometimes it may not be necessary to model the movements into and out of the machine at all. In any case, we will consider the modelling of movements of material by conveyor and robots in chapter 12.

9.2   CENTRAL OR COMMON STORAGE BUFFERS

Central or common storage buffers are positions where pallets can be stored between operations, while they wait either for the required workstation to become free, or until another component of the type whose fixture is on the pallet is to be loaded into the system. They may be all together in one central place in the system, or they may be distributed throughout the system. Cummins' Makino line is an example of centralised stores, and the Hattersley Newman Hender system is an example of distributed common storage.

As with load–unload positions, there are two ways in which central buffers can be used:

1. Each buffer position is dedicated to one pallet, in which case it is the pallet's "home" position, and there will be a buffer for every pallet.
2. Any buffer can be used by any pallet. The buffers are a general resource and there will normally be fewer buffer positions than pallets in the system.

Central buffer stations have a similar cycle to machine buffers; they are either occupied or empty, and they alternate from one condition to the other as pallets are moved to and from them, as in Figure 9.11.

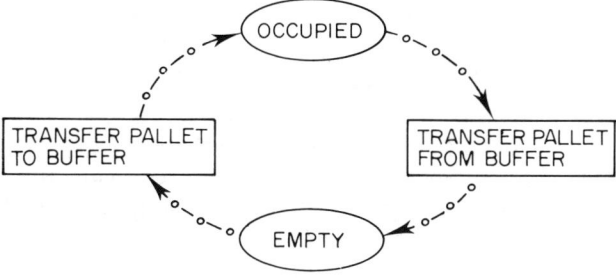

Figure 9.11   Activity cycle diagram for central or common storage buffer positions.

There are six principal forms of movement to or from the buffer positions, depending on whether the pallet is empty or loaded and on whether the component has been or is still to be machined. The movements might occur in a cycle as follows:

1. move an empty pallet from buffer to load–unload area,
2. move a loaded pallet from load–unload area to buffer,
3. move a loaded pallet from buffer to machine,
4. move a loaded pallet from machine to buffer,
5. move a loaded pallet from buffer to unload station,
6. move an empty pallet from unload station to buffer.

These, along with movements directly between load stations and machines, can all be shown individually in an activity cycle diagram, as in Figure 9.12. As with pallet shuttle logic, when the code is to be written, it may be more convenient to treat them all as just different forms of one general move activity.

This diagram does not indicate the circumstances under which an empty pallet will be taken away from the load–unload station to the store, or brought back from the store. As the diagram stands, the conditions for bringing a pallet back from store will exist as soon as it has been placed in store. To prevent a never-ending cycle of non-productive activities from occurring, other conditions must be laid down. These will normally relate to other entities such as the component, which is not shown in this diagram, but examples are given in Models Four and Five later in this chapter.

It is not normally necessary to include dedicated buffers as entities in the model. If a pallet requires to go to the buffer, the buffer must be available, as no other pallet can occupy it. Consequently, no test for the availability of a buffer is necessary. It is only necessary to record the state of the pallet as it moves to or from the buffer, and this can be done using attributes of the pallet in the normal way.

In the case of general buffers, it is normally necessary to test for the availability of a buffer before a pallet can be moved to it, since there are usually fewer general buffers than pallets. For this reason it may be desirable to include the buffers as entities in the model. On the other hand, it may be possible to avoid doing so by maintaining a count of the number of free buffer spaces. So long as this value is greater than zero a pallet can be moved to a buffer.

However, we may be interested in more than simply whether the buffer is occupied or not. If the simulation model is to provide a graphical display of the system, a further factor is introduced. Each buffer will have its own specific location on the screen, and might be displayed in one colour when empty and in another colour when occupied by a pallet. Depending on the characteristics of the software package being used, it may be necessary to include the buffers as entities, with the co-ordinates of the buffer being held as attributes. On the other hand it may be simpler to use a table of co-ordinates of buffer positions with an attribute of the pallet recording which buffer position it occupies. This would apply to both dedicated and general buffers.

The locations of buffers may also be important if movement times of vehicles are to be computed using accurate distances. In this case we will need to identify the co-ordinates of the start and finish points of each journey, compute the distance and hence the time to travel between them. Once again, depending on the facilities in the software package being used, it may be more convenient to include the buffers as entities or else to use look-up tables.

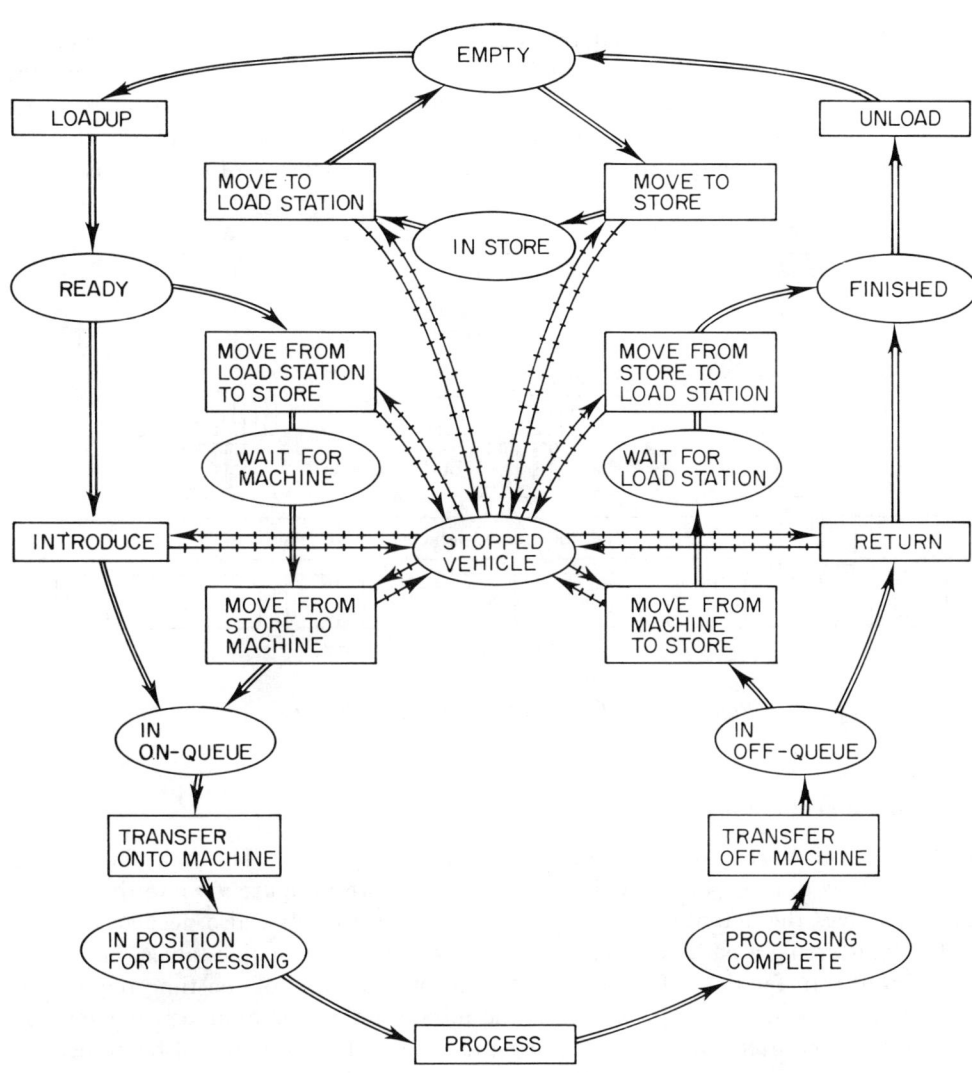

Figure 9.12   Activity cycle diagram for pallet and vehicle when central storage is included.

9.3   MODEL TWO—SYSTEM WITH MACHINE BUFFERS AND DEDICATED LOAD STATIONS

With Model One we observed that the machines have to be idle before a pallet can be brought to them, and that it should be possible to obtain more output from the system if the machine had buffers. Let us therefore amend the model to incorporate machine buffers. Since machining centres almost always have buffer positions, this is a much more practical configuration. The layout of the system for Model Two is shown in Figure 9.13.

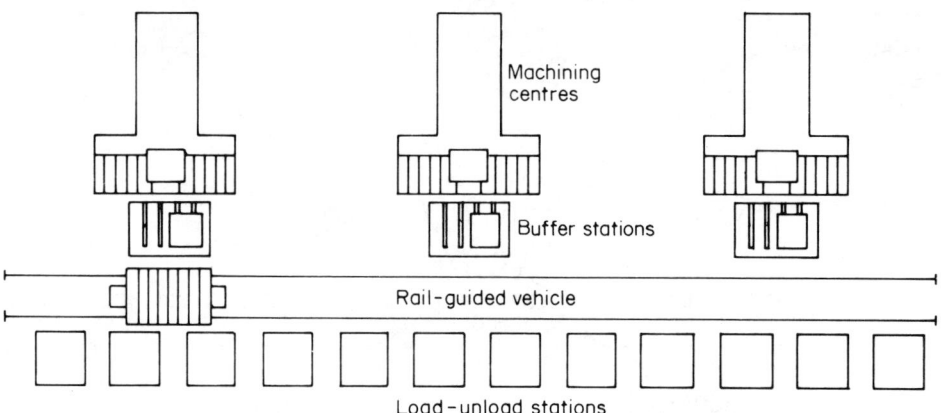

Figure 9.13   Layout of system for Model Two.

9.3.1   Amendments to model

The changes which are required to the logic are relatively minor. The main thing which we need to do is to incorporate the buffer cycles as shown in figure 9.4 into the activity cycle. This gives the activity cycle diagram of Figure 9.14. We also need to define a time for transferring a pallet between buffer stands and the machine tables. Let us use a one-minute transfer time. This is probably an overestimate of the time required in reality, however, since we are working in one minute units, one minute is the smallest time which can be used, without redefining the time unit. (There would be no need to consider this with a package whose clock is a real variable.)

The code of the model will be amended in various places. For example we should add statements to record the number of pallets waiting in machine buffers, the utilisation of the buffers and so on and to print out the values obtained. These statements will be of similar form to those given in Model One in the last chapter and are not presented here. There will also be a need to define the transfer time in the Data section. We can use a variable TRTIME to hold the value. The coding of only a few activities will be affected by adding machine buffers. INTRODUCE and RETURN will be concerned with the contents of the buffers instead of the machines directly. We will have two new activities, TRFRON and TRFROF, transferring a pallet to a machine from an in-queue buffer and transferring a pallet off a machine to an off-queue buffer. The ECSL code for these four activities is shown in Figure 9.15.

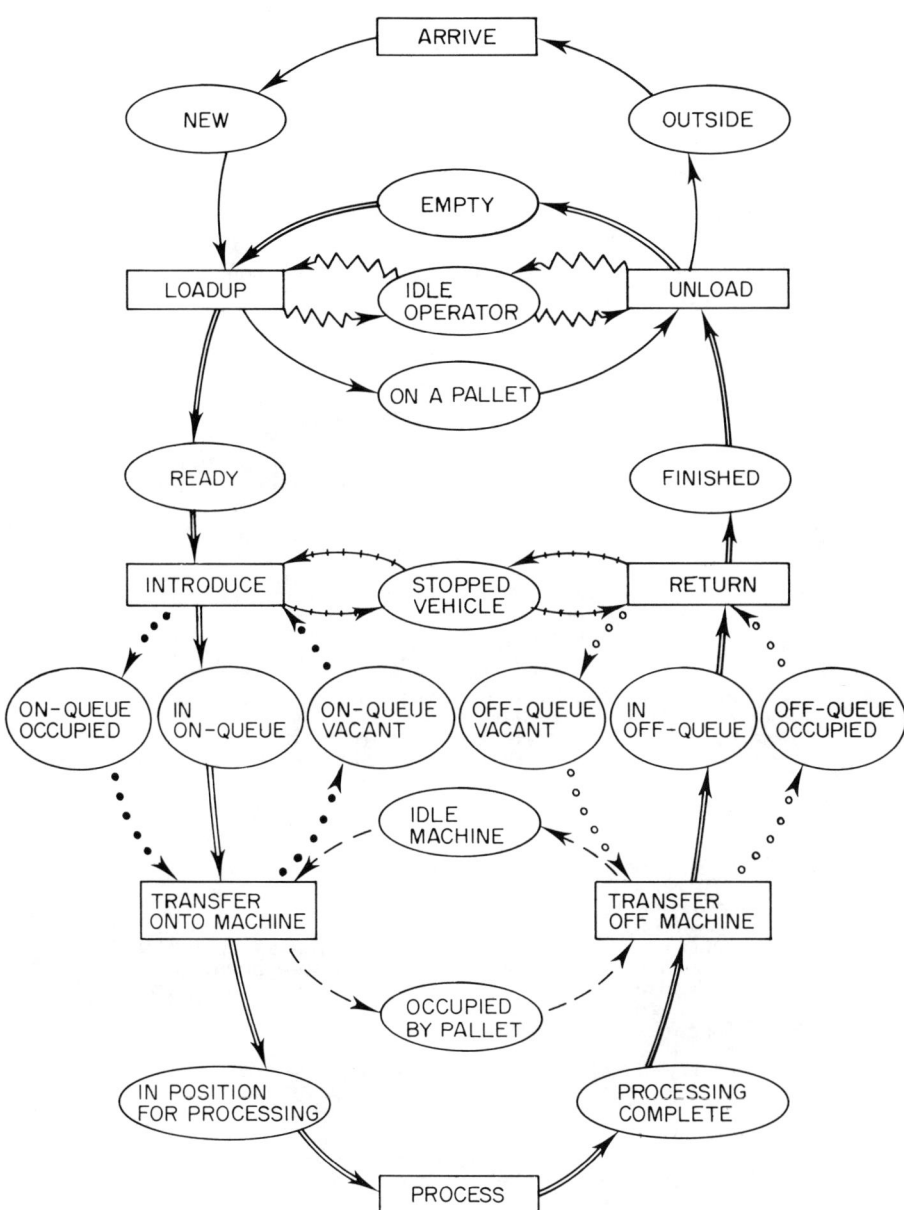

Figure 9.14   Activity cycle diagram for system of Model Two.

```
                    THERE ARE 3 ONBUFF SET ONOCCU ONVACA
                    THERE ARE 3 OFFBUF SET OFFOCC OFFVAC

                    ADD INONQ TO HIST ZJINONQ DURATION
                    ADD INOFFQ TO HIST ZKINOFFQ DURATION
                    ADD ONOCCU TO HIST ZMONOCCU DURATION
                    ADD ONVACA TO HIST ZNONVACA DURATION
                    ADD OFFOCC TO HIST ZOOFFOCC DURATION
                    ADD OFFVAC TO HIST ZPOFFVAC DURATION

                    BEGIN INTROD
                    ....
                    FIND FIRST ONBUFF B IN ONVACA
                    MCAT OF PALLET A = B
                    ....
                    PALLET A FROM READY INTO INONQ AFTER DURATION
                    ONBUFF B FROM ONVACA INTO ONOCCU AFTER DURATION

                    BEGIN RETURN
                    ....
                    FIND FIRST PALLET A IN INOFFQ
                    B = MCAT OF PALLET A
                    MCAT OF PALLET A= 0
                    ....
                    PALLET A FROM INOFFQ INTO FINSHD AFTER DURATION
                    OFFBUF B FROM OFFOCC INTO OFFVAC AFTER DURATION

                    BEGIN TRFRON
                    FIND FIRST PALLET A IN INONQ
                      B = MCAT OF PALLET A
                      MACHIN B IN IDLE
                    C = MCAT OF PALLET A
                    DURATION = TRTIME
                    AADURATION = DURATION + MACHTIME OF PALLET A
                    PALLET A FROM INONQ INTO COMPLE AFTER AADURATION
                    MACHIN B FROM IDLE INTO OCCUPI
                    ADD 1 TO MADE OF PALLET A
                    ADD MACHTIME OF PALLET A TO TMACHTIME
                    ONBUFF C FROM ONOCCU INTO ONVACA AFTER DURATION
                    ADD 1 TO TRFRON
                    REPEAT

                    BEGIN TRFROF
                    FIND FIRST PALLET A IN COMPLE
                      B = MCAT OF PALLET A
                      OFFBUF B IN OFFVAC
                    DURATION = TRTIME
                    PALLET A FROM COMPLE INTO INOFFQ AFTER DURATION
                    MACHIN B FROM OCCUPI INTO IDLE AFTER DURATION
                    OFFBUF B FROM OFFVAC INTO OFFOCC AFTER DURATION
                    ADD 1 TO TRFROF
                    REPEAT

                    TYPE 'TRFRON was started' TRFRON ' times'
                    TYPE 'TRFROF was started' TRFROF ' times'
                    ....
                    TYPE 'Average number in queue INONQ 'MEAN(ZJINONQ)
                    TYPE 'Average number in queue INOFFQ'MEAN(ZKINOFF)
                    TYPE 'Average number in queue ONOCCU'MEAN(ZMONOCC)
                    TYPE 'Average number in queue ONVACA'MEAN(ZNONVAC)
                    TYPE 'Average number in queue OFFOCC'MEAN(ZOOFFOC)
                    TYPE 'Average number in queue OFFVAC'MEAN(ZPOFFVA)

                    DATA
                    ONVACA 1 TO *
                    OFFVAC 1 TO *
                    TRTIME 1
```

Figure 9.15   Amendments to ECSL code for converting Model One into Model Two, which includes machine buffers.

9.3.2   Results and comparison with Model One

When the model is run, using the same data as in Model One for operation times, and so on, the results shown in Table 9.1 are obtained. The output achieved is similar to that obtained in Model One (Table 8.5), since, as before, the system's output is constrained by the rate of arrivals, not the machine capacity. However, while the machine utilisation remains at 68%, the machine occupancy is reduced from 77% to 72%, because the

Table 9.1   Results from Model Two

INTROD was started	442 times
RETURN was started	437 times
LOADUP was started	442 times
UNLOAD was started	437 times
TRFRON was started	440 times
TRFROF was started	437 times
ARRIVE was started	441 times
Utilization of OPERAT	.3662
Occupancy of MACHIN	.7201
Utilization of MACHIN	.6824
Utilization of VEHICL	.2440
Utilization of ARRIVE	1.0000
Average number in queue NEW	1.3804
Average number in queue ONPALL	2.9656
Average number in queue READY	.0221
Average number in queue INONQ	.4779
Average number in queue COMPLE	.0000
Average number in queue INOFFQ	.0192
Average number in queue FINSHD	.0418
Average number in queue EMTY	8.6682
Average number in queue OCCUPI	2.0998
Average number in queue IDLE	.8394
Average number in queue ONOCCU	.4779
Average number in queue ONVACA	2.3383
Average number in queue OFFOCC	.0191
Average number in queue OFFVAC	2.7988

Component Type	Number made	Average time in system
1	80	36.2500
2	62	73.6065
3	61	57.8688
4	54	57.5471
5	33	59.0909
6	39	40.7692
7	35	56.8571
8	25	47.5000
9	13	36.1538
10	14	38.5714
11	14	51.4285
12	10	54.0000
overall	440	52.4256

machines are usually able to discharge their pallet to the buffer immediately after the operation is complete, eliminating any delay until the vehicle arrives.

This indicates that the capacity of the system of Model Two is greater than that of Model One. To exploit this situation we would need to load the system up to closer to its capacity. The results which are obtained when the model is amended to assume that components are always available are given in Table 9.2. These results can be compared with those given in table 8.4. The output of the system, 624 components, is approximately 15% greater than that obtained when there were no machine buffers. The occupancy

Table 9.2   Results from Model Two when it is assumed that components are always available and the component entity is deleted from the model

INTROD was started	624 times
RETURN was started	624 times
LOADUP was started	624 times
TRFRON was started	624 times
TRFROF was started	624 times
Utilization of OPERAT	.5200
Occupancy of MACHIN	1.0000
Utilization of MACHIN	.9400
Utilization of VEHICL	.3466
Average number in queue READY	5.3847
Average number in queue INONQ	2.6417
Average number in queue COMPLE	.0000
Average number in queue INOFFQ	.0806
Average number in queue FINSHD	.0264
Average number in queue OCCUPI	2.9133
Average number in queue IDLE	.0000
Average number in queue ONOCCU	2.6417
Average number in queue ONVACA	.0983
Average number in queue OFFOCC	.0806
Average number in queue OFFVAC	2.6594

Component Type	Number made	Average time in system
1	92	79.6703
2	65	110.9230
3	82	86.8674
4	78	93.0769
5	65	113.1250
6	56	129.2857
7	45	159.3333
8	37	188.9474
9	34	189.4118
10	30	206.0000
11	22	210.0000
12	18	210.0000
overall	624	126.2500

of the machines has risen, as anticipated, to 100%, and the machine utilisation is 94% compared with 88% in Model One. Comparing the queue lengths in table 9.1 with 9.2 we can see that the on-queue buffers are much more heavily utilised at the higher rate of output.

The comparisons between Model One and Model Two are summarised in Table 9.3.

Table 9.3   Comparison of results from Model Two (with machine buffers) and Model One (no machine buffers)

Arrival pattern:	Components always available		Controlled arrival rate		
Mean time between arrivals, minutes:	—	—	17	17	12
Machine buffers:	No	Yes	No	Yes	Yes
Number of arrivals	—	—	441	441	607
Output	558	624	441	440	595
Average time in system	125.2	126.3	51.4	52.4	71.2
Occupancy of machines	1.00	1.00	0.77	0.72	0.96
Utilisation of machines	0.88	0.94	0.68	0.68	0.91

9.4   MODEL THREE—SYSTEM WITH GENERAL LOAD STATIONS AND MACHINE BUFFERS

We observed that in Model One and in Model Two, the utilisation of the load–unload stations was very small. If there were only one load–unload station it would have the same utilisation as the operator, or about 35%. For Model Three let us see the effect of combining all load–unload operations at a single load station. Since most systems which have general load stations have at least two, we will amend the logic in a way which will apply to any number of load stations. This model is representative of several real systems, and a layout diagram is given in Figure 9.16. The Anderson Strathclyde

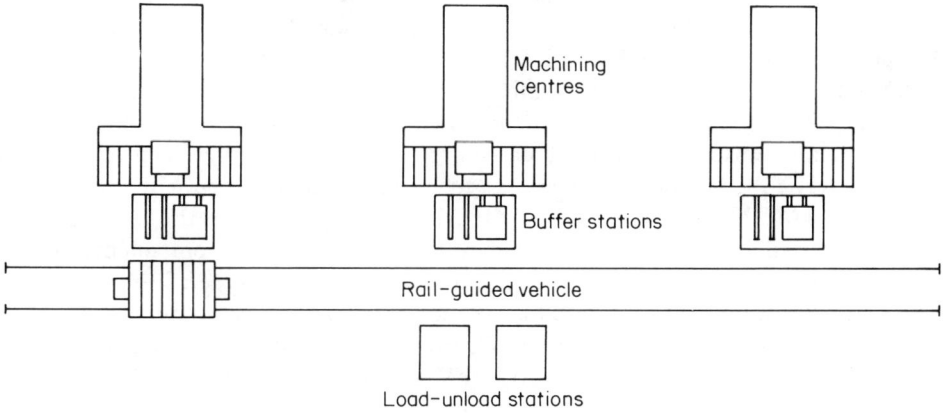

Figure 9.16   *Layout of system for Model Three.*

system, described in chapter 7, is of this type, although in that system the processing requirements of the components are rather complex, involving several settings. In the next chapter we will examine how the sequence of operations and fixtures can be added to the model. For the present, we will maintain the assumption that the components can be completed in a single operation.

9.4.1 Amendments to model

The basic difference between Model Three and Models One and Two is that the load–unload station must be included in the model, since a pallet cannot be RETURNed to a load–unload station unless one is available. This involves a simple modification to the activity cycle diagram, which is shown in Figure 9.17.

However, several consequential amendments are necessary. Since there are no dedicated load stations, there is no "home" position for each pallet. The number of positions where pallets can be placed is 11 (three at each machine, one at the load–unload station and one on the vehicle). Since this is less than the number of component types, it will no longer be practical to have dedicated pallets. Consequently, we have to consider the number of pallets in the system. The number of pallets must be less than 11, otherwise there would be no place to move any pallet because every other space was occupied. The absolute maximum number of pallets is therefore 10, but with ten pallets there might be considerable congestion. Strictly, since in our logic a movement requires that the vehicle as well as the destination is free, the maximum number is reduced to nine. We shall experiment with varying numbers of pallets, and assume that any pallet can take any type of component. Since there are three machines, three is the minimum number of pallets with which the system can be operated without a certain reduction in output. The number can be increased from three until the system appears saturated.

As a consequence of the pallets no longer being dedicated to a particular type of component, we will have to alter the structure of our attributes. We can no longer use the pallet type attribute, PTYPE. Instead, we will need to know the type of component which is on the pallet. A further complication is that it is conceivable that now there could be components of the same type on two or more pallets simultaneously. In the earlier models we were able to use the fact that there could be only one component of each type in the queue ON A PALLET at a time to identify the correct component to be taken out of that queue when the pallet was unloaded. Now we need some means of uniquely identifying which component is mounted on each pallet. An attribute CMPNO of the pallet can be used to record which component is on the pallet. The value of that component's CTYPE attribute defines the type of component on the pallet.

Similarly, MACHTIME, RATIO, REQD, LOADED and TIMHIST cannot be attributes of the pallet any more. Nor can they be attributes of the components, because they are associated not with the component directly, but with a type of component. We could define arrays to hold this data. However, since the data is associated with a type of component, we could define a new kind of entity called PARTTYPE whose attributes store this data. We can then use ECSL's FIND command to locate the component type with the lowest ratio for loading up, and for this purpose we need to define a set for the entity, which we can call PARTLIST. (PARTTYPE and PARTLIST are preferred to COMPONENTTYPE and COMPONENTLIST because they are shorter and

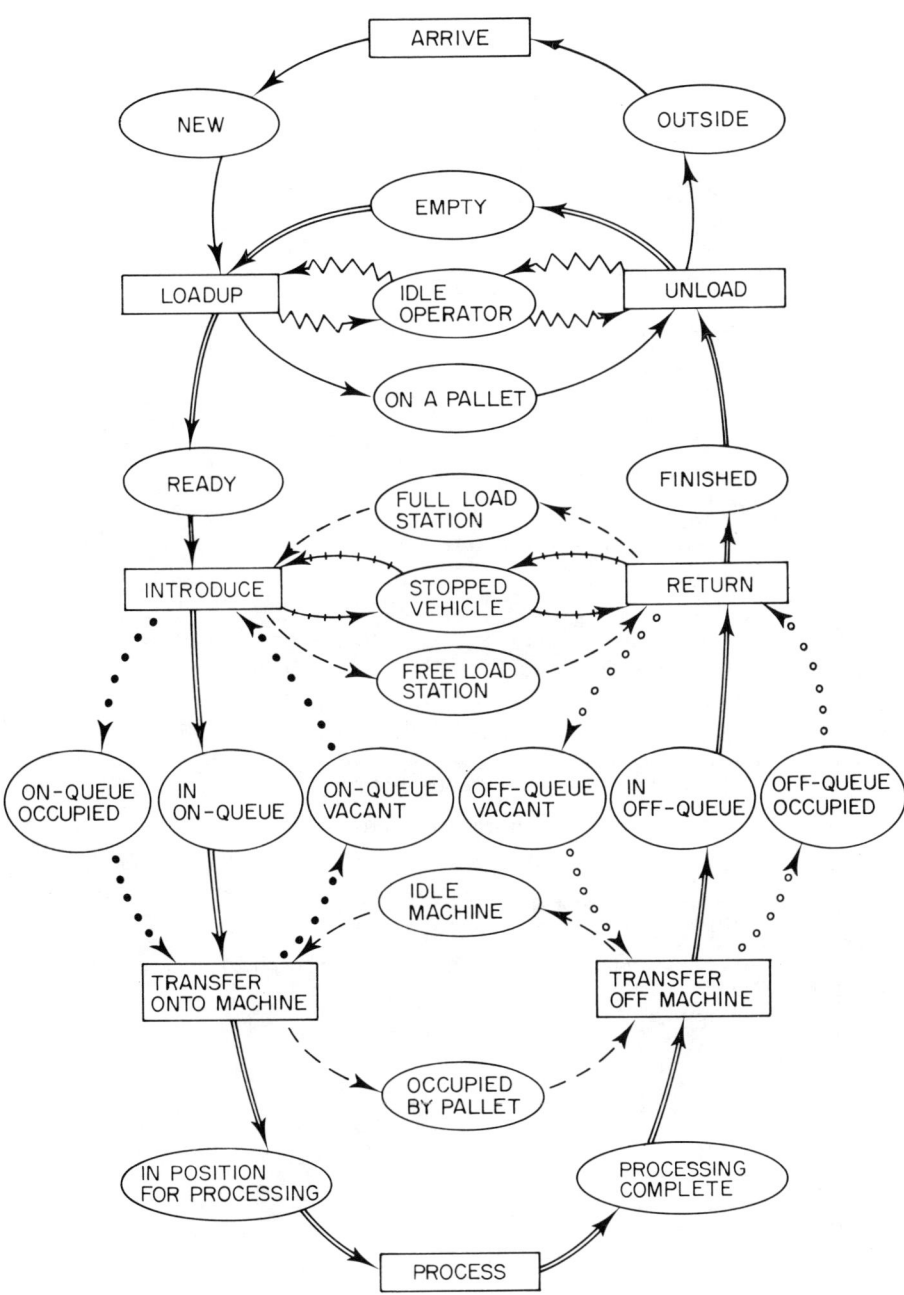

Figure 9.17   Activity cycle diagram for system of Model Three.

```
        THERE ARE 1 LOADSTN SET LDFULL LDFREE
        THERE ARE 5 PALLET SET READY INONQ COMPLE INOFFQ FINSHD EMTY WITH
       +LDAT MCAT CMPNO
        THERE ARE 12 PARTTYPE SET PARTLIST WITH
       +LOADED RATIO REQD MACHTIME MADE HIST TIMHIST(11 10 20)

        FOR LDFREE
          ADD DURATION TO EZLDST

        BEGIN INTROD
        ....
        C = LDAT OF PALLET A
        LOADSTN C FROM LDFULL INTO LDFREE AFTER DURATION
        LDAT OF PALLET A = O

        BEGIN RETURN
        ....
        FIND FIRST LOADSTN C IN LDFREE
        ....
        LOADSTN C FROM LDFREE INTO LDFULL
        LDAT OF PALLET A = C

        BEGIN LOADUP
        ....
        FIND PALLET B FROM EMTY
        FIND PARTTYPE T IN PARTLIST WITH MIN RATIO
          FIND FIRST COMPON A IN NEW
             CTYPE EQ T
        ....
        CMPNO OF PALLET B = A
        LOADED OF PARTTYPE D + 1
        X = 1000 / REQD OF PARTTYPE D
        RATIO OF PARTTYPE D = X * LOADED OF PARTTYPE D

        BEGIN UNLOAD
        ....
        A = CMPNO OF PALLET B
        D = CTYPE OF COMPON A
        ADD TIMIN OF COMPON A TO TIMHIST OF PARTTYPE D

        BEGIN TRFRON
        ....
        C = CMPNO OF PALLET A
        D = CTYPE OF COMPON C
        AADURATION= DURATION+ MACHTIME OF PARTTYPE D
        ADD 1 TO MADE OF PARTTYPE D
        ADD MACHTIME OF PARTTYPE D TO TMACHTIME

        TYPE'Occupancy of Load Stations',(1-EZLDST/(1.*(CLOCK-RUNINZ)))
        ...
        TYPE 'Component      Number     Average time'
        TYPE '  type          made       in system'
        FOR PARTTYPE I
           TYPE *4,I *13,MADE OF PARTTYPE I *14,+4,MEAN(TIMHIST)
        FOR PARTTYPE I SUM MADE
        TYPE 'Overall' I *14,+4,MEAN(TIHIST)
        DATA
CMPNO O 2 3 4 5
PARTLIST 1 TO *
NEW 1
ONPALL 2 3 4 5
OUTSID 6 TO *
CTYPE 1 2 3 1 2 45*0
LDFULL 1
OFFVAC 2 TO *
OFFOCC 1
ONOCCU 1 TO *
INONQ 2 3 4
INOFFQ 5
LDAT 1 4*0
MCAT O 1 2 3 1
EMTY 1
IDLE 1 TO *
```

Figure 9.18   Amendments to ECSL code for converting Model Two into Model Three, which contains a limited number of general load-stations.

avoid ambiguities, since ECSL treats only six characters as significant.) Other software packages may or may not involve such considerations and may introduce others.

Amendments to the ECSL coding of the Definitions section of the program to include the new entity type and the new use of attributes will be required. A new attribute of the pallet, LDAT, can be used to record which load station the pallet is at, even although we will start with only one. In the activities section, the coding of the INTRODUCE and RETURN activities will be amended to include the load stations. The statements referring to attributes of the pallet which have been made attributes of the part type will need to be amended. These are mainly used for recording results, such as TIMIN and TMACHTIME, or for updating RATIO. The coding of the selection of the type of component in LOADUP will also be amended. To illustrate these changes the parts of the new Definitions section and of the new versions of INTRODUCE, RETURN and LOADUP activities are given in Figure 9.18. There will also be new statements for recording and printing certain statistics.

Thought must be given to the initial conditions for Model Three. It is not possible to place all the pallets in EMTY as previously, because there is only one load station. The other pallets must be placed around the system. With five pallets and one load–unload station, we could place three pallets in the on-queue buffers, one in an off-queue buffer and one in EMTY at the load station. The pallets at the machine buffers will have components on them, and we could put one component in NEW with the rest in OUTSIDE. The values of CTYPE, CMPNO, MCAT, LDAT and so on must be assigned. Figure 9.18 includes the data section when there are five pallets. The initial conditions vary slightly with the number of pallets, since each pallet must be located somewhere within the system. Simulation runs were performed with varying numbers of pallets. In these runs, one pallet was placed initially at a load station, the next three were placed on the on-queue buffers, the fifth, sixth and seventh, if present, were placed on off-queue buffers, and in the run with eight pallets one was placed on the table of machine one.

9.4.2   Results with one load station and varying numbers of pallets

The results when Model Three is run with one load station and varying numbers of pallets are given in Table 9.4(a). They clearly show that the time in the system increases with the number of pallets, from 46.7 with only three pallets to 122.5 with eight pallets. The occupancy of the machine and load station increase in a similar way, reaching 100% when there are seven pallets. At this point the system is becoming congested and any additional pallets would merely extend the time in the system, and might possibly lead to a reduction in output. There is some evidence that only three pallets is not adequate since the output at 438 is somewhat below that for 4 to 8 pallets. With 4 to 8 pallets the observed output was 440 compared with 441 arrivals. The differences appear to be related to the number of pallets, but they are small and, strictly, more experiments with different random number streams should be performed to determine whether these effects are genuinely attributable to the number of pallets or are merely a coincidental effect of the sampling process. If there were nine pallets, it is conceivable that there would be lengthy delays at the machines with pallets having to wait on the machine table after the operation until an off-queue buffer becomes free.

Table 9.4   Performance of Model Three with varying numbers of pallets, load–unload stations and rates of arrival

(a) One load station, controlled rate of arrivals

Number of pallets	3	4	5	6	7	8
Number of arrivals	441	441	441	441	441	441
Output	438	440	440	440	440	440
Average time in system	46.7	58.2	73.4	89.1	106.0	122.5
Occupancy of machines	0.72	0.72	0.76	0.86	1.00	1.00
Utilisation of machines	0.68	0.68	0.68	0.68	0.68	0.68
Occupancy of load stations	0.63	0.85	0.94	0.97	1.00	1.00

(b) Two load stations, controlled rate of arrivals

Number of pallets	3	4	5	6	7	8
Number of arrivals	441	441	441	441	441	441
Output	441	441	441	441	440	441
Average time in system	45.9	54.0	64.9	78.0	93.0	107.8
Occupancy of machines	0.72	0.72	0.74	0.77	0.87	1.00
Utilisation of machines	0.68	0.68	0.68	0.68	0.68	0.68
Occupancy of load stations	0.37	0.62	0.78	0.88	0.95	1.00

(c) Two load stations, components always available

Number of pallets	3	4	5	6	7	8
Output	300	429	509	567	589	589
Average time in system	47.9	49.7	57.1	64.3	74.6	86.5
Occupancy of machines	0.51	0.73	0.86	0.96	1.00	1.00
Utilisation of machines	0.48	0.69	0.82	0.91	0.94	0.94
Occupancy of load stations	0.71	0.80	0.85	0.89	0.91	1.00

9.4.3   Results with two load stations

As observed in chapter 7, most real systems have at least two load stations. Table 9.4(b) gives the results when there are two load stations, although still only one operator. The system can clearly handle the rate of arrival of the components. The time in the system and the occupancy of the machines and load stations increase with the number of pallets, though at reduced levels compared to the single load station results. With eight pallets the machines and load stations are 100% occupied. Since there is only one operator it is evident that the value of the second load station is not in providing additional processing capacity, but merely in facilitating the movement of pallets around the system.

Table 9.5 gives more complete results for the case with two load stations and five pallets.

9.4.4   Results when components are assumed always available

To get a measure of the capacity of the system, we could amend the model so that whenever a pallet is returned to the load station and unloaded, a component of the type with the least value of RATIO is placed upon it. The results when this was done

Table 9.5   Results from Model Three with two
load stations and five pallets

INTROD was started	442 times
RETURN was started	441 times
LOADUP was started	442 times
UNLOAD was started	441 times
TRFRON was started	441 times
TRFROF was started	439 times
ARRIVE was started	441 times
Utilization of OPERAT	.3675
Occupancy of MACHIN	.7405
Utilization of MACHIN	.6835
Utilization of VEHICL	.2450
Occupancy of Load Stations	.7788
Utilization of ARRIVE	1.0000
Average number in queue NEW	.9136
Average number in queue ONPALL	3.7343
Average number in queue READY	.0098
Average number in queue INONQ	.4880
Average number in queue COMPLE	.0577
Average number in queue INOFFQ	.7326
Average number in queue FINSHD	.0370
Average number in queue EMTY	.8981
Average number in queue OCCUPI	2.1606
Average number in queue IDLE	.7783
Average number in queue ONOCCU	.4880
Average number in queue ONVACA	2.3281
Average number in queue OFFOCC	.7326
Average number in queue OFFVAC	2.0838

Component Type	Number made	Average time in system
1	80	58.8889
2	62	90.9523
3	61	62.7869
4	54	71.1111
5	33	61.5151
6	40	52.0512
7	35	63.7143
8	25	59.1666
9	13	39.2307
10	14	45.7142
11	14	62.8571
12	10	70.0000
overall	441	64.8752

are shown in Table 9.4(c). They clearly show that the capacity of the system increases with the number of pallets up to a point when an additional pallet does not give more output but merely lengthens the queues and delays. The figures show that the eighth pallet gives no more output. An interesting observation is that with 3 or 4 pallets the output is actually less than when the rate of arrivals was controlled. This seems an illogical result, since we would expect at least as many components produced. However, we have slightly changed the logic of the model, in joining UNLOAD to LOADUP, and eliminating the component entity and the queue EMTY. There is now no provision for pallets to accumulate at their dedicated load stations, as with Models One and Two. Having one large activity instead of two small ones reduces the flexibility of the operator. Although his utilisation is still below 50% even with eight pallets, it is conceivable that operator availability caused a small loss of output. Thus some reduction from the 624 observed in Model Two is not really surprising.

9.5   MODEL FOUR—A SYSTEM WITH A CENTRAL PALLET STORE

If we add central or common storage to Model Three we will have a model which is representative of many FMS cells, namely a system with general load–unload stations, machine buffers and also some central common storage for pallets. The Victor Products, Mirrlees Blackstone and Cummins' Makino line systems are all of this type. A fundamental point is whether there as many spaces in the common store as there are pallets, and whether there is a "home" position for each pallet, or pallet–fixture combination. The Cummins system has a location for each pallet, whereas the Victor and Mirrlees systems do not. If there is a location for each pallet, then there will always be a space for any pallet wishing to enter the store, and the spaces themselves may be omitted from the model, at least so far as the basic logic goes. If not, then they must be included, just as general load–unload stations must be included. Let us assume first that we are dealing with the simpler case, in which there is a home position for each pallet, as in the Cummins' Makino line. Figure 9.19 shows the layout of the system for Model Four.

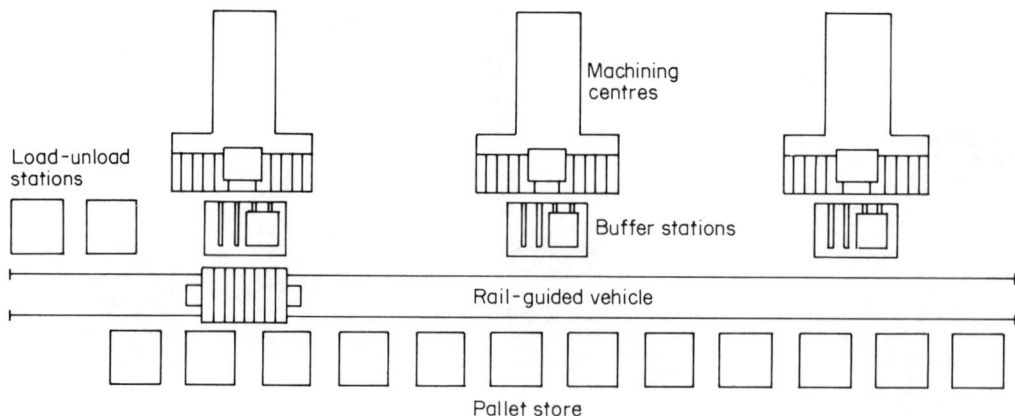

Figure 9.19   Layout of system for Model Four.

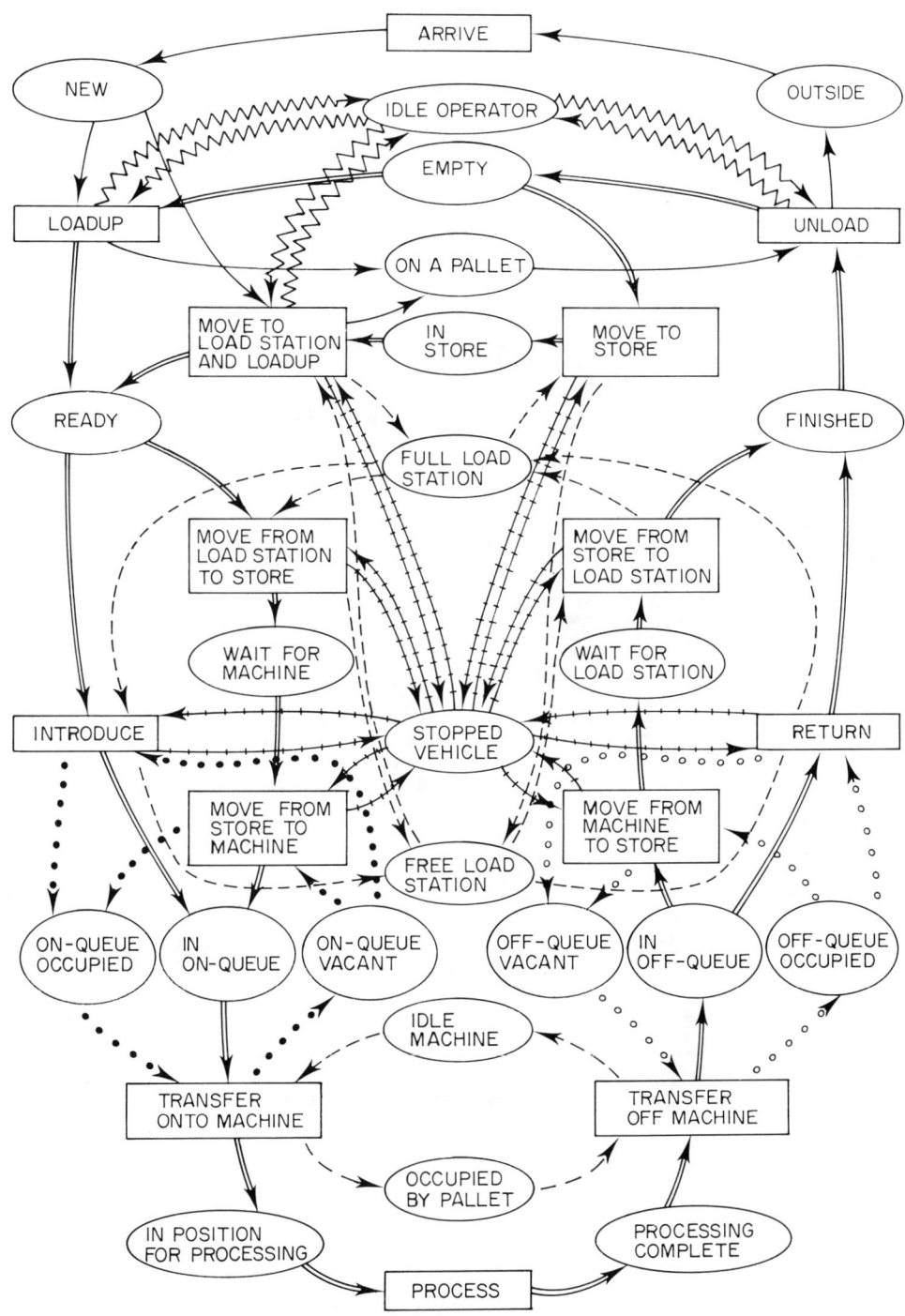

Figure 9.20   Activity cycle diagram for system of Model Four.

```
  THERE ARE 12 PALLET SET READY INONQ COMPLE INOFFQ FINSHD EMTY
+INSTORE LDWAIT MCWAIT WITH LDAT MCAT CMPNO PTYPE
  THERE ARE 12 PARTTYPE SET PARTLIST WITH
+LOADED RATIO REQD MACHTIME MADE HIST TIMHIST(11 10 20)

  ADD INSTORE TO HIST ZRINST DURATION
  ADD MCWAIT TO HIST ZSMCWA DURATION
  ADD LDWAIT TO HIST ZTLDWA DURATION

  BEGIN LOADUP
  . . . .
  ZERO PARTLIST
  FOR PALLET B IN EMTY
     EXISTS COMPON IN NEW WITH CTYPE EQ PTYPE
     PARTTYPE B INTO PARTLIST
  FIND PARTTYPE T IN PARTLIST WITH MIN RATIO
  FIND FIRST PALLET B IN EMTY WITH PTYPE EQ T
  FIND FIRST COMPON A FROM NEW WITH CTYPE EQ T
  . . . .

  BEGIN MVSTLU
  TIME OF VEHICL LE 0
  TIME OF OPERAT LE 0
  FIND FIRST LOADSTN C IN LDFREE
  ZERO PARTLIST
  FOR PALLET A IN INSTORE
     EXISTS COMPON IN NEW WITH CTYPE EQ PTYPE
     PARTTYPE A INTO PARTLIST
  FIND PARTTYPE T IN PARTLIST WITH MIN RATIO
  FIND FIRST PALLET A IN INSTORE WITH PTYPE EQ T
  FIND FIRST COMPON B FROM NEW WITH CTYPE EQ T
  DURATION = MOVETIME + LOADTIME
  COMPON B FROM NEW INTO ONPALL AFTER LOADTIME
  PALLET A FROM INSTORE INTO READY AFTER DURATION
  LOADSTN C FROM LDFREE INTO LDFULL
  TIME OF VEHICL = MOVETIME
  TIME OF OPERAT = DURATION
  LDAT OF PALLET A = C
  CMPNO OF PALLET A = B
  LOADED OF PARTTYPE T + 1
  X = 1000 / REQD OF PARTTYPE T
  RATIO OF PARTTYPE T = X * LOADED OF PARTTYPE T
  TIMIN OF COMPON B= CLOCK
  ADD 1 TO MVSTLU

  BEGIN MVSTUL
  TIME OF VEHICL LE 0
  FIND FIRST LOADSTN B IN LDFREE
  FIND FIRST PALLET A IN LDWAIT
  DURATION = MOVETIME
  TIME OF VEHICL = DURATION
  PALLET A FROM LDWAIT INTO FINSHD AFTER DURATION
  LOADSTN B FROM LDFREE INTO LDFULL
  ADD 1 TO MVSTUL
  LDAT OF PALLET A = B
```

Figure 9.21   Amendments to ECSL code for converting Model Three into Model Four, which contains a central pallet store.

```
BEGIN MVSTMC
TIME OF VEHICL LE 0
FIND FIRST ONBUFF A IN ONVACA
FIND FIRST PALLET B IN MCWAIT
DURATION = MOVETIME
PALLET B FROM MCWAIT INTO INONQ AFTER DURATION
ONBUFF A FROM ONVACA INTO ONOCCU
MCAT OF PALLET B = A
TIME OF VEHICL = DURATION
ADD 1 TO MVSTMC

BEGIN MVLUST
TIME OF VEHICL LE 0
FIND FIRST PALLET A FROM READY
DURATION = MOVETIME
PALLET A FROM READY INTO MCWAIT AFTER DURATION
TIME OF VEHICL = DURATION
B = LDAT OF PALLET A
LOADSTN B FROM LDFULL INTO LDFREE AFTER DURATION
LDAT OF PALLET A = 0
ADD 1 TO MVLUST

BEGIN MVMCST
TIME OF VEHICL LE 0
FIND FIRST PALLET A IN INOFFQ
B = MCAT OF PALLET A
DURATION = MOVETIME
OFFBUFF B FROM OFFOCC INTO OFFVAC AFTER DURATION
PALLET A FROM INOFFQ INTO LDWAIT AFTER DURATION
TIME OF VEHICL = DURATION
MCAT OF PALLET A = 0
ADD 1 TO MVMCST

BEGIN MVULST
TIME OF VEHICL LE 0
FIND FIRST PALLET A IN EMTY
DURATION = MOVETIME
PALLET A FROM EMTY INTO INSTORE AFTER DURATION
B = LDAT OF PALLET A
LOADSTN B FROM LDFULL INTO LDFREE AFTER DURATION
TIME OF VEHICL = DURATION
LDAT OF PALLET A = 0
ADD 1 TO MVULST

TYPE 'MVULST was started' MVULST ' times'
TYPE 'MVSTLU was started' MVSTLU ' times'
TYPE 'MVMCST was started' MVMCST ' times'
TYPE 'MVSTMC was started' MVSTMC ' times'
TYPE 'MVLUST was started' MVLUST ' times'
TYPE 'MVSTUL was started' MVSTUL ' times'
....
TYPE 'Average number in store empty'+4,MEAN(ZRINST)
TYPE 'Average number in store WAITING FOR MACHINE'+4,MEAN(ZSMCWA)
TYPE 'Average number in store WAITING FOR LOADSTN'+4,MEAN(ZTLDWA)
```

Figure 9.21   (continued).

9.5.1 Amendments to the model

To amend Model Three to add common storage, we have to introduce the more complex pallet and vehicle cycles from figure 9.12. This adds six new activities to the model. We must consider the conditions under which we should move an empty pallet from the load–unload station to the central store or from the central store back to the load station. As figure 9.12 stands, an empty pallet would be brought from the store to the load–unload area as soon as there is one in the store, the vehicle is stopped, and the load station is free. These conditions will exist immediately after a pallet has been brought to the store. Unless we add some other condition, the vehicle would take the pallet back to the load station immediately after it has brought it to the store, in a never-ending cycle! To handle this problem we could introduce a dummy entity to represent the need to bring a pallet from the store. It could be generated by the arrival of a component, and destroyed by the component being loaded on a pallet. However, we can avoid this by defining a new activity which depends on the existence of a NEW component and a pallet of the right type in IN STORE. Since, as soon as the movement of the pallet back to the load–unload station is complete the LOADUP of the component can begin, we can combine these two activities into one and define a new activity MOVE TO STORE AND LOADUP. This will take the pallet from INSTORE and put it into READY, and take the component from NEW and place it in ON A PALLET.

Similarly, figure 9.12 implies that an empty pallet would be removed from the load–unload station and placed in store immediately it has been unloaded if there were no component waiting to be loaded on it. In practice, it might be permissible to leave it there until some other pallet requires to use the load-station. We could use similar logic to that just described, except that it would be a pallet in the IN OFF-QUEUE or WAITING FOR LOAD STATION which would trigger the removal of an empty pallet, followed immediately by the moving of a loaded one to the load–unload area. However, the diagram is already quite complex, and this will be omitted. We will therefore place an empty pallet in store if no component is waiting to be placed on it.

The complete activity cycle diagram for Model Four is shown in Figure 9.20. The code of the model will be amended by defining 12 pallets as in Models One and Two, and by adding the six new activities, which can be named MVSTLU, MVSTUL, MVSTMC, MVLUST, MVMCST and MVULST, where LU, UL, MC and ST represent movements to or from a loading or unloading operation or a machine or storage location. Thus MVSTUL is moving from the store to the load–unload area for an unloading operation. The code for these activities is given in Figure 9.21. The relative priorities of the vehicle's activities have been implied by entering them in the model in the order INTRODUCE, RETURN, MVSTLU, MVSTUL, MVSTMC, MVLUST, MVMCST, MVULST. The initial conditions can be simply specified if we assume that all the pallets start empty in the pallet store.

9.5.2 Results

Table 9.6 gives the results obtained with either one or two load stations, again using the same data for component requirements and processing and movement times. The total numbers of components made are 439 and 440 components, compared with 441 arrivals. Further tests would be necessary to establish whether this represents a reduction

Table 9.6   Results from Model Four with one and two load stations

Number of load stations	One	Two
Times INTROD was started	439	433
Times RETURN was started	234	419
Times LOADUP was started	176	164
Times UNLOAD was started	437	437
Times TRFRON was started	439	440
Times TRFROF was started	437	437
Times ARRIVE was started	441	441
Times MVULST was started	262	273
Times MVSTLU was started	265	277
Times MVMCST was started	203	18
Times MVSTMC was started	1	8
Times MVLUST was started	1	8
Times MVSTUL was started	203	18
Utilization of OPERAT	.4384	.4426
Occupancy of MACHIN	.7200	.7203
Utilization of MACHIN	.6805	.6824
Utilization of VEHICL	.4463	.4037
Occupancy of Load Stations	.7666	.4068
Utilization of ARRIVE	1.0000	1.0000
Average number in queue NEW	1.5861	1.4140
Average number in queue ONPALL	3.1977	2.9270
Average number in queue READY	.0033	.0059
Average number in queue INONQ	.3119	.3823
Average number in queue COMPLE	.0000	.0000
Average number in queue INOFFQ	.0258	.0190
Average number in queue FINSHD	.0000	.0151
Average number in queue EMTY	.0080	.0301
Average number in store empty	8.3558	8.6012
Average number in store MCWAIT	.0006	.0070
Average number in store LDWAIT	.3222	.0084
Average Time in system—overall	55.7208	51.5560
TOTAL MADE	439	440

in capacity of the system, or is a sampling effect due merely to the interactions of the events over the run. There are some clear effects. The vehicle utilisation figures have risen to 40% when there are two load stations and 45% when there is one, as against 24% in Models One and Two. This reflects the additional journeys to and from the central pallet store.

A comparison of the results with one and two load stations reveals some interesting observations. In the single load station case, we find that of the 437 pallets unloaded, 260 were moved to store and only 177 loaded with a new component without being moved to the store. Similarly, of the 437 pallets transferred to the off-queue buffers, 231 were returned directly to the load–unload station and 206 taken to the store en route. When a second load station is provided the number of pallets moved to store before unloading is reduced to 14. These figures reflect the occupancy of the load stations. With only one, the occupancy was 77%, a high level probably constraining system operation, but only 41% when there are two load stations. The utilisation of the on-queue buffers is very low in both runs and only a few pallets are placed in store after loading up, 1 and 8 respectively.

A substantial number of pallets, 260 and 273, were moved into store after unloading. As discussed above, the logic of the model does not allow pallets to wait at the load station or machine off-queue buffers, but places them into store if the principal destination is occupied, if there is no component of the right type waiting at the load–unload area. Particularly in the case when there are two load stations, when their occupancy is low, it might be sensible to allow the pallet to wait until either a NEW component of the right type arrives, or until some other pallet needs a load station. This would involve rewriting the logic of some of the activities. We could also experiment with different priorities for the vehicle's activities.

These results again highlight the value of an extra load–unload station, even although the work load on the load–unload operator is low.

9.6   MODEL FIVE—A SYSTEM WITH LIMITED CENTRAL STORE

In many systems there are only a limited number of pallet storage locations, instead of one for each pallet. Examples are the Mirrlees and Victor Products systems. Figure 9.22 gives the layout of the system. If there are only a limited number of locations in the central store then the model must include testing for the availability of a space before a pallet can be put into store. Let us amend the model to include this feature. One of the parameters we may wish to investigate is the number of spaces which should be provided in the central store.

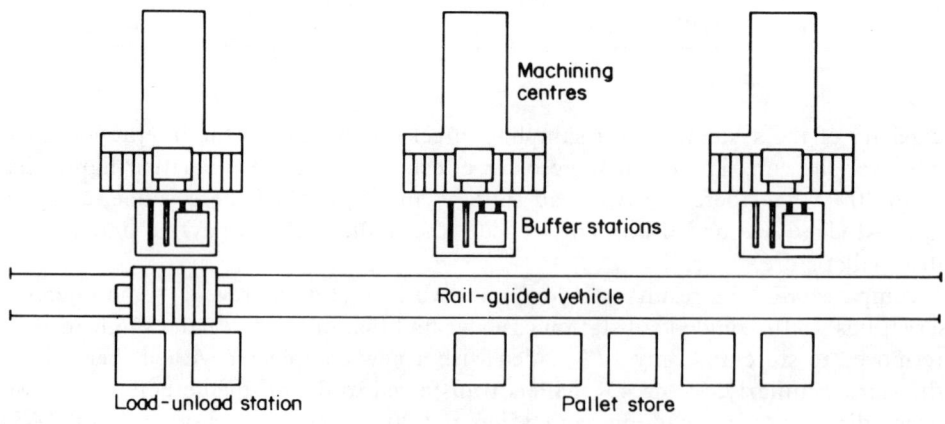

Figure 9.22   Layout of system for Model Five.

9.6.1 Amendments to the model

To introduce the restriction on the capacity of the central store we must include the spaces in store as a new type of entity, SPACE, with an activity cycle similar to those of the machines and load stations. Movements into store will fill a previously free space, and movements out of store will make a full space free again. The queues representing the free and full states can be called SPFREE and SPFULL. The coding of the MV—ST activities will have to include a test of the form FIND FIRST SPACE S IN SPFREE. We can record which space the pallet is placed in via an attribute SPAT and

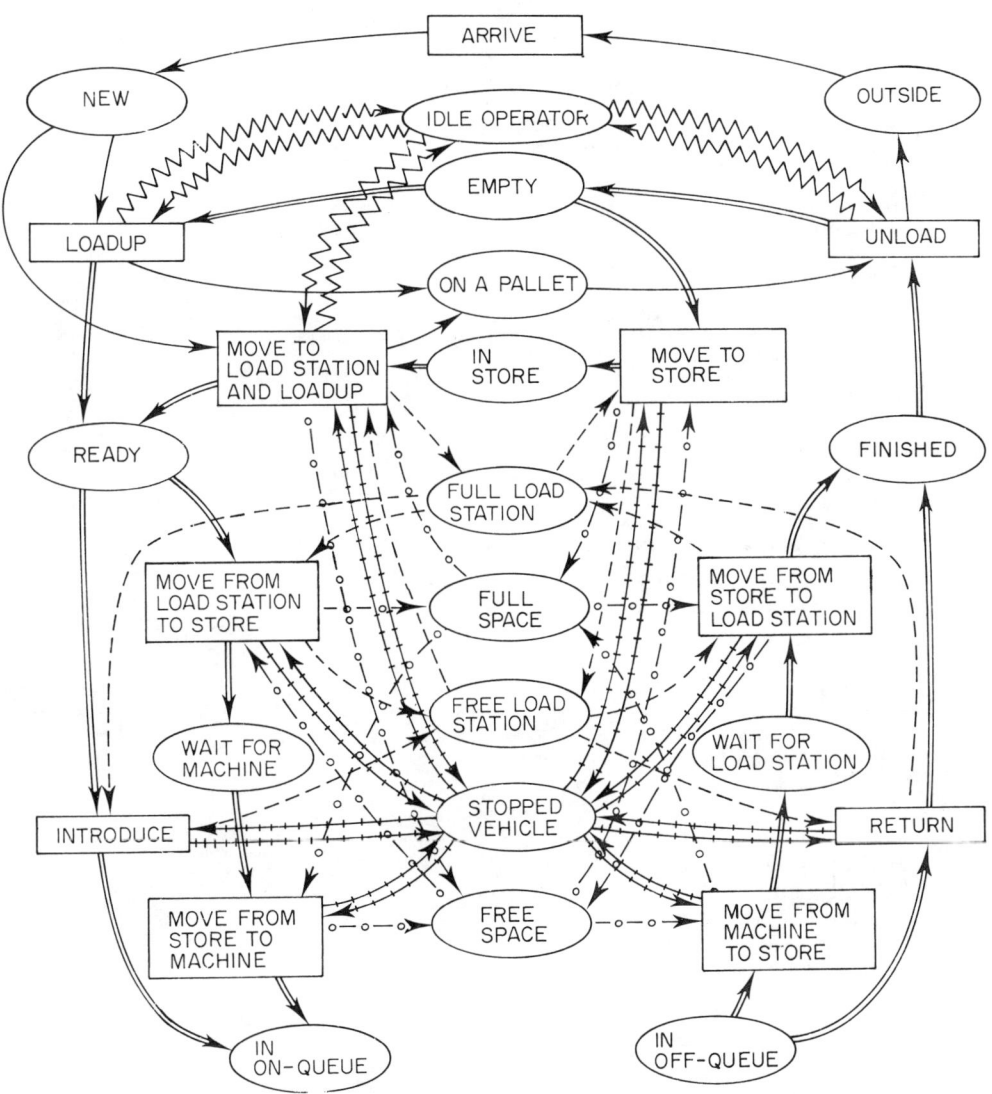

Figure 9.23   Activity cycle diagram for system of Model Five.

a statement of the form SPAT OF PALLET A = S. When the pallet is taken out of store, in MVST— activities, we will identify which space is freed by statements such as S = SPAT OF PALLET A and SPACE S FROM SPFULL INTO SPFREE AFTER DURATION. We will also wish to record information on the occupancy of the spaces. Figure 9.23 gives most of the activity cycle diagram for the system of Model Five, and Figure 9.24 the coding which must be added to the model.

Some thought must be given to the initial conditions, because as in Model Three the pallets must be distributed through the system. The membership of the sets and the

```
THERE ARE 5 SPACE SET SPFULL SPFREE
THERE ARE 12 PALLET SET READY INONQ COMPLE INOFFQ FINSHD EMTY
+INSTORE LDWAIT MCWAIT WITH LDAT MCAT SPAT CMPNO PTYPE

ADD INSTORE TO HIST ZRINST DURATION
ADD MCWAIT TO HIST ZSMCWA DURATION
ADD LDWAIT TO HIST ZTLDWA DURATION
....
FOR SPFREE
   ADD DURATION TO FZSPAC

BEGIN MVSTLU ( or MVSTUL or MVSTMC)
....
S = SPAT OF PALLET A
SPACE S FROM SPFULL INTO SPFREE AFTER MOVETIME
SPAT OF PALLET A = 0

BEGIN MVLUST ( or MVMCST or MVULST)
....
FIND FIRST SPACE S IN SPFREE
....
SPACE S FROM SPFREE INTO SPFULL
SPAT OF PALLET A = S

TYPE 'Occupancy of CENTRAL STORE'+4,(1-FZSPAC/(5.*(CLOCK-RUNINZ)))

   DATA
PTYPE 1 2 3 4 5 6 7 8 9 10 11 12
PARTLIST 1 TO *
MCAT 1 2 3 1 2 3 6*0
SPAT 6 * 0 1 2 3 4 0 0
LDAT 10 * 0 1 2
CMPNO 1 2 3 4 5 6 6 * 0
CTYPE 1 2 3 4 5 6 1 2 3 4 5 6 7 8 9 10 11 12 32*0
INSTORE 7 TO 10
INONQ 1 TO 3
INOFFQ 4 TO 6
EMTY 11 12
LDFULL 1 2
SPFULL 1 TO 4
ONPALL 1 TO 6
NEW 7 TO 18
OUTSID 19 TO *
OFFOCC 1 TO *
ONOCCU 1 TO *
IDLE 1 TO *
....
 END
```

Figure 9.24   Amendments to ECSL code for converting Model Four into Model Five, which contains only a limited number of common pallet storage locations.

values of the attributes must be consistent. With two load stations and at least four spaces we can place three pallets in on-queue buffers, three in off-queue buffers, two at the load stations and four in the store. This gives a total of 12 pallets, which as before can be dedicated to component types.

9.6.2 Results—blockage of the system

When the model was run it came to a premature halt. The system had become blocked, in the sense that no pallet could be moved because there was no free location for it. Suppose that pallets 11 and 12, which have the lowest requirement, are at the load stations, and that there are no components of types 11 or 12 waiting for loading. Then the logic of the model would seek to move each pallet away from the load stations into store, so that a pallet can be brought from store for loading up. However, if there are pallets in all the spaces of the store, it will not be possible to move either pallet 11 or 12 from the load station. Similarly, pallet 1, with the highest demand, may be in store and there may be components of type 1 waiting at the load area, but pallet 1 cannot be moved to a load station because both are occupied, by pallets 11 and 12. When all spaces are full and all load stations are occupied, no pallets can be moved away from the machines and they become blocked, and the whole system shuts down.

To overcome this problem we need to be able to move a pallet from either a load station or storage space to an unoccupied buffer position, so that the other pallets can be moved round. In real life this is obvious and the system's real control software may include suitable logic, but we have not put the necessary logic into the model. To solve the problem in the model we need an activity, which we could call UNBLOCK, which would take one pallet from the load station to the store, and another from the store to the load station, temporarily parking one elsewhere in the system, and then load a component on the pallet now at the load station. Figure 9.25 gives the code for this activity. Notice that we do not need to identify a free buffer position, because there must be at least one available somewhere. Of course, if the duration of the movements is to be calculated precisely on the distance travelled, then we would need to locate a vacant buffer. In the models so far we are using an average movement time and can overlook this detail.

An observation concerning the UNBLOCK activity can be made in passing. In discussing rotary pallet shuttles earlier in the chapter, it was suggested that it was in some ways more difficult to express the desired logic in an activity cycle diagram than in actual code. This activity is an example. It is quite straightforward to write the code to identify two entities of the same type and act on them in separate ways, whereas in the activity cycle diagram ambiguity may be introduced. Rotary pallet shuttles could be handled in a similar way. Unfortunately, the appropriate method and coding is language-specific.

9.6.3 Results with unblocking routine

Table 9.7 gives the results obtained when the system with 12 pallets and 2 load stations is run, with between 4 and 8 spaces in the central pallet store. They show several important effects. Firstly we see that the output of the system is lower than in the previous models, reaching, when there are 8 spaces, the level in Model Four with only one load station.

```
BEGIN UNBLOCK
LDFREE EMPTY
SPFREE EMPTY
TIME OF VEHICLE LE 0
TIME OF OPERAT LE 0
ZERO PARTLIST
FOR PALLET B IN INSTORE
   EXISTS COMPON IN NEW WITH CTYPE EQ PTYPE
   PARTTYPE B INTO PARTLIST
FIND PARTTYPE T IN PARTLIST WITH MIN RATIO
FIND FIRST PALLET B IN INSTORE WITH PTYPE EQ T
FIND FIRST COMPON A FROM NEW WITH CTYPE EQ T
FIND FIRST PALLET C IN EMTY
DURATION = 2 * MOVETIME + LOADTIME
PALLET C FROM EMTY INTO INSTORE AFTER 3 * MOVETIME
PALLET B FROM INSTORE INTO READY AFTER DURATION
COMPON A FROM NEW INTO ONPALL AFTER DURATION
TIME OF VEHICLE = 3 * MOVETIME
TIME OF OPERAT = DURATION
LDAT OF PALLET B = LDAT OF PALLET C
SPAT OF PALLET C = SPAT OF PALLET B
LDAT OF PALLET C = 0
SPAT OF PALLET B = 0
CMPNO OF PALLET B = A
LOADED OF PARTTYPE B + 1
X = 1000 / REQD OF PARTTYPE B
RATIO OF PARTTYPE B = X * LOADED OF PARTTYPE B
TIMIN OF COMPON A = CLOCK
ADD 1 TO UNBLOCK

TYPE 'UNBLOCK was started' UNBLOCK ' times'
```

Figure 9.25   Activity added to Model Five, to prevent blockage of the system occurring.

In other words, there would have to be more than 8 spaces if the output is not to be constrained. It would therefore appear to be a false economy to have fewer storage positions than pallets, at least with the component variety, quantities and processing times used in the experiments. The length of the queue NEW varies dramatically with the number of spaces, from below 2 on average with 8 spaces, to nearly 40 with only 4 spaces. This is clear evidence that the system cannot cope with the rate of arrivals if there are insufficient spaces, even although the occupancy and utilisation of the machines are quite low. The average time in the system also increases dramatically as the number of spaces is reduced, again reflecting the effects of congestion in the system. The occupancy of the central store is over 95% in each case, and reaches 100%, as does the occupancy of the machines and load stations, as the number of spaces is reduced to 4 or 5.

Other results are given in the table to show, for example, how the queue lengths vary with the number of spaces. Some statistics vary erratically, indicating complex interactions among the various entities and conditions in the model.

SUMMARY

In this chapter we have presented the logical considerations which are associated with buffers at machines and pallet storage locations within the system. We have presented four models which introduce these features and generally increase in complexity.

The most important general result from the experiments is that the performance of a system deteriorates significantly if the system becomes congested, either because the number of pallets in use is too great, or the number of storage spaces becomes too small.

Table 9.7 Results from Model Five, for varying numbers of spaces in common storage, with twelve pallets and two load stations

Number of Spaces	4	5	6	7	8
Times INTROD was started	394	417	387	417	427
Times RETURN was started	394	423	417	424	414
Times LOADUP was started	316	335	307	259	226
Times UNLOAD was started	394	436	438	437	437
Times TRFRON was started	393	437	437	437	439
Times TRFROF was started	393	437	435	434	437
Times ARRIVE was started	441	441	441	441	441
Times MVULST was started	0	29	59	78	148
Times MVSTLU was started	0	29	58	78	149
Times MVMCST was started	0	13	20	13	24
Times MVSTMC was started	0	19	50	20	12
Times MVLUST was started	0	19	50	20	12
Times MVSTUL was started	0	13	21	13	24
Times UNBLOCK was started	79	72	72	101	65
Utilization of OPERAT	.3725	.4109	.4202	.4422	.4429
Occupancy of MACHIN	1.0000	.8806	.7856	.7201	.7200
Utilization of MACHIN	.6049	.6714	.6754	.6750	.6808
Utilization of VEHICL	.2847	.3272	.3550	.3794	.3901
Occupancy of Load Stations	.9972	.9676	.8872	.8731	.7404
Occupancy of central store	1.0000	.9992	.9885	.9804	.9577
Utilization of ARRIVE	1.0000	1.0000	1.0000	1.0000	1.0000
Average number in NEW	39.2068	15.1237	9.1079	4.3458	1.9212
Average number in ONPALL	6.3270	5.5413	5.0202	3.8520	3.3319
Average number in READY	.0909	.0306	.0069	.0050	.0054
Average number in INONQ	.8823	.7968	.8197	.5052	.4737
Average number in COMPLE	1.0766	.5036	.2209	.0186	.0009
Average number in INOFFQ	2.1231	1.6550	1.1701	.7769	.3154
Average number in FINSHD	.0116	.0084	.0194	.0156	.0154
Average number in EMTY	1.3004	1.2348	1.0683	1.0190	.7325
Average number in INSTORE	3.9342	4.7528	5.4308	6.6025	7.4388
Average number in MCWAIT	.0000	.0613	.0762	.0279	.0198
Average number in LDWAIT	.0000	.0880	.2923	.0872	.0469
Average Time in system	113.0965	94.3120	83.1964	66.6133	58.6957
Total made	393	437	437	437	439

We have also illustrated that the capacity of a system may depend more on the interactions between entities and circumstances than on the direct work load on machines, operators or vehicles. It is for precisely this reason that simulation is important.

EXERCISES

1. Add the pallet's and machine's cycles to figure 9.8.

2. Develop an activity cycle diagram for a system in which some machines have buffers and some do not.

3. Develop an activity cycle diagram for a system with rotary pallet shuttles, omitting the transfer between the shuttle and the machine.

CHAPTER 10

Building FMS models—
3: Operation sequences, fixtures and tools

In all the models so far we have assumed that the components would be completed in a single operation. Although this is quite a common arrangement, if suitable fixturing methods are exploited, in general, components will require several operations, and possibly several different fixtures. The more complicated the component the more operations will be required, and usually more fixtures and tools. In this chapter we will examine the logic associated with these aspects, and generalise the models already presented, to include them. Figure 10.1 gives the notation for the activity diagrams for this chapter.

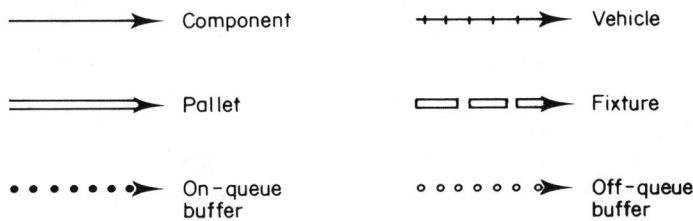

Figure 10.1   Arrow notation for activity cycle diagrams.

10.1   DEFINING THE SEQUENCE AND DURATIONS OF OPERATIONS

First let us consider the methods which can be used to specify the sequence and times of operations which are to be performed on the components. By sequence, or routing, we mean the number of operations and the workstation at which each must be done. Even if the components do not appear as entities in the model, which might be feasible in the circumstances discussed earlier, the processing requirements must be defined. In general there are three possibilities:

— using random sampling,
— using look-up tables, or
— using some combination of these.

These have already been illustrated. In the simple job shop in chapter 1 each job had a predetermined routing through the shop, the information being held in a table. In chapter 4 the use of attributes to hold this information was discussed. When the model was turned into a computer program using CAPS in chapter 5, instead of a table, random sampling was used, as described in chapter 3. As was explained in chapter 5, this was done for simplicity, because CAPS is not designed to use look-up tables. When we come to build an FMS model we have to decide which approach to adopt. In "classical" simulation, observed data from the system being modelled is usually summarised in a probability distribution from which values are sampled. However in FMS simulation we are almost certainly interested in the effects of the actual sequences of operations of the components as they travel through the system, and the actual or estimated cycle times. This means that we would read data from look-up tables rather than sample randomly. The term "emulation" is sometimes used to describe the process of modelling an actual system in a great deal of detail—as much detail as can sensibly be included in the model—using actual cycle times and routings.

10.1.1   Selecting operation times and workstations by sampling

In order that operation times can be sampled from a distribution, the type of distribution must first be determined, and its parameters calculated. The actual operation times would be studied, and plotted as a frequency distribution. If the shape of the distribution suggests that it conforms to one of the standard distributions, such as normal, exponential, binomial, Poisson, etc, then the goodness of fit of the observed data can be assessed and the parameters for that distribution computed. In the case of a normal distribution the mean and standard deviation would be calculated. The cycle times used in the model can then be sampled using the functions provided in the simulation package. Packages generally provide functions for the principal distributions.

If the frequency distribution of the actual times does not conform to a standard distribution, then the data can be expressed as a histogram and samples drawn from that, again using the functions provided.

If work stations are also to be sampled randomly, and this would be unusual in FMS simulation, we could sample from a histogram giving the probability of an operation being performed at each work station.

10.1.2   Selecting operation times and workstations by look-up tables

As stated above, in FMS simulation we normally require to model actual rather than randomly sampled operation sequences. If the system manufactures only a small number of types of component, then a look-up table can be used. Table 10.1 illustrates the layout of such a table, containing a pair of rows for every component type. As was suggested in an earlier section, load–unload operations can be included in such a table without difficulty.

If a large number of component types are manufactured the tables would be very large and it may not be possible to accommodate the data. Some method of condensing the data must be found. Many FMSs produce distinct families of components. For example, Cummins' Makino line produces several dozen component types, but they fall into three families, exhaust manifolds, water manifolds and thermostat housings.

Table 10.1   Layout of data table for operation sequences and times defining each component's routing

Component number		Operation number				
		1	2	3	4	5
1	Workstation	A	B	C	D	E
	Operation time	15	24	10	30	20
2	Workstation	B	C	B	A	0
	Operation time	25	80	30	15	0
3	Workstation	C	D	E	0	0
	Operation time	20	30	5	0	0

Within each type of part the routing is the same, and the cycle times are reasonably consistent. Thus it is possible to maintain modelling accuracy by using an attribute of each component defining its type and using that to extract information from a small look-up table. Assuming the operation times conform to a normal distribution, samples can be drawn from the data using the mean and standard deviation for the appropriate part family. Table 10.2 illustrates how the information might be held. Using this approach

Table 10.2   Layout of data table for routings and times when components can be arranged in families

Component family		Operation number				
		1	2	3	4	5
1	Workstation	A	B	C	D	E
	Mean operation time	15	24	10	30	20
	Standard deviation of operation time	4	8	2	10	5
2	Workstation	B	C	B	A	0
	Mean operation time	25	80	30	15	0
	Standard deviation of operation time	3	4	2	2	0
3	Workstation	C	D	E	0	0
	Mean operation time	20	30	5	0	0
	Standard deviation of operation time	4	8	1	0	0

any number of components can be held without any increase in the amount of data to be stored, since the amount of data depends on the number of families, not on the number of components.

We must also specify the number of operations. In chapters 1 and 4 this was defined by a separate attribute, NOOPS, or table entry, such as table 4.2. An alternative method can be deduced from tables 10.1 and 10.2, which does not require an attribute or particular table entry. Notice that in the second and third component families some of the operations appear to be done on machine 0 and have a duration of 0. The test could then be that if the next operation is to be done on a machine of type zero then there are no further operations. To apply this method, the length of the operation list has to be one greater than the maximum number of operations. Although this method may simplify the test, it adds to the volume of data just as much as adding a separate attribute. However, if the data were being held in a file instead of in memory, and memory were in short supply, this might be advantageous.

10.1.3   Transferring component data to pallets

In a previous section it was suggested that the components might not be included in the model. The data on the number, sequence and times of operations must still be provided. Similarly, even if the component is modelled it might not be modelled in detail while it is on a pallet. Thus we need to know about the pallet, where it must be sent and how long it must be held there, rather than about the component. This can be done by assigning an attribute to the pallet which specifies which type of component it holds, thus defining which entry in a table, such as table 10.2, should be interrogated. The value of this attribute would normally be set in the LOADUP activity. If, instead of referring to look-up tables, the data is held as attributes of the component, then a similar number of attributes of the pallet can be defined and the data written from the component to the pallet during the LOADUP activity.

10.1.4   Several operations

Figure 10.2 shows how the activity cycle diagram for the pallet given in figure 9.4 would be amended when several operations are required. The logic is very similar to that in the diagrams presented in chapter 5, although in this case the machines have on-queue and off-queue buffers. After each operation it is necessary to test whether there are more operations, to determine whether the pallet should be RETURNed to the load–unload area, or should be MOVEd to another machine.

10.1.5   Alternative operations or workstations

In most FMSs there is a choice of machine at which operations can be done. For example, it is normal to provide the tools for each operation at more than one machining centre, so that there is greater flexibility when scheduling components through the system. This was touched upon in chapter 7. For other operations, such as washing or gauging operations, there is less likely to be any choice of machine. To deal with the case of alternative workstations for operations we need a more complex data structure for holding the routing data. Instead of specifying which machine each operation is to be

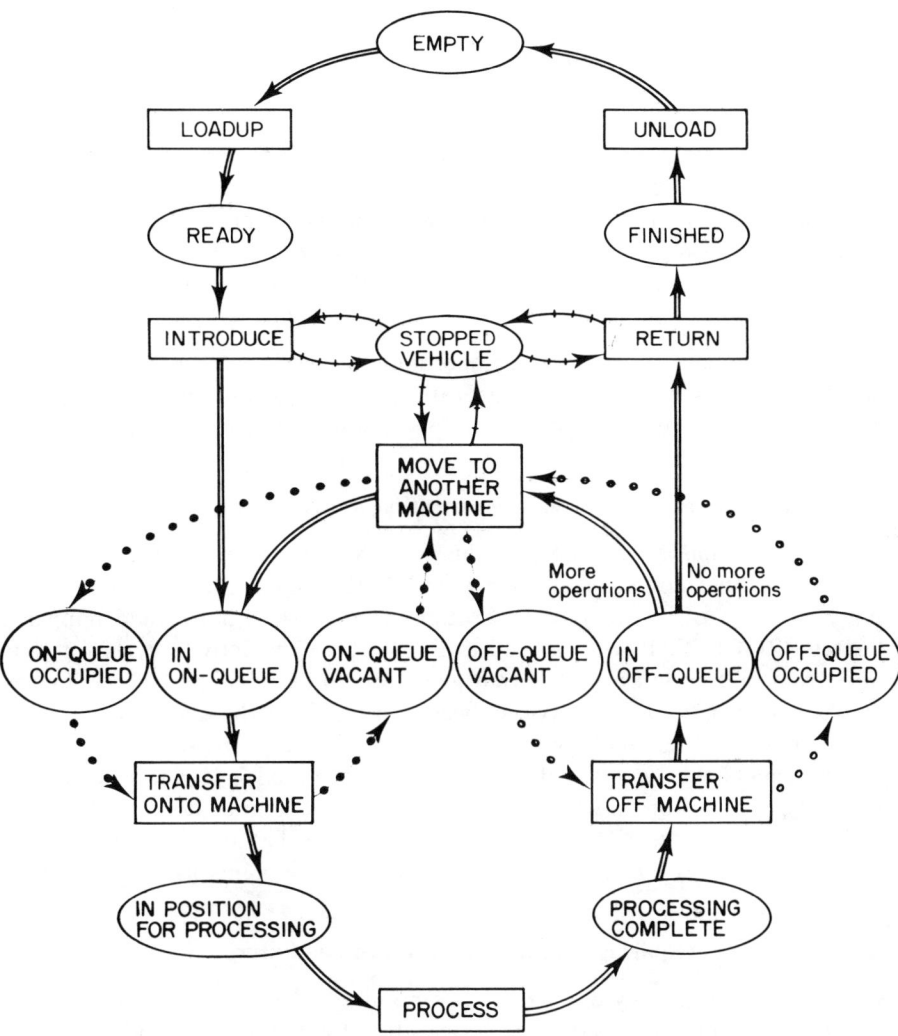

Figure 10.2   Activity cycle diagram for pallet, vehicle and machine buffers when several operations on a component are required.

done on, we have to define a group of machines, and the group, or groups, to which each workstation belongs. These groups need not be mutually exclusive groups. For example, a machine might be a member of group 1, which could be the list of machines which can perform the first operation on a component 1, and also be a member of group 17, the list of machines for the fourth operation on component 10.

Doing this presents no problem, but requires additional attributes and variables to be defined. One of the models at the end of this chapter will give an example of how this can be done.

10.2   FIXTURES

The fixturing arrangements of the components are among the most important aspects of component processing in determining the complexity of the model, or more specifically, of the rules of operation of the system. In many systems, components can be completely machined in a single operation. In others, several operations may be required. In the most complex situation the components may need to be mounted in a succession of different fixtures so that different faces may be presented to the cutting tools, with several operations while in each fixture. Although these remarks imply that we have machining of prismatic parts in mind, the principle applies to any type of operation. To consider the simplifications or complications introduced to the model by these possibilities, let us examine three cases:

— components which require only one operation in one fixture,
— components which require several operations in only one fixture, or
— components which require one or more operations in two or more fixtures.

10.2.1   Components requiring only one operation in one fixture

The simplest situation is when components need only one operation in one fixture. As demonstrated in the models of chapter 8 and 9, the model can be simplified because it is not necessary to test for more operations to discover what should happen to the pallet after PROCESS. It will proceed to RETURN. The activity cycle for the pallet simplifies from that of figure 10.2 to that given for Models Two and Three, figures 9.14 or 9.16. Although only one activity has been omitted from the diagram, a fairly trivial simplification, in addition, it is not required to define or store the sequence of operations on the components. This can significantly simplify the model. It will only be necessary to specify the workstation on which the single operation is to be done, and the duration of the operation. In Models Two and Three we assumed that any machine could process any component, and as a result no consideration of these points had to be given.

10.2.2   Components requiring several operations in one fixture

If several operations in only one fixture are required, then the diagram of figure 10.2 applies, which includes a test after each operation for further operations. The number and sequence of operations and their durations will be defined as described above.

10.2.3   Components requiring operations in two or more fixtures

If more than one fixture is needed in processing a component, then the fixture changing operation must be defined, by some means. Normally, the component will return to the load–unload area for the fixture to be changed, but in some cases there are special re-fixturing stations, especially in systems where the normal loading and unloading are done by a robot and refixturing is the only normal manual intervention. Cell 1 of Cummins' Con-rod line is an example.

   If the load–unload stations are used for refixturing, the routing of the component will specify an operation at the load–unload area whenever a fixture change is required. Then, when the component arrives at the load station, we can check whether it has had all the necessary operations, in which case it will be unloaded, or if not, it will need a fixture change before more operations are carried out. The component's activity cycle would be amended as shown in Figure 10.3. To conform to the rules of activity cycle diagrams, the NEW and PARTLY PROCESSED components have to be placed in a common queue before LOADUP. It would not be correct to have separate queues because the rules of ACDs would take that to mean that it was necessary to have a NEW component as well as a PARTLY PROCESSED one (and any other entities involved) before LOADUP could begin. In fact, LOADUP requires only one or other, so that there should be a single queue and attributes used to distinguish between NEW and PART-PROCESSED components.

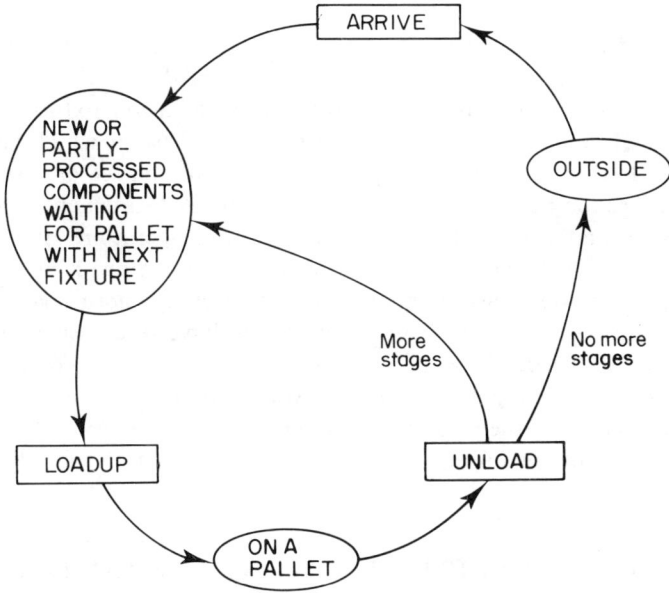

Figure 10.3 Activity cycle for component when several stages of processing in different fixtures are required.

We do not need to define the physical implications of the queue of partly-processed components at present. What actually happens to the component, such as being placed into a store or laid down on the floor nearby, will vary from system to system. If it were placed in a store, such as an Automated Storage and Retrieval System, then other activities and entities would have to be added to the model. Conceivably, this might re-introduce a LEAVE activity after each stage of processing. In fact, this raises an important point in building the model. Two methods of modelling successive fixture stages may be found in the literature:

— treating each stage of machining as if it were a different component;
— modelling the fixture changing operations in detail.

The first method permits the model to be simplified, even although some modelling accuracy may be lost, because the components would all go to OUTSIDE after UN-

LOADing. Several general FMS simulators do not provide facilities for modelling fixture stages in detail and use this method. For example, MAST does not handle fixture stages directly. It does allow different types of pallet to be defined, and for each type of pallet a list of the types of component which use it. By relating the pallet types to the different fixtures used, each stage of machining can be treated as if it were a separate component. However this method does not enable the processing stages of any specific component to be linked together, and consequently the total elapsed time to completely machine a component cannot be obtained. This may be no great loss, especially if the components may wait in a marshalling area for some time between successive stages of machining. To minimise the effects of the loss of control due to this approach, MAST permits the user to specify that one type of component cannot be started until another one is complete.

With the second method, when fixture changing is to be modelled in detail, several points must be considered. Is the component taken out of the fixture at the load station? Is there another component waiting to go into the now vacant fixture on the pallet at the load station? If not, what happens to the fixture and pallet which are now empty? Are the part and fixture lifted from the pallet together and placed on a setting up bench of some kind, where the part is removed from the fixture? If a bench is used, how many of them are there, and will their availability constrain operations at the load–unload area? Is the fixture, which the component needs next, already waiting for it at another load station? Is it already in use on another pallet carrying another component around the system? Is it in a fixture store from which it will have to be retrieved, and perhaps mounted on a pallet before the component can be placed in it? Will the component be placed back on the same pallet after the fixture has been changed? Where will the component be placed while the fixtures are being changed over? Will it be a marshalling area where it may wait for quite some time, or on a bench of some kind at the load station?

10.3   RELATIONSHIP BETWEEN FIXTURES AND PALLETS

Most of these questions concern the relationships between pallets and fixtures. Two common practices are:

— the fixture and pallet are never separated;
— the fixtures are mounted on pallets when required, and dismounted when not required.

Let us consider briefly the effect these two methods would have on the activity cycle diagram.

10.3.1   Fixtures permanently assigned to pallets

There are good engineering reasons why fixtures should be kept mounted on pallets all the time. Of course this requires having as many pallets as there are fixtures, which could add considerable cost. From the modelling standpoint, however, it simplifies things somewhat. If the fixtures are never removed from the pallets then there will be no need to define a cycle for the fixtures, since they have no existence independent of the pallet.

Of course, when the system is being commissioned the fixtures would be separate from the pallets and would need to be placed on them, but we may not need to model that stage of the system's operation.

There will be no effect on the ACD of the pallet. The two pallets concerned act independently of each other. The pallet for one stage will become EMPTY as soon as the component has been UNLOADed from it, and the pallet for the next stage must be EMPTY before the component can be LOADed. However, we do need to ask what happens to the pallet when no more components are waiting for it. If we have dedicated load stations it will merely wait at its own load station until required. On the other hand, if we do not have dedicated load stations it may have to be moved away so as to free the load station for use by another pallet. If there is a pallet store in the system, then there should be no problem. However, with a system like Anderson Strathclyde's there is no pallet store and the pallet must be moved to a vacant buffer somewhere in the system, where it may obstruct the movement of other pallets. The case when there is a fixture store in the system will be discussed shortly.

10.3.2 Fixtures removed from pallets when not required

If the fixtures will become separated from the pallets a somewhat different situation will apply. The main effect is that it will be necessary to define the fixtures as entities, independently of the pallets. There are two possible methods of handling components, fixtures and pallets:

— remove the component from the fixture, and then, if a change is required, remove the fixture from the pallet;
— remove the fixture and component together from the pallet and then remove the component from the fixture.

The first method is probably the more common, especially with systems in which a pallet holds several components. It might also be used in systems in which loading and unloading are done at one location, perhaps automatically by a robot, and fixture changing is done at another location, perhaps manually, such as in Cell 1 of Cummins' Con-rod line. With this method, the cycle for the fixtures might be as in Figure 10.4.

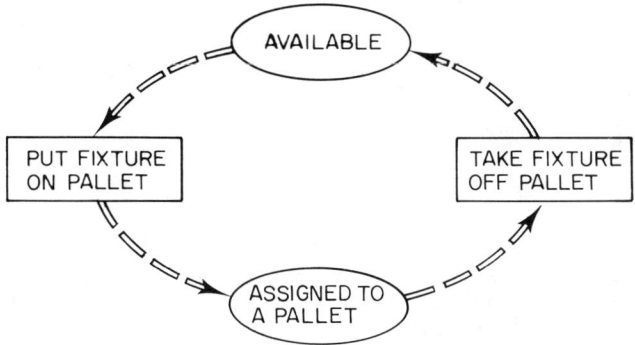

Figure 10.4   Activity cycle diagram for fixtures.

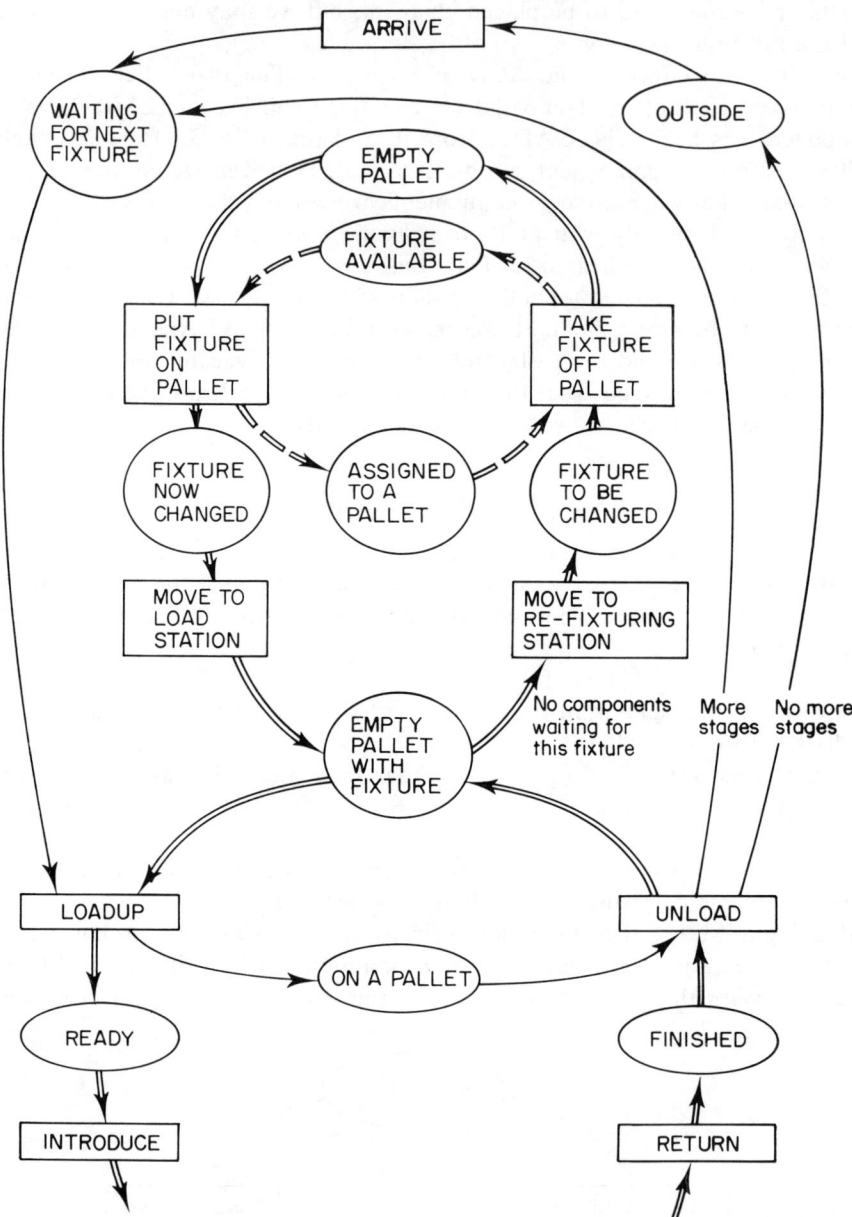

Figure 10.5   Activity cycle diagram showing cycle of component and fixture and part of pallet's cycle when fixtures are removable from pallets, and the component is taken out of the fixture while the fixture remains on the pallet.

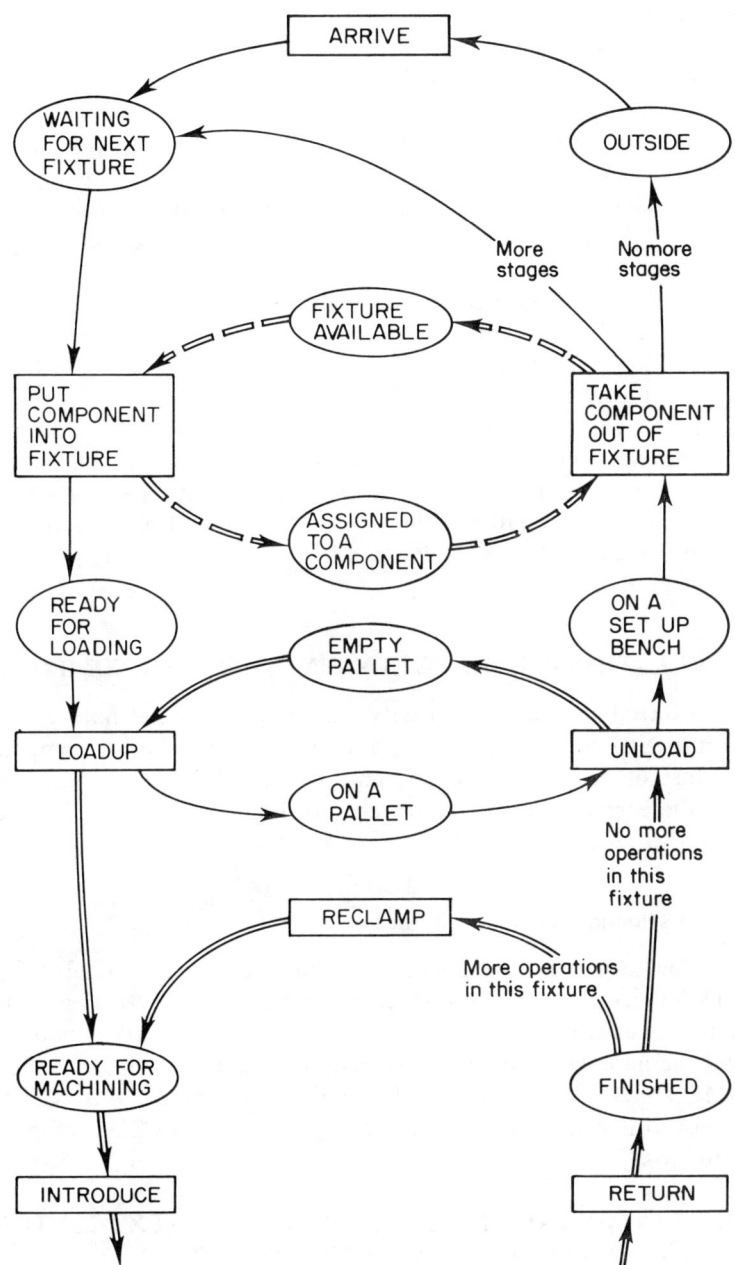

Figure 10.6   Activity cycle diagram showing cycle of component and fixture and part of pallet's cycle when fixtures are removable from pallets, and the fixture and component are taken off the pallet before the component is removed from the fixture.

Although entities with such a simple cycle can often be replaced by a variable counting the number of available entities, in this case we must distinguish between different types of fixtures. It may be convenient, therefore, to include them as entities, with an attribute defining their type. This cycle implies that the life of a fixture is not modelled in detail, but all the time it is on a pallet is absorbed into the queue ASSIGNED TO A PALLET. This cycle would also apply to the case where fixtures are normally permanently assigned to pallets but it is required to model initial fixture setting up. The pallet cycle needs to be amended to include extra activities for mounting and dismounting fixtures. Figure 10.5 shows part of the pallet's cycle and the cycles for the component and fixture under this system of operation.

The second method, which is used in the Anderson Strathclyde system, due to the size of the components and the fact that there are many more fixtures than pallets, involves slightly different logic from that in figure 10.5. The pallet cycle will generally not be affected, since the activities PUT FIXTURE ON PALLET and TAKE FIXTURE OFF PALLET would be represented by LOADUP and UNLOAD, while the component cycle involves the extra activities of PUT COMPONENT INTO FIXTURE and TAKE COMPONENT OUT OF FIXTURE. Figure 10.6 presents the cycles for components and fixtures and part of the pallet cycle. It includes a RECLAMP activity for the case when components may need to be reclamped for further machining in the same fixture, as occurs in the Anderson system.

10.4   RELATIONSHIP BETWEEN FIXTURES AND COMPONENTS

These remarks introduce the relation between components and fixtures, and there are several important aspects of this subject, quite apart from whether components are put in fixtures before or after the fixture is put on the pallet. These relate principally to ensuring that the correct fixture is used, and to the number of components which can be held in a fixture.

10.4.1   Fixture selection queue disciplines

The diagram does not show the queue disciplines, but clearly PUT COMPONENT INTO FIXTURE requires checking that the correct fixture is available. If we assume that each fixture is used by only one type of component then an attribute of the component can hold the number of the next fixture required. Then, when attempting to begin a PUT COMPONENT INTO FIXTURE activity, the model will need to match the fixture numbers with that attribute of the component. The ECSL coding for this check could be as follows:

```
FIND FIRST COMPONENT A IN WAITING-FOR-NEXT-FIXTURE
   I = NEXT-FIXTURE-NUMBER OF COMPONENT A
      FIXTURE I IN AVAILABLE
put component into fixture .....
```

The indentation defines a loop, so that the processor would examine each WAITING component in turn until it finds one for which the required fixture is in AVAILABLE. "A" is a variable used to identify the component being considered. In practice the hyphenated names would need to be shortened. Notice that this method can be used to identify the fixtures used for successive stages of machining easily.

This method will only work if there is only one fixture of each type. If there are several fixtures which would satisfy the component then we have to define an attribute of the fixture as well. This attribute would define a fixture type or a group of similar fixtures. If we use FTYPE of the component to define the type of fixture it requires and CTYPE to define the type of component for which a fixture is suitable, then the ECSL coding might be:

```
FIND FIRST COMPONENT A IN WAITING-FOR-NEXT-FIXTURE
   FIND FIRST FIXTURE B IN AVAILABLE
      FTYPE OF COMPONENT A EQ CTYPE OF FIXTURE B
   put component into fixture .....
```

This would cause the processor to examine each component in turn and for each component examine each fixture until it finds a matching pair.

10.4.2 Fixtures common to several types of component

Many systems manufacture families of similar components. In these systems it is probable that several types of components can be held by a common fixture. It might be required to identify all the components which could be placed into an AVAILABLE fixture. In this case the queue discipline logic has to be generalised further. Rather than give each fixture a type attribute we might want to define a list of components which the fixture could accommodate. There are several ways in which this can be done. The simplest method is to define a table of components for each fixture, but often the software package provides some facility which simplifies the necessary logic. Most packages handle sets. While so far we have used sets only for holding the contents of queues, they are in fact quite general. Thus we could define a set containing the type number of each component which could be placed in each fixture. The logic would then be something like:

```
FIND FIRST COMPONENT A IN WAITING-FOR-NEXT-FIXTURE
   FIND FIRST FIXTURE B IN AVAILABLE
      COMPONENT A IN COMPONENT-LIST OF FIXTURE B
   put component into fixture .....
```

As before, the processor will examine each component which is WAITING-FOR-NEXT-FIXTURE in turn and, while considering each component, will examine every fixture in AVAILABLE until it finds a matching pair. The components suitable for each fixture would have to be loaded into the appropriate sets when the model is initialised.

To handle the case when components require several fixtures in succession the type numbers inserted in each fixture's list would need to indicate the stage of machining as well as the component identity. This would permit the fixtures to be examined first and then the components which could be placed in them identified. However it is probably easier to consider each component in turn and find whether a suitable fixture is available, by the method described above.

10.4.3　Fixtures which can hold several components simultaneously

When components are small, fixtures will normally be designed to hold several components. These may be several components all of the same type, one component of each of several different types, or any combination of these. Some companies try to place all the components for an assembly on one fixture. This has several implications for the model-building process. Depending on how important it is to model operations in great detail, we may adopt one of two tactics:

— treat the entire set of components as a single composite entity, or
— treat each individual component separately.

It will obviously save a lot of detail if the entire set of parts can be treated as a single entity. Whether this can be done will depend on such factors as likelihood of the fixture going through the system with a partial load, whether the components are all going to a common assembly, whether assembly operations are to be included in the model, and whether change-over and start-up aspects of system operation are to be modelled. Even if the fixture might go through with a partial load it may still be possible to treat the set of components as a single entity by establishing the value of those parameters which would depend on the number of components actually present, such as the processing time, when the LOADUP operation is performed.

10.4.4　Multi-stage fixtures

A method commonly used in systems where the components are small but require operations on two (or more) fixtures in succession is to mount components at both stages on the same pallet. A four-sided fixture arrangement might be used with components at the first stage on two faces, and components at the second stage on the two other faces. Whenever a pallet visits the load–unload station, the two components on which the second operation has been done are unloaded, the two on which operation one has been done are moved to the locations for operation two, and two new components are mounted in the locations for the first stage of machining. The Cummins' Makino line is an example of a system in which this method is used. This effectively reduces a sequence of two or more operations into a single operation. The activity cyle diagram would simplify from that of figure 10.2 back to the simpler ones of chapter 8 and 9.

　If the model were to be simplified in this way it would be less easy to model start-up or change-over phases of system operation. If these phases are to be included, the model must treat the components on a pallet in more detail. An attribute could be defined which specifies the number of components on a pallet, and used to calculate the duration of load–unload and machining operations. The value of the attribute would be set or reset every time a load–unload operation was performed.

10.5　FIXTURE STORES

Another question must be considered before we can pass on from fixtures, and that relates to their storage when not in use. There must be some means of storing fixtures when not in use. This may either be:

— within the system proper, with movement of fixtures between the store and the fixture setting or load–unload area by the vehicles of the system, or

— near the load–unload or fixture setting area, with handling being done manually.

In the former case, for movement to be possible, the fixture must be mounted on a pallet, and therefore the store is not merely a fixture store but also a pallet store. This therefore comes within the case of the central or common storage for pallets described in chapter 9, especially in figure 9.12.

The second method will usually apply in the case of fixtures removed from the pallets when not in use. Although above we stated that handling would be done manually there could be other entities involved, such as cranes, trucks or hand carts. These might have to be included in the model if their availability can constrain the operation of the system.

10.6   TOOLS

The allocation of tools to machines was raised in the discussion in chapter 7 on the planning problems in FMS. We will now review some aspects of tool supply and their effect on model building. Obviously if an operation requires tools, such as an operation on a machining centre, the tools must be present before the operation is carried out. Depending on the complexity of the tool management problem, this may require us to model the movement and allocation to machines of tools in great detail. Fortunately, this is seldom necessary. Indeed it will generally only be necessary if the model aims specifically at investigating the tool supply part of the system. We will consider this case shortly.

10.6.1   Tool changing

In the discussion about the routing of components through the system, we referred to the methods of defining the machine or group of machines on which each operation is to be done. If a system consists of a number of identical machining centres, then they are only identical, so far as the operations they can perform are concerned, provided identical sets of tools are placed in their magazines. In other words, it is the allocation of tools to machines which determines the flexibility in the system. We also referred above to the number of fixtures which might be required in processing a component. Different fixtures are used when different faces of the component must be presented to the machine, and, depending on the nature of the component, different faces may involve different types of operation, and hence different tools sets. Thus the fixturing arrangements are also closely related.

When we consider the problems of modelling tooling aspects, the method of tool supply is a major factor. There are many variants in practice, but we can readily conceive of two regimes:

— in which all the tools for an operation must be present in the magazine before the operation can be allocated to the machine, the component taken to it, and the operation commenced,

— in which tools can be supplied at any time to any machine and an operation can be initiated in anticipation of all the tools being brought to the machine before the individual tool is required.

An important physical aspect of tool supply is whether:

— tools can be placed into or removed from the magazine while the machine is in operation, or
— tools must be loaded before the operation commences.

If tools are changed manually, for safety reasons it will probably be necessary to change the tools needed for an operation before the operation commences. On the other hand, if an automatic method is provided, as in the Mirrlees Blackstone system, then the first method is possible. The control software can co-ordinate the movements of the machine and the tool supply robot and ensure no collision will occur. Many recent designs of machines have arranged the tool chain so that it may be accessed easily from outside. Some earlier designs enclosed the tool magazine within a cabinet thereby preventing access for tool changing, even if a mechanical system had been supplied. If tools can be changed automatically at random during machine operation, many of the tool management problems disappear. We will continue to discuss the case where tools will have to be changed between operations.

Tools will have to be changed for two reasons:

— because tools wear out, and
— because the variety of tools is too great for all the tools to be mounted in the magazines permanently.

Whether the second reason occurs in a particular system depends on the variety and complexity of components. The more complex the components the greater the number of tools involved. In addition to loading one of each tool which will be needed, it is normal to load extra, or sister, tools so that when a tool wears out the sister tool can be used, and the time between tool changes extended.

Taking these factors together three situations can be envisaged:

— when all the tools can be mounted in all the machines all the time,
— when each of the tools can be mounted in at least one machine all the time,
— when it is not possible to mount all the tools (in any machine) at the same time.

In the first case, it is unlikely that tool supply problems will constrain the performance of the system. The simulation model will probably not have to take tooling into account. Only tool changes due to wear will occur, and, especially if the tool lives are good, it may not be necessary to change sets of tools more than once per day. The tool changing may therefore be accommodated outside the normal running of the system.

In the last case however it will be necessary to schedule the tools and workpieces jointly. Depending on the work content and the tool lives, decisions will be made at say daily intervals on which parts are to be produced and which tools should be loaded in the magazines. This can be a complex problem. In the Anderson Strathclyde system this is a major consideration in the operation of the system, as discussed by Carrie and Perera.[5]

10.6.2   Tool supply methods

Another factor concerning tools and how they can be modelled is that of tool supply methods. Three methods are obvious possibilities:

— manual tool supply,
— automatic tool supply by a dedicated handling system,
— automatic tool supply by the same transport system as handles components.

In the first case, an example of which is the Anderson Strathclyde system, there will be little effect on the model, since no new entities may be required. If the operators who supply the tools are the same personnel as do loading and unloading operations then the frequency and duration of tool supply actions has to be allowed for. Possibly, this can be handled under the heading of "other duties" as discussed in chapter 8. Alternatively, a small amount of time each shift may be "deducted" from the model to represent such activities.

In the second case, an example of which is the Mirrlees Blackstone system, the system effectively subdivides into two parts which have little interaction, except at the machine tools. Once again, unless the tool supply side is specifically to be modelled, it may not be necessary to treat the tool supply in detail. Similarly, if tool supply is to be modelled then much of what happens in the component supply side of the system will have little direct impact on tool supply, and effectively a separate model may be produced.

In the third case, the JCB system is an example, it is unfortunately necessary to model tool movements in some detail. Considerable thought may well be necessary to define all the activities involved. It may be possible to simplify the modelling of the tool supply side by generating "tool requests" on a random basis at the necessary frequency. These tool requests initiate demands on the material handling system. There are two effects. One is simply to make a vehicle unavailable for handling workpiece pallets. The other effect is that movements of tools obviously alter the location of the vehicle concerned, and this may have to be monitored for calculating movement durations, as will be described in the next chapter.

10.6.3   Modelling tool flow

These remarks serve to indicate that it will often be possible to model tool movements without going into the full detail of tracking the location and life of each individual tool. Let us now consider briefly the approach to the worst possible case, namely that in which the presence of tools at machines determines where and if operations can be carried out.

Typically several hundred tools will be involved. The data storage required to handle this number is considerable. The processor time required to check the status of this number of tools and the contents of the magazines will also be cosiderable, and may slow down the operation of the model to such an extent that visual interactive modelling is impracticable.

We can still reduce the complexity of the task by thinking in terms of sets of tools required for each operation. This can often be done by exploiting the facilities provided by the language used, provided it is powerful enough. Sets of tools are just like sets of any other kind of entity. Many packages provide functions to compare the content of sets, and in fact perform all the basic operations of set mathematics. Suppose that we define a set PRESENT for each machine tool, and a set REQUIRED for each operation. We can then test whether all members of REQUIRED are within PRESENT. ECSL provides a keyword, WITHIN, for this purpose. Thus

REQUIRED WITHIN PRESENT

performs the test and appropriate action can then be taken. We can also discover which tools are not present and add them to a list of tools to be moved to the machine:

ZERO NEEDED
FOR TOOL T IN REQUIRED
    TOOL T NOTIN PRESENT
    TOOL T INTO NEEDED

NEEDED is the list of tools not in the machine which are needed for the operation. NOTIN is a test which succeeds if the entity specified is not a member of the specified set.

In this way we can check whether operations can be done, or if not which tools are short. There could be complications in applying this approach, since if it were discovered that an operation could not be performed we might find that the system blocked up with pallets containing components which could not be processed. As was shown with Model Five, apparently straighforward methods of running a system can lead to unforeseen congestion.

A more practical approach would be to set up the sets of tools at the start of each shift. We could compare the PRESENT and NEEDED sets and create a list of the machines at which each operation on each part could be performed. If any operation on a component required a set of tools not present on any machine that component would not be loaded onto a pallet until the next re-allocation of tools had been made. This is effectively what would be done in real life in any case, unless the system has a random tool supply method. An interesting question is how frequent should the re-allocation of tools to machines be? What period should elapse before the next assignment? In each period a set of components to be produced is determined and an assignment of tools and operations to machines decided. The longer the period the fewer the tool changes, but the less flexible the arrangement, and potentially the lesser the utilisation of machines. The shorter the period, the more frequent the tool changes, but the system can be kept topped up with work.

If the life of the tools must also be included, then we at least double the amount of data and testing which has to be coped with. Rather than just test for the presence of a tool we would have to check whether it has sufficient life remaining before we could assign the pallet to the machine. Of course, this is what happens in real life, and is what generates the need to change tools. When we get to this stage we need a model which is virtually as complex as the real system software. As was suggested by Carrie,[1,2] rather than develop a model of the system we would probably be better using the real control software in "simulation mode"; system suppliers should provide this mode of operation of their software.

We will not pursue this aspect further. These remarks should serve to point the way, which must depend very much on the situation in each individual system.

10.7 MODEL SIX—A SYSTEM PROCESSING COMPLEX COMPONENTS

To illustrate the topics introduced in this chapter let us now build a model of a system which processes complex components. It can be notionally based on the Anderson Strathclyde system, described in chapter 7. The components typically require three different fixtures. There are three or four operations in each fixture, in one of which a re-clamping operation is necessary. The system has one vehicle, two load–unload stations, thirteen pallets and six machining centres, which have been grouped according to the type of operation being done. The machines are grouped as follows:

Group 1: machine 1 (the facing-head machine) for special operations
Group 2: machines 2 and 3 for roughing operations
Group 3: machines 4 and 5 for semi-finishing operations
Group 4: machine 6 for finishing operations

The breakdown into semi-finishing operations and finishing operations is somewhat arbitrary, because there are some shared tools between machines 4, 5 and 6. Some operations can be done only at one of these machines, others can be done at two of them. This means that we have to have a way of defining the grouping of machines in terms of the machines which can perform each operation. We will assume that the allocation of tools to machines is permanent and that tools either do not wear down, or can be replaced in a way which does not affect the operation of the system. Operation times vary from about 15 minutes up to about two hours. Although the number of types of components produced on the system has built up considerably over the years, initially it produced six different components.

10.7.1 Data used

Figure 10.7 gives the layout of our notional system for Model Six. Artificial data has been created, as given in Table 10.3. If an operation has to be done at the load–unload area, zero has been given for the machine group, and for the operation time. Times

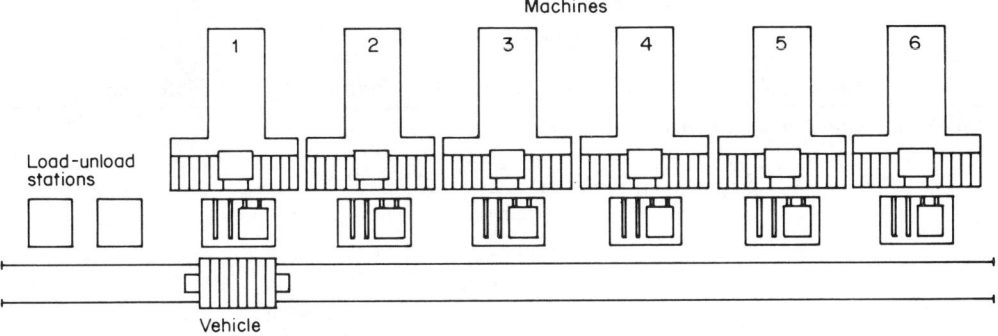

Figure 10.7   Layout of system for Model Six.

Table 10.3   Data used in Model Six, defining component processing requirements

Component type:						1	2	3	4	5	6			
Number of operations:						15	11	15	15	13	12			
Monthly requirement:						10	8	6	4	6	2			

Component type	Operation number														
	1	2	3	4	5	6	7	8	9	10	11	12	13	14	15
Machine group required for each operation															
1	0	2	3	4	0	2	3	0	2	4	1	0	2	7	0
2	0	2	1	0	2	5	4	0	2	1	0	0	0	0	0
3	0	2	4	1	0	2	3	0	2	4	1	0	3	4	0
4	0	2	3	4	1	0	2	3	4	1	0	2	4	1	0
5	0	2	1	0	2	6	7	0	2	3	4	1	0	0	0
6	0	2	3	0	2	1	0	2	5	6	1	0	0	0	0
Type of fixture require for each operation															
1	11	11	11	11	21	21	21	21	21	21	21	31	31	31	0
2	12	12	12	22	22	22	22	22	22	22	0	0	0	0	0
3	13	13	13	13	23	23	23	23	23	23	23	33	33	33	0
4	14	14	14	14	14	24	24	24	24	24	34	34	34	34	0
5	15	15	15	25	25	25	25	35	35	35	35	35	0	0	0
6	16	16	16	16	16	16	26	26	26	26	26	0	0	0	0
Duration of each operation (minutes)															
1	0	156	120	65	0	130	140	0	203	60	67	0	67	168	0
2	0	101	57	0	73	80	80	0	152	87	0	0	0	0	0
3	0	89	40	60	0	80	120	0	150	70	80	0	120	80	0
4	0	183	120	40	100	0	80	120	60	100	0	20	50	60	0
5	0	57	100	0	80	90	40	0	100	120	90	80	0	0	0
6	0	98	120	0	90	100	0	60	80	80	80	0	0	0	0

Time to place component in fixture	15 minutes
Time to place fixtured component on pallet	10 minutes
Time to reclamp a component	10 minutes
Time to unload fixtured component from pallet	5 minutes
Time to take component out of fixture	10 minutes
Average time between arrivals	270 minutes

Table 10.4   Analysis of work loads resulting from the data in Table 10.3

(a) Work content per component

Componenent type	Work content	
	per piece	minutes per month
1	1251	12510
2	680	5440
3	964	5784
4	998	3992
5	822	4932
6	758	1516

(b) Work content per machine group

Machine group	Work content (minutes per month)
1	5142
2	13132
3	5960
4	4170
5	800
6	700
7	1920

(c) Expected load per machine, operator, vehicle

Facility	Expected work load (minutes per month)
Machine 1	5142
Machine 2	6566
Machine 3	6566
Machine 4	4740
Machine 5	3680
Machine 6	5130
Operator	2350
Vehicle	2450

(d) Work load per fixture

Fixture type	Component type	Fixture stage	Work load per component	Work load minutes per month
11	1	1	351	3510
12	2	1	168	1344
13	3	1	199	1194
14	4	1	453	1812
15	5	1	167	1002
16	6	1	428	856
21	1	2	635	6350
22	2	2	507	4056
23	3	2	535	3210
24	4	2	385	1540
25	5	2	235	1410
26	6	2	325	650
31	1	3	260	2600
33	3	3	225	1350
34	4	3	155	620
35	5	3	415	2490

(e) System capacity by number of shifts per day

Working time available	shifts/day 1	shifts/day 2	shifts/day 3
Per day	480	960	1440
Per week	2400	4800	7200
Per month	9600	19200	28800

for load and unload operations, reclamping and fixturing operations have been given separately. The fixture type is given as (xy) where x is the processing stage and y is the component type. As with Model One it would be wise to assess the overall workload on machines, vehicle and other facilities, and check that the system design is sound, or in this case that the artificial data is sensible. We can aggregate the operation times to get the total work content per component, and multiply that by the average monthly requirements to get the monthly work content. These figures are given in Table 10.4(a). The work content can be allocated to the machine groups, as in Table 10.4(b). Although the allocation of work to machines will depend on the situation in the model when the decision is made, if we assume that the work load on a group is allocated equally among the machines in the group, we obtain the figures Table 10.4(c) which also gives the load on the operator and the vehicle. The work content of the operations while the component is in each fixture can be calculated, producing the figures in Table 10.4(d). There is a total of 16 fixture types. There are wide variations in the work load on the fixtures.

We need to compare these work loads to the time available. If we omit losses due to breakdowns or other causes, the time available, assuming 480 minutes per shift, five days per week and 4 weeks per month, would be as in Table 10.4(e). Comparing the machine loads with these figures it is clear that single shift operation should be sufficient. (Whether building a system like this for single shift operation would be economical is questionable, but beyond the present discussion.) Since there are 9600 minutes in a month, the work content figures are approximately 100 times the expected percentage utilisation of the facilities concerned. The machine utilisations will be between about 40% and 70%. One fixture of each type should be sufficient, giving a total of 16, although some will have very low utilisation. However, since a component cannot be loaded on a fixture until the previous component has finished with it, there is a risk of delays due to fixture availability, especially in the case of the components with the higher production requirements. The benefits of having extra fixtures of certain types could be investigated. The total average monthly requirement is 36 components, which gives a mean time between arrivals of approximately 270 minutes.

10.7.2 Amendments to the model

Since the system has no central pallet store, Model Three is the closest of the previous models to what we require. Several amendments will be necessary, some minor, some complete new activities. Rather than provide a full activity cycle diagram for Model Six, which would be quite complex, we can refer to figures 10.2, 10.6 and 9.17. New activities, as indicated in the activity cycle diagrams figures 10.2 and 10.6, are MOVE, RECLAMP, TAKEOFF and PUTON.

First we can consider the data structure. The data in table 10.3 can be held as arrays, such as NOOPS(J), NREQD(J), OPSEQ(I,J), FIX(I,J) and OPTIME(I,J) for the five sets of data, where I is the operation number and J is the type of component. The times of the operations at the load–unload area can be defined by new constants, RECLAMPTIME, PUTONTIME and UNFIXTIME, or by new values for LOAD-TIME and ULOADTIME. MOVETIME can be used for the time to move from one machine to another, as well as for movements to or from the load–unload area.

We need to define a new class of entity, namely FIXTURE with its sets AVAILABLE and ASSIGNED. We also need to associate each fixture with its component. This could be done by numbering the fixtures from 1 to 16, and giving a table of these numbers to indicate which fixture is needed for each operation. However, we can use a more informative numbering system by defining an attribute for the type of each fixture, say FTYPE, and using the values in the table 10.3 directly. This would also facilitate the logic if there were duplicate fixtures of any type.

We also need to define some new sets for the components. We can use FXWAIT for components waiting for a fixture and LDWAIT for components ready for loading. NEW no longer exists, since FXWAIT serves the same purpose. ONBENCH can be the name for the queue of components waiting to be taken off a fixture.

Several new attributes of the components will be required. We will need to record which fixture a component is currently using, and we will need to know the type of that fixture and which type of fixture the component needs for its next operation, so that we can check whether a fixture change is needed. For these three attributes we can use FNO, FTHS and FNXT. For keeping track of the progress of the component through its sequence of operations we can use OPNO and NEXT as in chapters 4 and 5. Now NEXT defines the group of machines which can perform each operation. We must therefore have a means of defining which machines are within each group. In ECSL an easy way to do this is to use a group of sets all with the same name, called in this case GROUP. Above, 4 groups were defined. Due to the fact that some operations can be done at one or two of the semi- and finishing machines, three other groups must be defined, numbered 5, 6 and 7 in table 10.3. The complete definition of the machine groups is:

Group 1: machine 1
Group 2: machines 2 and 3
Group 3: machines 4 and 5
Group 4: machine 6
Group 5: machine 4
Group 6: machine 5
Group 7: machines 4 and 6

The three additional groups indicate that some operations can be done only at machine 4, some only at machine 5 and some at either machine 4 or 6. These sets have to be sets of the on-queue buffers, since it is one of these entities which is searched for in the INTRODUCE activity. Thus we add "GROUP 7" to the list of sets for the ONBUFF.

We must consider in which activities the values of the new attributes are set and tested. As just stated we need to select an on-queue buffer which is vacant and in the group required for the next operation. Therefore NEXT will be tested in INTRODUCE and MOVE, and must be set beforehand in ARRIVE, TRFRON, RECLAMP and PUTON. Similarly, the fixture required will be tested in PUTON. It will also be tested in RECLAMP and UNLOAD so as to determine which of these operations is required. Thus FNXT must be set up in advance of a component coming to the load–unload area, either in ARRIVE or in RETURN.

The system has 13 pallets, so careful consideration has to be given to their initial conditions. We can place six at on-queues and six at off-queues and the remaining one at a load station. The attributes for operation number, fixture in use, and next fixture required, machine currently at and the group required for the next operation must all be consistent.

In addition to the results statistics, which were obtained in the earlier model, we should record the utilisation of each machine separately, since we expect different values for each machine.

Figure 10.8 gives the code for the new activities and for the other amendments relative to Model Three.

```
THERE ARE 50 COMPON SET LDWAIT FXWAIT ONBENCH ONPALL OUTSID WITH TIMIN
+CTYPE FNO FNXT FTHS NEXT OPNO
THERE ARE 16 FIXTURE SET AVAILABLE ASSIGNED WITH FTYPE GZFIXT
THERE ARE 6 ONBUFF SET ONOCCU ONVACA GROUP 7
THERE ARE 6 MACHINE SET OCCUPI IDLE WITH TMACH TIMIDLE
REAL RT R
ARRAY OPSEQ(15,6) OPTIME(15,6) FIX(15,6) NOOPS(6)
HIST TIHIST ( 30 1000 200 ) TYPEDIST ( 6 1 1 )
FOR FIXTURE I IN AVAILABLE
   ADD DURATION TO GZFIXT OF FIXTIURE I
FOR MACHIN I IN IDLE
   ADD DURATION TO TIMIDLE OF MACHIN I

BEGIN INTROD
....
FIND FIRST PALLET A IN READY
   C = CMPNO OF PALLET A
   N = NEXT OF COMPON C
   FIND FIRST ONBUFF B IN ONVACA
      ONBUFF B IN GROUP N
....

BEGIN TRFRON
....
C = CMPNO OF PALLET A
D = CTYPE OF COMPON C
OT = OPTIME(OPNO OF COMPON C, CTYPE OF COMPON C)
AADURATION = DURATION + OT
CHAIN
   OPNO OF COMPON C EQ 2
   ADD 1 TO MADE OF PARTTYPE D
   OR CONTINUE
OPNO OF COMPON C + 1
NEXT OF COMPON C = OPSEQ(OPNO OF COMPON C, CTYPE OF COMPON C)
ADD OT TO TMACHTIME
ADD OT TO TMACH OF MACHIN B
....

BEGIN MOVE
TIME OF VEHICL LE O
FIND FIRST PALLET A IN INOFFQ
   B = CMPNO OF PALLET A
   N = NEXT OF COMPON B
   N GT O
   FIND FIRST ONBUFF D IN ONVACA
      ONBUFF D IN GROUP N
DURATION = MOVETIME
TIME OF VEHICL = DURATION
PALLET A FROM INOFFQ INTO INONQ AFTER DURATION
ONBUFF D FROM ONVACA INTO ONOCCU
E = MCAT OF PALLET A
OFFBUF E FROM OFFOCC INTO OFFVAC AFTER DURATION
MCAT OF PALLET A = D
ADD 1 TO MOVE
```

Figure 10.8   Additional and amended code for converting Model Three into Model Six.

```
BEGIN RETURN
. . . .
FIND FIRST PALLET A IN INOFFQ
  B = CMPNO OF PALLET A
  NEXT OF COMPON B EQ 0
. . . .
FNXT OF COMPON B = FIX(OPNO OF COMPON B, CTYPE OF COMPON B)
. . . .

BEGIN RECLAMP
TIME OF OPERAT LE 0
FIND FIRST PALLET A IN FINSHD
  B = CMPNO OF PALLET A
  FNXT OF COMPON B EQ FTHS OF COMPON B
DURATION = RECLMPTIME
PALLET A FROM FINSHD INTO READY AFTER DURATION
OPNO OF COMPON B + 1
NEXT OF COMPON B = OPSEQ(OPNO OF COMPON B, CTYPE OF COMPON B)
TIME OF OPERAT = DURATION
ADD 1 TO RECLAMP

BEGIN UNLOAD
TIME OF OPERAT LE 0
FIND FIRST PALLET A IN FINSHD
  B = CMPNO OF PALLET A
  CHAIN
    OPNO OF COMPON B LT NOOPS(CTYPE  OF COMPON B)
    FNXT OF COMPON B NE FTHS OF COMPON B
    OR OPNO OF COMPON B EQ NOOPS(CTYPE OF COMPON B)
DURATION = ULOADTIME
PALLET A FROM FINSHD INTO EMTY AFTER DURATION
COMPON B FROM ONPALL INTO ON BENCH AFTER DURATION
TIME OF OPERAT = DURATION
CMPNO OF PALLET A = 0
ADD 1 TO UNLOAD

BEGIN TAKEOFF
TIME OF OPERAT LE 0
FIND FIRST COMPON A IN ONBENCH
DURATION = UNFIXTIME
CHAIN
    OPNO OF COMPON A EQ NOOPS(CTYPE OF COMPON A)
    COMPON A FROM ONBENCH INTO OUTSIDE AFTER DURATION
    TIMIN OF COMPON A = CLOCK - TIMIN OF COMPON A
    ADD TIMIN OF COMPON A TO HIST TIHIST
    T = CTYPE OF COMPON A
    ADD TIMIN OF COMPON A TO HIST TIMHIST OF PARTTYPE T
    OR COMPON A FROM ONBENCH INTO FXWAIT AFTER DURATION
F = FNO OF COMPON A
FIXTURE F FROM ASSIGNED INTO AVAILABLE AFTER DURATION
TIME OF OPERAT = DURATION
ADD 1 TO TAKEOFF


BEGIN PUTON
TIME OF OPERAT LE 0
ZERO PARTLIST
FOR COMPON A IN FXWAIT
  EXISTS FIXTUR IN AVAILABLE WITH FTYPE EQ FNXT
  T = CTYPE OF COMPON A
  PARTTYPE T INTO PARTLIST
```

Figure 10.8   (continued).

```
FIND PARTTYPE T IN PARTLIST WITH MIN RATIO
FIND FIRST COMPON A IN FXWAIT WITH CTYPE EQ T
   FIND FIRST FIXTURE F IN AVAILABLE WITH FTYPE EQ FNXT
DURATION = PUTONTIME
TIME OF OPERAT = DURATION
FIXTUR F FROM AVAILABLE INTO ASSIGNED
COMPON A FROM FXWAIT INTO LDWAIT AFTER DURATION
FTHS OF COMPON A = FNXT OF COMPON A
FNO OF COMPON A = F
OPNO OF COMPON A + 1
NEXT OF COMPON A = OPSEQ(OPNO OF COMPON A, CTYPE OF COMPON A)
CHAIN
   OPNO OF COMPON A EQ 2
   LOADED OF PARTTYPE T + 1
   X = 1000 / REQD OF PARTTYPE T
   RATIO OF PARTTYPE = X * LOADED OF PARTTYPE T
   OR CONTINUE
ADD 1 TO PUTON

BEGIN LOADUP
TIME OF OPERAT LE 0
FIND FIRST PALLET A IN EMTY
FIND FIRST COMPON B IN LDWAIT
DURATION = LOADTIME
PALLET A FROM EMTY INTO READY AFTER DURATION
CMPNO OF PALLET A = B
COMPON B FROM LDWAIT INTO ONPALL AFTER DURATION
TIME OF OPERAT = DURATION
ADD 1 TO LOADUP

BEGIN ARRIVE
....
OPNO OF COMPON A = 1
FNXT OF COMPON A = FIX(OPNO OF COMPON C, CTYPE OF COMPON C)

TYPE 'PUTON was started' PUTON ' times'
TYPE 'TAKEOFF was started' TAKEOFF' times'
TYPE 'RECLAMP was started' RECLAMP' times'
TYPE 'TRFROF was started' TRFROF ' times'
TYPE 'MOVE was started' MOVE' times'


FOR MACHIN I
   TYPE 'Occupancy of MACHIN'+4,I (1-TIMIDLE/( 1. *(CLOCK -RUNINZ)))
FOR MACHIN I
   TYPE 'Utilization of MACHIN'+4,I,(TMACH/( 1. *(CLOCK -RUNINZ)))
RT = 0.
FOR I = 1 TO 16
   R = 1 - (GZFIXT OF FIXTURE I)/(1.*(CLOCK-RUNINZ))
   RT = RT + R
   TYPE 'Utilization of fixture'+4,I,FTYPE OF FIXTURE I,R
 TYPE 'Overall utilization of fixtures'+4,(RT/16)

 DATA
CMPNO 1 2 3 4 5 6 7 8 9 10 11 12 0
LDAT 12*0 1
MCAT 2 3 6 5 1 4 2 1 3 4 6 5 0
INONQ 1 TO 6
```

Figure 10.8   (continued).

```
INOFFQ 7 TO 12
EMTY 13
NEXT 2 2 4 3 1 3 3 0 4 4 0 1 38*0
FNO  1 2 3 4 5 6 7 8 9 10 11 12 38*0
FNXT 12*0 31 34 33 35*0
FTHS 11 12 13 14 15 16 21 22 23 24 25 26 21 24 23 35*0
OPNO 2 2 3 3 3 3 7 11 10 9 8 1 12 11 13 35*0
CTYPE 1 2 3 4 5 6 1 2 3 4 5 6 1 4 3 35*0
ONPALL 1 TO 12
FXWAIT 13 TO 15
OUTSID 16 TO *
FTYPE 11 12 13 14 15 16 21 22 23 24 25 26 31 33 34 35
ASSIGNED 1 TO 12
AVAILABLE 13 TO *
PARTLIST 1 TO *
LDFULL 1
LDFREE 2
OFFOCC 1 TO *
ONOCCU 1 TO *
IDLE 1 TO *
NOOPS  15 11 15 15 13 12
OPSEQ  0 2 3 4 0 2 3 0 2 4 1 0 2 7 0
       0 2 1 0 2 5 4 0 2 1 0 0 0 0 0
       0 2 4 1 0 2 3 0 2 4 1 0 3 4 0
       0 2 3 4 1 0 2 3 4 1 0 2 4 1 0
       0 2 1 0 2 6 7 0 2 3 4 1 0 0 0
       0 2 3 0 2 1 0 2 5 6 1 0 0 0 0
FIX     11 11 11 11 21 21 21 21 21 21 21 31 31 31 0
        12 12 12 22 22 22 22 22 22 22  0  0  0  0 0
        13 13 13 13 23 23 23 23 23 23 33 33 33 0
        14 14 14 14 14 24 24 24 24 24 24 34 34 34 0
        15 15 15 25 25 .25 25 35 35 35 35 35  0  0 0
        16 16 16 16 16 16 26 26 26 26 26  0  0  0 0
OPTIME 0 156 120  65   0 130 140   0 203  60  67   0  67 168   0
       0 101  57   0  73  80  80   0 152  87   0   0   0   0   0
       0  89  40  60   0  80 120   0 150  70  80   0 120  80   0
       0 183 120  40 100   0  80 120  60 100   0  20  50  60   0
       0  57 100   0  80  90  40   0 100 120  90  80   0   0   0
       0  98 120   0  90 100   0  60  80  80  80   0   0   0   0
GROUP 1 1
        2 2 3
        3 4 5
        4 6
        5 4
        6 5
        7 4 6
REQD 10 8 6 4 6 2
TYPEDIST 36 10 8 6 4 6 2
SA 111
SB 333
ONEDAY    2400
ONEWEEK   9600
OVRLART   270
TRTIME 1
ULOADTIME 5
LOADTIME 10
MOVETIME 5
PUTONTIME 15
UNFIXTIME 10
   END
```

Figure 10.8   (continued).

10.7.3 Results—blockage of the system

When this model was run the system ground to a halt half way through the month, when no further pallet movement was possible. This result is to be expected from our findings with some of the earlier models, and was observed by Carrie, et al.[3,4] in presenting a case study based on the Anderson system. The problem is, as with Model Five, that it may be possible for both load stations to be occupied by pallets which want to go to a machine whose on-queue is full, whose machining table is occupied by a pallet, and whose off-queue buffer is occupied by a pallet which should go to a load station. Alternatively, the pallets at the load stations may be empty because there are no waiting components and no available fixtures for them. Figure 10.9 gives the status of the system when the blockage occurred.

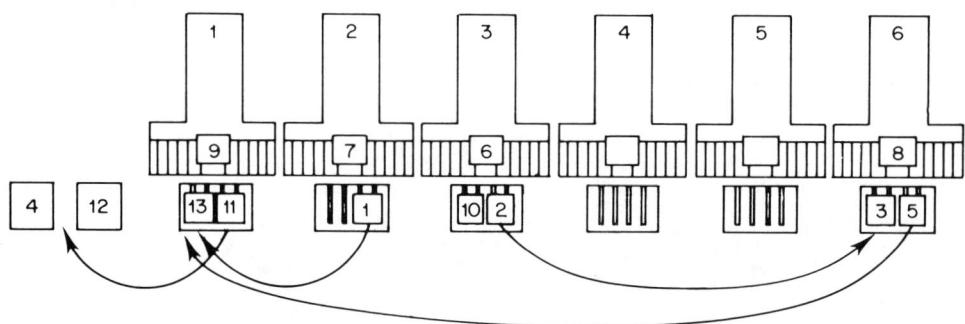

Figure 10.9   Location and destination of pallets in Model Six when blocking occurred.

10.8   MODEL SEVEN—INCORPORATING UNBLOCKING MOVEMENTS
OF EMPTY PALLETS

One method of solving this problem would be to lift the empty pallet off the pallet stand at the load–unload area and lay it aside somewhere. However, although this would be easy to program, it is unrealistic in practice, as continual handling in this way would soon damage the locating surfaces on the underside of the pallet.

Instead, we need to be able to move empty pallets to vacant pallet stands elsewhere in the system, so that loaded pallets can be moved to the load–unload area and to machine on-queue pallet stands or away from off-queue buffers. We can do this by defining new activities similar to activity UNBLOCK in Model Five, which switches round two pallets, one loaded and an empty one which occupies the location to which the loaded pallet needs to be taken. The new activities will be as follows:

BINTROD: Move an empty pallet away from an on-queue pallet stand to any vacant pallet stand, then move a loaded pallet from a load station to the on-queue buffer, and finally move the empty pallet to the load station.

BRETURN: Move an empty pallet away from a load station, bring a loaded pallet from a machine off-queue to the load station for unloading, and finally place the empty pallet stand at the off-queue from which the loaded pallet came.

BLOADUP: Bring an empty pallet from a pallet stand at a machine to a vacant load station, and load a component (already in its fixture) on the pallet.

BMOVE: Move an empty pallet away from the on-queue of a machine to which a loaded pallet requires to be moved, then move the loaded pallet and finally move the empty pallet at the off-queue stand from which the loaded one came.

BTRFROF: Move an empty pallet away from the off-queue of a machine at which a loaded pallet is waiting to be transferred off the machine table. The empty pallet will be left at a vacant stand located elsewhere in the system.

As in Model Five, with the exception of BTRFROF, we do not need to identify the vacant buffer where the empty pallet is temporarily parked, since, due to the number of pallets and locations, there must be one, and the movement durations are based on average times.

One other additional activity, which does not involve an empty pallet, may be useful:

BUNBLOCK: When a loaded pallet is on an off-queue pallet stand waiting to go to a load–unload station, and the operator is available, but the load stations are occupied, and the machine where the pallet is waiting is itself waiting to transfer off the pallet on its machine table, then move the pallet on the off-queue to a vacant off-queue stand elsewhere in the system.

To amend the model to include these extra movements we need to define some new sets for the pallet and pallet stand entities. We need to be able to distinguish between buffers which are occupied by loaded pallets, as before, and those which are occupied by empty pallets. A pair of new queues are required. We can call them ONPARK and OFPARK for the on-queue and off-queue buffers respectively. Similarly we need to distinguish between loaded pallets at on- or off-queue buffers. The new queues we can call PARKON and PARKOF. Figure 10.10 gives the code for the new activities and other amendments.

With these routines empty pallets may be on either an on-queue or an off-queue pallet stand, but loaded pallets coming off a machine may only be located at off-queues, and pallets going to machines may only be placed at on-queue pallet stands. Empty pallets are kept at the load stations where possible. These restrictions keep the model logic fairly simple. Other rules could be incorporated, for example to allow loaded pallets waiting to go to a load station to be parked at an on-queue buffer, but to do so will add greatly to the complexity of the model.

If we wished to measure movement times accurately, then we would need to identify the vacant pallet stands where pallets are parked temporarily. This case will be considered in the next chapter.

```
THERE ARE 13 PALLET SET READY INONQ COMPLE INOFFQ FINSHD EMTY
+PARKON PARKOF WITH LDAT MCAT CMPNO
THERE ARE 6 ONBUFF SET ONOCCU ONVACA ONPARK GROUP 7
THERE ARE 6 OFFBUF SET OFFOCC OFFVAC OFPARK
ADD PARKON TO HIST ZUPKON DURATION
ADD PARKOF TO HIST ZVPKOF DURATION


BEGIN BINTROD
TIME OF VEHICL LE O
FIND FIRST PALLET A IN READY
   C = CMPNO OF PALLET A
   N = NEXT OF COMPON C
   FIND FIRST ONBUFF B IN GROUP N
      ONBUFF B IN ONPARK
FIND PALLET E IN PARKON
   MCAT OF PALLET E EQ B
D = LDAT OF PALLET A
DURATION = 2 * MOVETIME
PALLET A FROM READY INTO INONQ AFTER DURATION
MCAT OF PALLET A = B
LDAT OF PALLET A = O
ONBUFF B FROM ONPARK INTO ONOCCU
DURATION = 3 * MOVETIME
PALLET E FROM PARKON INTO EMTY AFTER DURATION
MCAT OF PALLET E = O
LDAT OF PALLET E = D
LOADSTN D FROM LDFULL INTO LDFULL AFTER DURATION
TIME OF VEHICL = DURATION
ADD 1 TO BINTROD


BEGIN BMOVE
TIME OF VEHICL LE O
FIND FIRST PALLET A IN INOFFQ
   B = CMPNO OF PALLET A
   N = NEXT OF COMPON B
   N GT O
   FIND FIRST ONBUFF D IN GROUP N
      ONBUFF D IN ONPARK
FIND PALLET C IN PARKON
   MCAT OF PALLET C EQ D
DURATION = 2 * MOVETIME
PALLET A FROM INOFFQ INTO INONQ AFTER DURATION
ONBUFF D FROM ONPARK INTO ONOCCU
E = MCAT OF PALLET A
MCAT OF PALLET A = D
DURATION = 3 * MOVETIME
OFFBUF E FROM OFFOCC INTO OFPARK
PALLET C FROM PARKON INTO PARKOF AFTER DURATION
MCAT OF PALLET C = E
TIME OF VEHICL = DURATION
ADD 1 TO BMOVE
```

*Figure 10.10   Additional and amended code for converting Model Six into Model Seven,
including movements of empty pallets.*

```
BEGIN BTRFROF
TIME OF VEHICL LE 0
FIND FIRST PALLET A IN COMPLE
  B = MCAT OF PALLET A
  OFFBUF B IN OFPARK
FIND PALLET C IN PARKOF
  MCAT OF PALLET C EQ B
DURATION = MOVETIME
CHAIN
  FIND OFFBUF D IN OFFVAC
  OFFBUF D FROM OFFVAC INTO OFPARK
  PALLET C FROM PARKOF INTO PARKOF AFTER DURATION
  OR FIND ONBUFF D IN ONVACA
  ONBUFF D FROM ONVACA INTO ONPARK
  PALLET C FROM PARKOF INTO PARKON AFTER DURATION
MCAT OF PALLET C = D
TIME OF VEHICL = DURATION
DURATION = MOVETIME + TRTIME
PALLET A FROM COMPLE INTO INOFFQ AFTER DURATION
MACHIN B FROM OCCUPI INTO IDLE AFTER DURATION
OFFBUF B FROM OFPARK INTO OFFOCC
ADD 1 TO BTRFROF
REPEAT

BEGIN BRETURN
TIME OF VEHICL LE 0
FIND FIRST PALLET A IN INOFFQ
  B = CMPNO OF PALLET A
  NEXT OF COMPON B EQ 0
FIND PALLET E IN EMTY
C = LDAT OF PALLET E
D = MCAT OF PALLET A
DURATION = 2 * MOVETIME
PALLET A FROM INOFFQ INTO FINSHD AFTER DURATION
MCAT OF PALLET A = 0
LDAT OF PALLET A = C
FNXT OF COMPON B = FIX(OPNO OF COMPON B, CTYPE OF COMPON B)
DURATION = 3 * MOVETIME
PALLET E FROM EMTY INTO PARKOF AFTER DURATION
MCAT OF PALLET E = D
LDAT OF PALLET E = 0
OFFBUF D FROM OFFOCC INTO OFPARK
LOADSTN C FROM LDFULL INTO LDFULL AFTER DURATION
TIME OF VEHICL = DURATION
ADD 1 TO BRETURN
```

Figure 10.10   (continued).

```
BEGIN BLOADUP
TIME OF VEHICL LE 0
TIME OF OPERAT LE 0
FIND FIRST COMPON B IN LDWAIT
EMTY EMPTY
FIND LOADSTN C IN LDFREE
VDURATION = MOVETIME
DURATION = VDURATION + LOADTIME
CHAIN
  FIND FIRST PALLET A IN PARKON
  E = MCAT OF PALLET A
  ONBUFF E FROM ONPARK INTO ONVACA AFTER VDURATION
  PALLET A FROM PARKON INTO READY AFTER DURATION
  OR FIND PALLET A IN PARKOF
  E = MCAT OF PALLET A
  OFFBUF E FROM OFPARK INTO  OFFVAC AFTER VDURATION
  PALLET A FROM PARKOF INTO READY AFTER DURATION
CMPNO OF PALLET A = B
MCAT OF PALLET A = 0
LDAT OF PALLET A = C
LOADSTN C FROM LDFREE INTO LDFULL
COMPON B FROM LDWAIT INTO ONPALL AFTER DURATION
TIME OF OPERAT = DURATION
TIME OF VEHICL = VDURATION
ADD 1 TO BLOADUP

BEGIN BUNBLOCK
TIME OF VEHICL LE 0
LDFREE EMPTY
TIME OF OPERAT LE 0
FIND FIRST OFFBUF A IN OFFVAC
FIND FIRST PALLET B IN INOFFQ
    M = MCAT OF PALLET B
    MACHIN M IN OCCUPI
    FIND FIRST PALLET D IN COMPLE WITH MCAT EQ M
    C = CMPNO OF PALLET B
    NEXT OF COMPON C EQ 0
OFFBUF A FROM OFFVAC INTO OFFOCC
MCAT OF PALLET B = A
DURATION = MOVETIME
OFFBUF M FROM OFFOCC INTO OFFVAC AFTER DURATION
TIME OF VEHICL = DURATION
ADD 1 TO BUNBLOCK

TYPE 'BINTROD was started' BINTROD ' times'
TYPE 'BRETURN was started' BRETURN ' times'
TYPE 'BLOADUP was started' BLOADUP ' times'
TYPE 'BTRFROF was started' BTRFROF ' times'
TYPE 'BMOVE   was started' BMOVE   ' times'
TYPE 'BUNBLOCK was started' BUNBLOCK' times'
TYPE 'Histogram of length of queue PARKON'/PICTURE(ZUPKON)
TYPE 'Histogram of length of queue PARKOF'/PICTURE(ZVPKOF)
```

Figure 10.10   (continued).

10.8.1   Results—effects of random number seeds

When Model Seven was run, the blocking problem appeared to be solved. However the output of the system was 29 components instead of the planned 36, and the mix of the components was not as close to the specified proportions as expected. In examining the results it was observed that only 27 components had arrived during the sampling period, and the output was constrained by the number of arrivals rather than the capacity of the system. Output greater than this was possible due to a drop in work in progress during the sampling period. In chapter 8, when the results were being discussed, the deviation between the number of arrivals and the proportions of the components to the expected values was commented upon. It was suggested that additional runs with alternative random number seeds would be wise. It was decided to see what the effects would be of using other seeds for the random number streams. There are two streams, one for the time between arrivals and one for the type of component.

Three different values for the seed for the time between arrivals were tried:

Random seed:	1	2	3
Number of arrivals:	27	41	35
Number expected:	36	36	36
Output achieved:	29	41	35

The output appears clearly to be constrained by the number of arrivals rather than by the system capacity. The third case gives a number of arrivals close to the expected value. With this value for the arrival time seed, other values were examined for the component mix stream:

Component type	Number expected	Number obtained with seed		
		1	2	3
1	10	7	6	10
2	8	13	9	9
3	6	7	10	7
4	4	3	3	3
5	6	3	5	5
6	2	2	2	1
total	36	35	35	35

It will be observed that with the third seed the proportions obtained were close to those expected, while for the other two seeds there was a substantial bias. Although doing so is of dubious validity from the statistical point of view (see discussion below), the third value of each seed was used for experiments with varying numbers of pallets and fixtures.

10.8.2   Results—varying numbers of pallets and fixtures

Table 10.5 gives some of the results obtained, including activity counts, facility utilisa-tions, fixture utilisations, average queue lengths and numbers of components produced, when there are 13 pallets and 16 fixtures. Various comments can be made. The activity counts suggest that the system is in a reasonably steady, balanced state. The frequency of the TRANSFER ON activity (313) is approximately the same as the TRANSFER OFF activities (12 + 299), as are the frequencies of INTRODUCE and RETURN, PUTON and TAKEOFF, and so on.

The fixture utilisations were expected to vary widely amongst themselves, but they also vary widely from the expected values in table 10.4(d). These utilisations are the proportion of time a fixture is not available, so include all the time when it is holding a component and when the pallet on which it is placed is being moved or is in a queue within the system. They are not directly comparable with the figures in table 10.4 and should exceed them substantially. However, the ratios of the two sets of figures also show considerable differences, indicating that the pallets holding different component types at different stages of manufacture had varying proportions of delays in production. On average 4.4 components were waiting for a fixture, although the fixtures were only 37% utilised on average. This suggests that the components were waiting for the few particular fixtures with high utilisation. We could investigate the performance of the system with additional fixtures of certain types.

The machine utilisations are roughly in line with the expected values, although ma-chine 2 is considerably more heavily utilised than machine 3, the other member of its group. On the other hand, the vehicle utilisation is more than double that expected. We also observe that on average only 5.2 components were on a pallet, which is small in relation to the number of pallets available, 13. From the activity counts we see that, when pallets were being transferred from a machine to its output buffer, on only 12 times could this be done without first moving an empty pallet out of the way, which was done 299 times. These figures all point to a high level of congestion in the system, and the large number of movements of empty pallets.

The congestion might be reduced by having fewer pallets. We might gain more by reducing congestion than we lose by lack of availablilty of pallets. Experiments were run with 9 and 11 pallets as well as 13. Selected results are given in Table 10.6. These show the characteristic reduction in time to put components through the system, with fewer pallets, and a slight increase in output. The reduction in congestion is highlighted by the greatly reduced vehicle utilisation and the much improved number of normal movements in relation to the number of movements involving empty pallets.

The model was run with an extra fixture of those whose predicted monthly work load was greater than 3000 minutes. Four fixtures fall into this category, namely 11, 21, 22 and 23. None of the others comes near this load, as can be seen from table 10.4. Some of the results obtained with these 20 fixtures and 9, 11 or 13 pallets are given in Table 10.7. The output is the same, since it is limited by the number of arrivals. The average times in the system have been reduced by about a quarter, and the maximum observed times in the system have been halved. The queue of components waiting for fixtures has almost disappeared. Figure 10.11 gives a graph of these results.

Table 10.5   Results from Model Seven (13 pallets, 16 fixtures)

Activity counts

INTROD	88 times	BINTROD	33 times
RETURN	74 times	BRETURN	45 times
LOADUP	85 times	BLOADUP	13 times
TRFROF	12 times	BTRFROF	299 times
MOVE	129 times	BMOVE	65 times
PUTON	97 times	BUNBLOCK	0 times
TAKEOFF	96 times		
RECLAMP	24 times		
UNLOAD	96 times		
TRFRON	313 times		
ARRIVE	35 times		

Facility utilisations

Facility	Occupancy	Utilisation
Machine 1	.5500	.4801
Machine 2	.8051	.7448
Machine 3	.5975	.5601
Machine 4	.5971	.5516
Machine 5	.3447	.3177
Machine 6	.5650	.4891
All machines	.5765	.5256
Vehicle		.5379
Operator		.4357
Load Stations	.6948	

Fixture utilisations

Fixture number	Fixture type	Utilisation
1	11	.5863
2	12	.3242
3	13	.3351
4	14	.1558
5	15	.2215
6	16	.0633
7	21	.9989
8	22	.5527
9	23	.6532
10	24	.1242
11	25	.3230
12	26	.0625
13	31	.4907
14	33	.3032
15	34	.0697
16	35	.6217
Overall		.3679

Table 10.5   (*continued*)

Average queue lengths

Components:	LDWAIT	0.275	FXWAIT	4.416		
	ONPALL	5.199	ONBENC	0.002		
Pallets:	READY	0.090	INONQ	1.122	COMPLE	0.096
	INOFFQ	0.173	FINSHD	0.020	EMTY	0.903
	PARKON	1.626	PARKOF	4.734		
On-queue buffers:	ONOCCU	1.291	ONVACA	2.943		
Off-queue buffers:	OFFOCC	0.359	OFFVAC	0.526		
Machines:	OCCUPI	3.271	IDLE	2.540		

Component production

Component type	Number required	Number arrived	Number made	Average time in system
1	10	10	9	4422
2	8	9	9	2040
3	6	7	7	2050
4	4	3	3	1600
5	6	5	6	2075
6	2	1	1	1200
overall	36	35	35	2648

Table 10.6   Results from Model Seven, with 16 fixtures and varying numbers of pallets

Number of pallets	9	11	13
Total number produced	36	36	35
Average time in system	2216	2451	2648
Vehicle utilisation	.38	.43	.54
Load stations	.53	.59	.69
Activity counts:			
INTROD	121	112	88
BINTROD	2	8	33
TRFROF	116	54	12
BTRFROF	203	264	299

Table 10.7   Results from Model Seven with extra fixtures

	Number of fixtures	Number of pallets		
		9	11	13
Total number produced	16	36	36	35
	20	36	36	35
Average time in system	16	2216	2451	2648
	20	1647	1821	1916
Maximum time in system	16	4000	5000	5200
	20	2400	2600	2600
Average number of components waiting for fixtures	16	3.09	3.81	4.42
	20	0.48	0.64	0.64

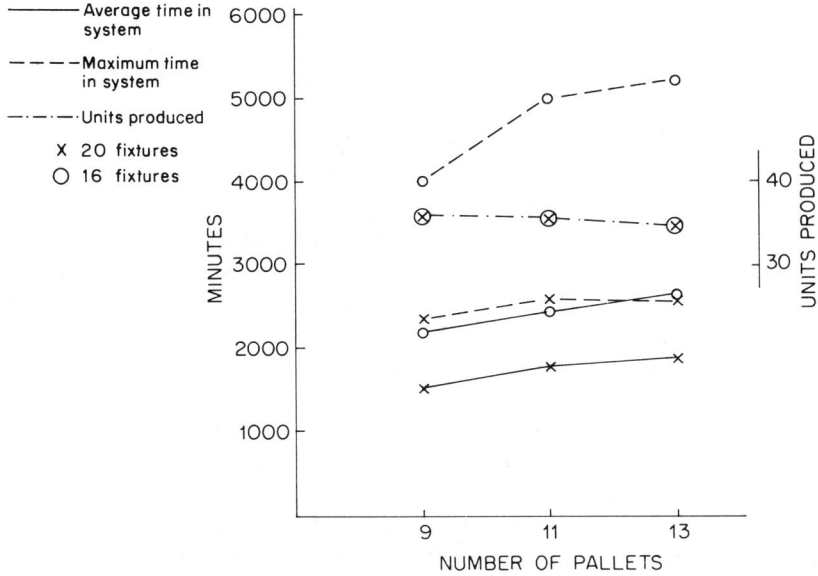

Figure 10.11   Graph of results with varying numbers of fixtures and pallets.

10.8.3   Discussion

These results highlight the assertion made in the opening chapter that with complex systems the interactions of the components are also complex, and that the performance of the system is difficult to predict without simulation.

As mentioned above, we used random number seed values which appeared to give unbiased results. This raises interesting theoretical questions. The underlying principle of using random numbers is that a sufficient number of runs, or runs of sufficient length, should be performed so that any bias in the random numbers obtained is removed. Instead, we selected seeds which seem to give no appreciable bias. Strictly, this is unjustifiable from the statistical validity standpoint. The run should have been increased to six months or so to ensure that the run was long enough. On the other hand, we need to know how the manufacturing system will perform on a monthly or weekly time frame. The runs would have to be short (from the statistical point of view), so many replications should be done.

Another view is that we should eliminate the random effect altogether. We could do this by placing the entire month's schedule of parts in the WAITING FOR FIXTURE QUEUE at the start of the run. We would then let the system work its way through this pile of work, and find out how long it took to complete the schedule, without any run-in period. On the other hand, this would eliminate from the model all randomness, such as would exist with the supply of material to the system. The method used, with random seeds which generate the correct total and mix of components, at least includes this effect. It is probably more representative of the on-going situation in the system than loading the entire monthly requirement at the start of the run.

This is largely a matter of definition of objectives. Frequently, we have to temper the theoretical principles with the practicalities of the system we wish to model.

A further comment, is that it was found that several of the statistics obtained from the model varied greatly with the different random number seeds. For example, the utilisation of fixtures is obviously highly dependent on the mix of components. The time in the system for individual types of component also varied. For example, with the original seeds the average time in the system for type 4 components was 2667 minutes, while with the third seeds it was 1600. This was the largest proportional difference. These differences arise because the components arrive in the system at random intervals, the fixtures they require may or may not be in use, and so on. This serves to illustrate, that it is very difficult to know whether the results which are obtained are generally true, or arose from a specific set of circumstances in a particular run. If they were due to a specific set of circumstances, then we would want to know what those circumstances were, so that they can be avoided, if adverse, or created, if beneficial. To seek this level of detail we need to study the model throughout the run, not merely observe the results at the end. This contradicts the notion of repeating experiments and averaging end results to eliminate bias. The theory treats the system as a black box. For a proper understanding of a manufacturing system we need to look inside. This re-emphasises the fact that complex systems have complex interactions, and the value of visual modelling techniques.

The output of the system was limited by the supply rather than by the capacity of the system. As a result, the proportional mix scheduling rule has played little part in controlling the operations. Had it been necessary to control the mix by this rule there might have been difficulties. One reason is that in this model we have simply copied the method of calculating the priority of each part type. For example, as in the earlier models, MADE was incremented when a component was transferred on to the machine for its first operation, in TRFRON. However, in this case there are several operations, but the component is not actually made until after the final operation. It would have been more sensible to move the updating of MADE to TAKEOFF. More significantly, every time a component was loaded on its first fixture, LOADED was incremented and RATIO was re-calculated. Perhaps we should have established separate values of RATIO for each fixture stage of each component.

Various forms of decision rule could have been examined, and some simple changes could have been made to the model. When a machine was needed we selected the one with the on-queue buffer which had been vacant longest. Perhaps the load on machines 2 and 3 would have been more balanced if the pallet were directed to the machine which was due to complete its current operation soonest. Many variations could have been evaluated, but we will not go in to this topic at present. It is a vast subject area, and goes right back to the design of the system, to the allocation of operations to machines and indeed to the process planning and part programming of the components. In chapter 7, a brief reference to this was made in the discussion on the planning problems in FMS, see for example Reference 5.

In this model we did not adopt the approach suggested earlier in the chapter about copying component data to attributes of the pallet. As a result, the coding of the model frequently includes statements to identify the component which is on a pallet and then to observe and set its attributes. The reader is left to form his or her own conclusion about the relative merits of the approaches.

SUMMARY

In this chapter we have discussed the modelling of components with more than merely one operation, and the fixturing and tooling aspects of the subject.

A model was constructed in which the components were complex, requiring several operations in a succession of fixtures. The results again highlighted the effects of congestion in an FMS system. Adding additional pallets does not guarantee extra output. We have also observed that the output of the system may be constrained by fixture availability rather than machine capacity.

No attempt has been made to present a model which includes tool management, since the quantity of data involved would expand the book beyond acceptable limits. Instead possible approaches have been suggested.

EXERCISES

1. In Models Six and Seven we have traced the component through all the successive stages of its processing, and were able to obtain figures for the total time to process them. How would the model be simplified if each fixturing stage were treated as a separate component?

2. Develop the complete activity cycle diagram for Models Six and Seven, ie without and with movements of empty pallets.

3. How would the coding of the model be altered if we wished to calculate a priority ratio for each fixture stage for each component?

4. We observed that the system operates more efficiently if there is a stock of partly machined components waiting for each fixture. How would the model simplify if we assumed that there was at least one component at each stage at all times?

5. In general terms, how would the model be amended if, instead of referring to attributes of the component, the data had been copied to attributes of the pallet?

REFERENCES

1. Carrie, A.S., The Role of Simulation in FMS, in *Flexible Manufacturing Systems: Methods and Studies*, edited by A. Kusiak, pp191–208, North-Holland, Amsterdam, 1986.

2. Carrie A.S., FMS simulation: needs, experience, facilities, *Proc. SIM-1, 1st International Conference on Simulation in Manufacturing*, Stratford-upon-Avon, March 1985, pp205–215, IFS (Publications) Ltd, Bedford, England (1985).

3. Carrie, A.S. and Adhami, E., Introducing FMS by simulation, *Proc. FMS-2, 2nd Int Conf on Flexible Manufacturing Systems*, and in *Simulation*, edited by R.D. Hurrion, pp173–184, IFS (Publications) Ltd, Bedford, England, 1986.

4. Carrie A.S., Adhami, E., Stephens, A. and Murdoch, I.C., Introducing a Flexible Manufacturing System, *Int. J. Prod. Res.*, 22, 6, 907–916, 1984.

5. Carrie, A.S. and Perera, D.T.S., Work Scheduling in FMS under tool availability constraints, *Int. J. Prod. Res.*, 24, 6, 1299–1308, 1986.

CHAPTER 11

Building FMS models—
4: Vehicles and movement durations

So far, we have had a very simple activity cycle for the vehicle, in that it was either stationary, in queue STOPPED, or it was moving, in a MOVE, INTRODUCE or RETURN activity. For many purposes this simple view of movements is satisfactory, but in general, if we are to accurately assess its utilisation and the duration of movements, we have to look more closely at the behaviour of the vehicle, just as we did with the machines in chapter 8. Figure 11.1 gives the notation which we will use in the diagrams which we develop.

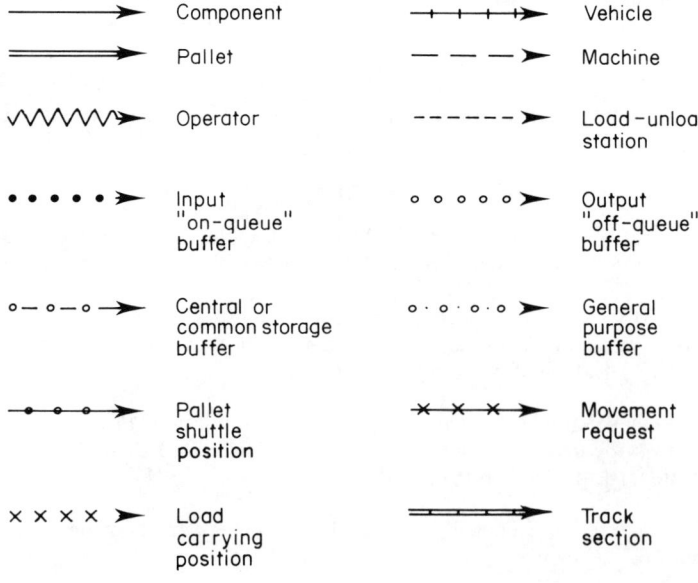

⟶	Component	⊢⊢⊢⊢⇥	Vehicle
⟹	Pallet	— — — ➤	Machine
∿∿∿➤	Operator	- - - - - ➤	Load–unload station
• • • • • ➤	Input "on-queue" buffer	○ ○ ○ ○ ○ ➤	Output "off-queue" buffer
○— ○— ○ ➤	Central or common storage buffer	○ · ○ · ○ · ○ ➤	General purpose buffer
—•—•—•—➤	Pallet shuttle position	✕ ✕ ✕ ➤	Movement request
✕ ✕ ✕ ✕ ➤	Load carrying position	⟹	Track section

Figure 11.1 Arrow notation for activity cycle diagrams.

11.1 EXPANDED ACTIVITY CYCLE FOR VEHICLES

In the discussion about machine buffers we introduced the transfer of pallets between a vehicle and a buffer. Therefore the simple movement activity has to be broken down in more detail, as in Figure 11.2.

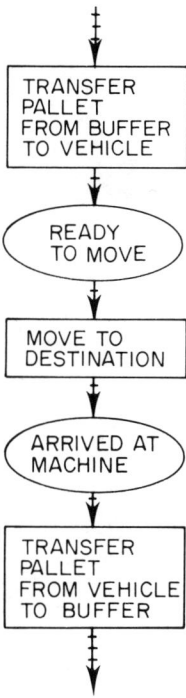

Figure 11.2 Expanded version of MOVE when transfers between vehicle and buffers are introduced.

The idle vehicle queue, STOPPED, also needs closer attention. Will the vehicle merely stop? Is there a parking area for idle vehicles? How does it get to the location of the next pallet to be moved? Are other activities involved? What conditions are necessary for the activities to occur? Which other entities are involved? Probably the vehicle will wait for instructions to move another pallet somewhere else. Thus the cycle of a movement consists of several independent elements:

1. Receive movement instruction.
2. Travel empty to location of pallet to be moved.
3. Pick up pallet.
4. Move to destination of pallet.
5. Deposit pallet.
6. Signal completion of movement.

Since the first and last of these would normally be instantaneous or virtually so, they can probably be safely deleted from the list. We will examine later on how the movement

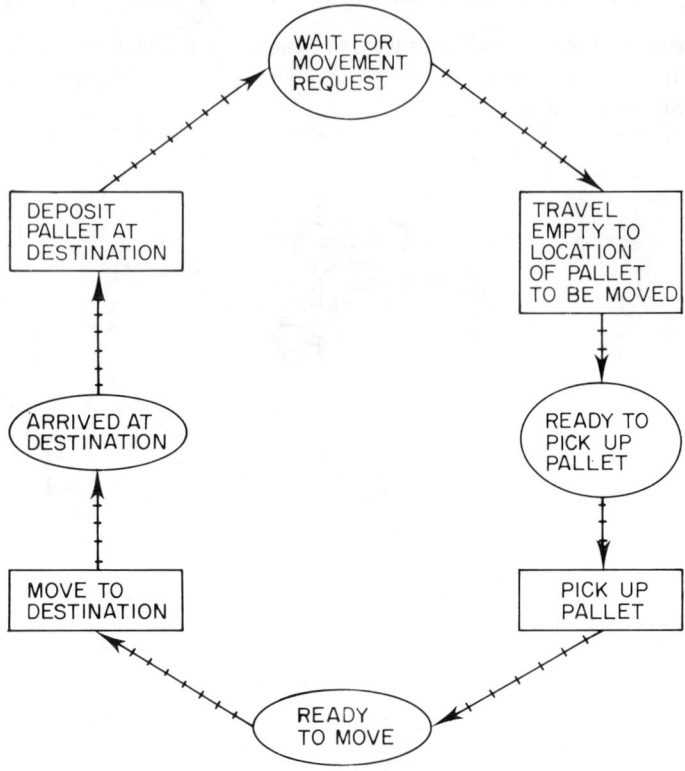

Figure 11.3   Activity cycle diagram for vehicle.

instruction is generated. The activity cycle diagram for the vehicle would be as Figure 11.3.

The destination could be any workstation or buffer. We will discuss the case which has already been mentioned, namely, the destination is an input buffer to a machine. We have already shown that picking up a pallet frees an occupied buffer, and depositing a pallet occupies a previously free buffer position. Thus we can add buffers to this cycle to obtain Figure 11.4. This indicates that the activity PICK UP PALLET FROM BUFFER requires that the vehicle is in queue READY TO PICK UP PALLET, and the appropriate buffer is in queue OCCUPIED OFF-QUEUE BUFFER. Similarly, DEPOSIT PALLET AT BUFFER requires that the vehicle has ARRIVED AT DES-TINATION and that the buffer there is in VACANT ON-QUEUE BUFFER. If the on-queue is not vacant then the vehicle will not be able to discharge its pallet, and will have to wait until the machine has finished its current operation and transferred the pallet from the on-queue position to the machine table. This could cause serious delays in the system, and it would be wise to check that the destination buffer is available before the movement begins.

Figure 11.5 shows the activity cycle diagram when MOVE TO DESTINATION depends on the availability of a VACANT ON-QUEUE BUFFER. Notice that it is

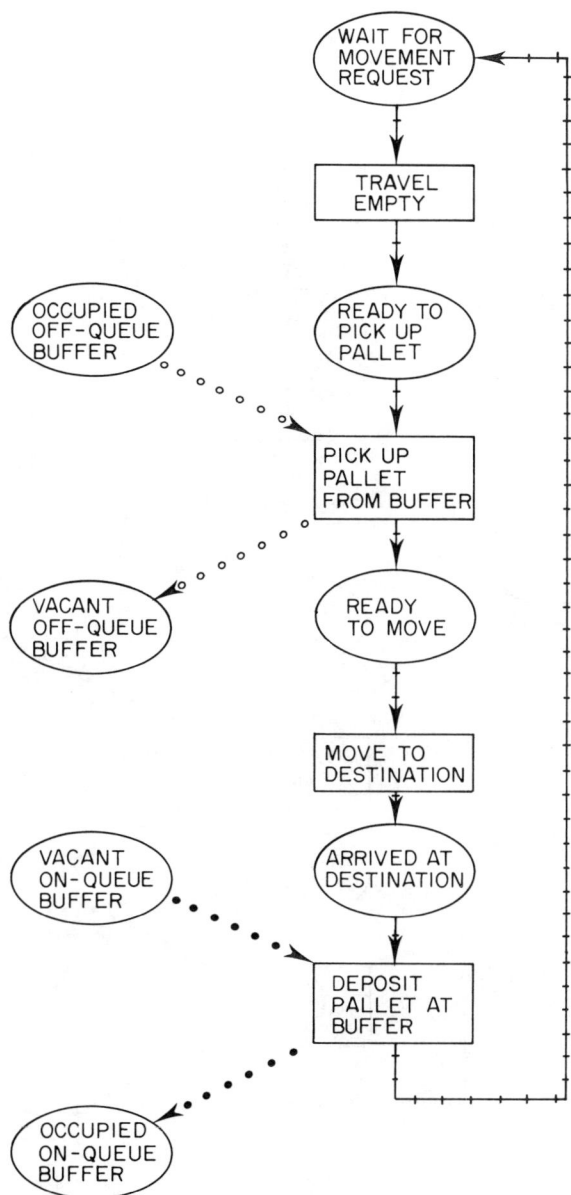

Figure 11.4   Interactions between cycles of vehicle and buffers.

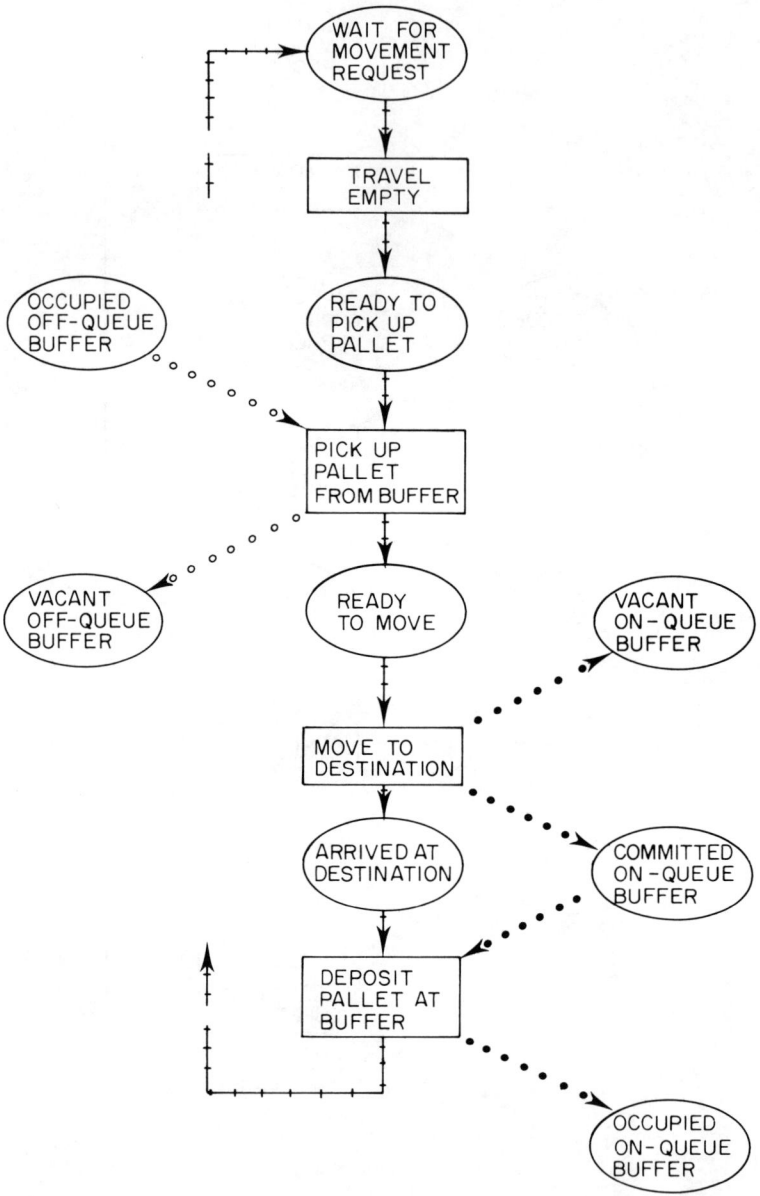

Figure 11.5   Revised interactions between cycles of vehicle and buffers when destination on-queue buffer is reserved before starting to move.

necessary to introduce a new queue for the on-queue buffer. This is for two reasons. One is the basic rule of activity cycle diagrams, that each entity involved in an activity must go into a queue before going to the next activity. The other reason is rather important for the correct operation of the model. We must not let the on-queue buffer remain in the queue VACANT ON-QUEUE BUFFER once the move begins, even though it will be vacant in real life until the pallet is placed upon it, because we must prevent another vehicle deciding to bring some other pallet to this same buffer.

We could extend this approach to require that the destination should be vacant before the vehicle picks up the pallet, Figure 11.6, or even before travelling empty to the pick up position, Figure 11.7. In figure 11.7 we have introduced an extra queue for the off-queue buffer to conform with the rules of ACDs to show that, although it is still occupied, it is no longer available for another PICK UP PALLET FROM BUFFER activity. In both figures 11.6 and 11.7 we have progressively extended the on-queue buffer's stay in the COMMITTED queue.

The purpose of placing the test for the availability of a vacant on-queue buffer to which the pallet could be moved, was to ensure that there would be no delay between the vehicle arriving at its destination and depositing the pallet. In other words the vehicle would move through the queue ARRIVED AT DESTINATION instantaneously. Figure 11.7 reflects this because it clearly shows that the conditions necessary for DEPOSIT PALLET AT BUFFER will be satisfied immediately MOVE TO DESTINATION has been completed. This means that DEPOSIT PALLET AT DESTINATION is a "bound" activity, and ARRIVED AT DESTINATION is a "dummy" queue, in that the preceding and succeeding activities can be combined into a single activity. Similarly, by advancing the test for an occupied off-queue to before TRAVEL EMPTY, the conditions for PICK UP PALLET FROM BUFFER will be satisfied as soon as TRAVEL EMPTY is complete, and these two activities may also be combined. This also enables the STILL OCCUPIED queue to be omitted, as shown in Figure 11.8.

Figures 11.5, 11.6, 11.7 and 11.8 represent slightly different statements of the conditions under which a movement may be initiated. Which version should be adopted? When building models it is necessary to identify the actions of which the system is mechanically capable, and also the rules incorporated into the control software. These rules may well be more restrictive than the physical arrangements. For example, figure 11.5 may be physically correct, but the software may require the model to be based on figure 11.8. In modelling most FMSs such factors allow the model to be simplified.

Some models include a further simplification, with a only marginal reduction in validity. Consider Figure 11.9. We have now reduced the diagram to the equivalent of the original activity cycle for the vehicle, as used in Model One of chapter 8, with WAIT FOR MOVEMENT REQUEST replacing STOPPED. This greatly simplifies the logic, but can it be justified? The definition of an activity dictates that all the entities involved in it are active from the start of the activity until the completion of the activity. Therefore figure 11.9 implies that the off-queue buffer will not be vacant until the vehicle has deposited its pallet elsewhere, whereas in fact it will be free as soon as the pallet has been placed on the vehicle. If the movements are rapid the loss of accuracy may be acceptable. Especially at the start of a simulation project it may be important to get working an approximate model which can be refined later, when more accurate and complete information becomes available.

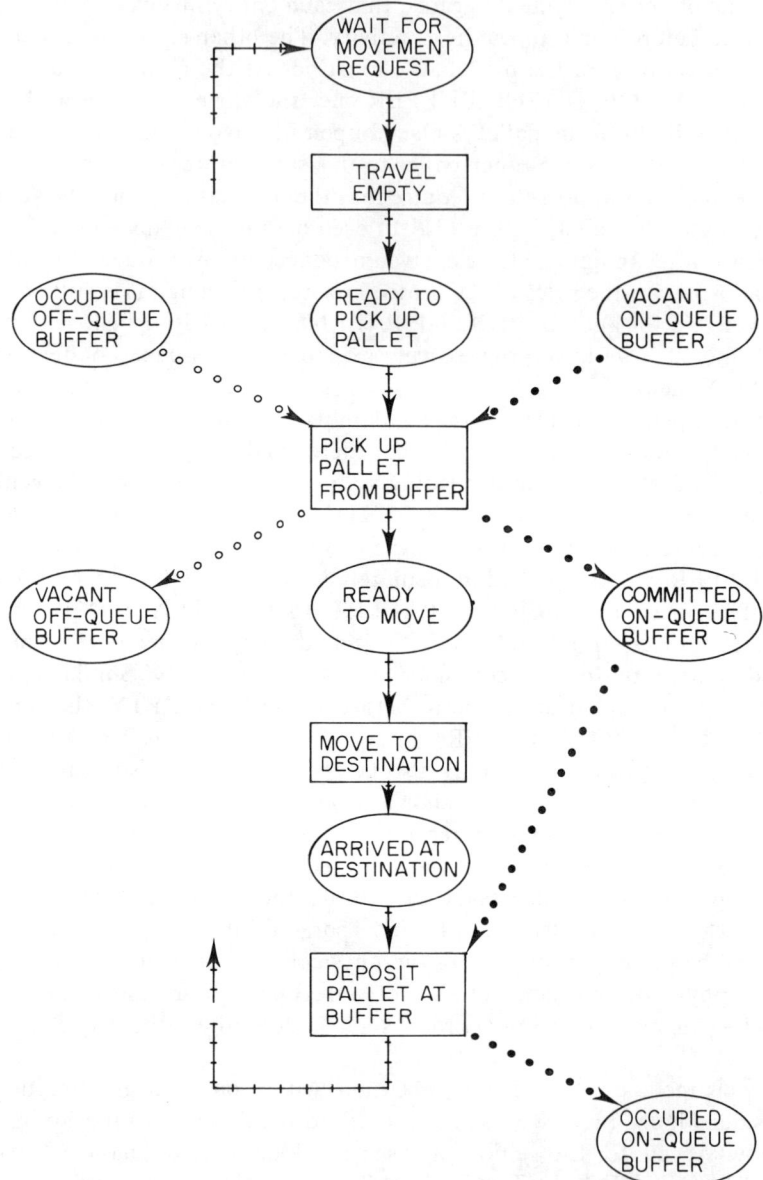

Figure 11.6   Revised interactions between cycles of vehicle and buffers when destination on-queue buffer is reserved before picking up pallet.

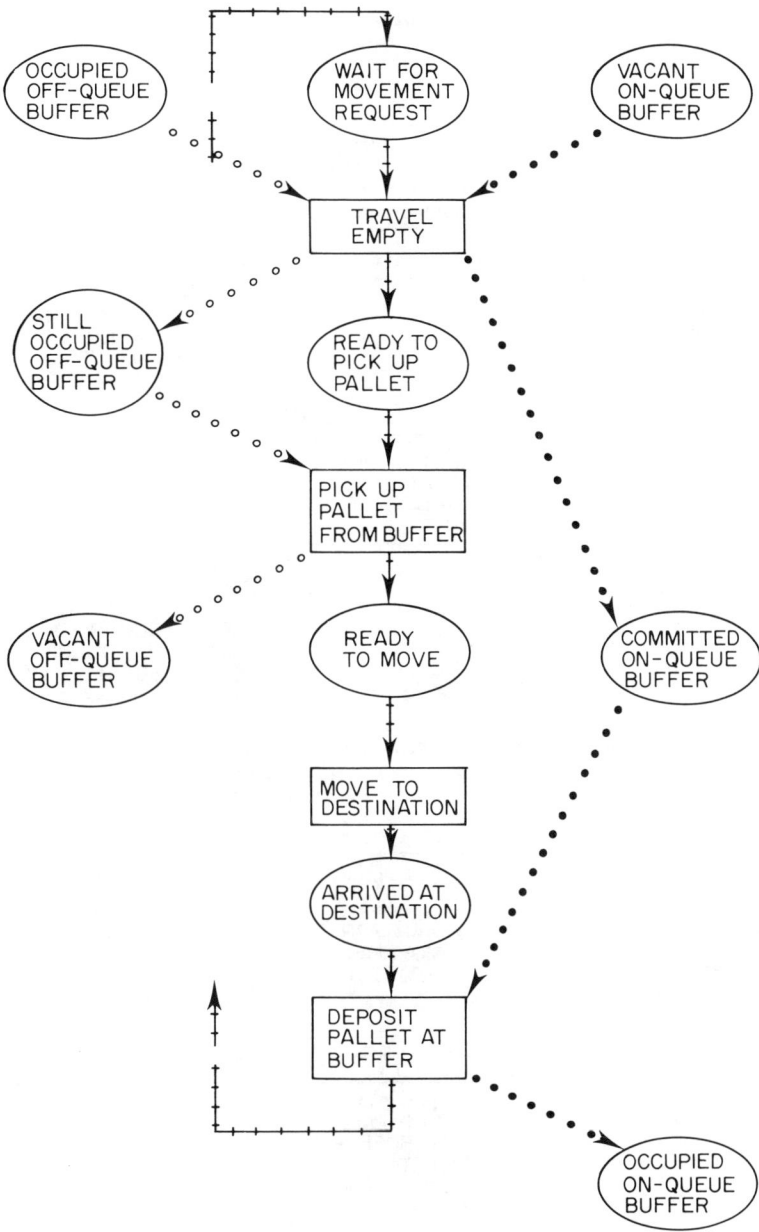

Figure 11.7   Revised interactions between cycles of vehicle and buffers when destination is reserved before vehicle accepts movement request.

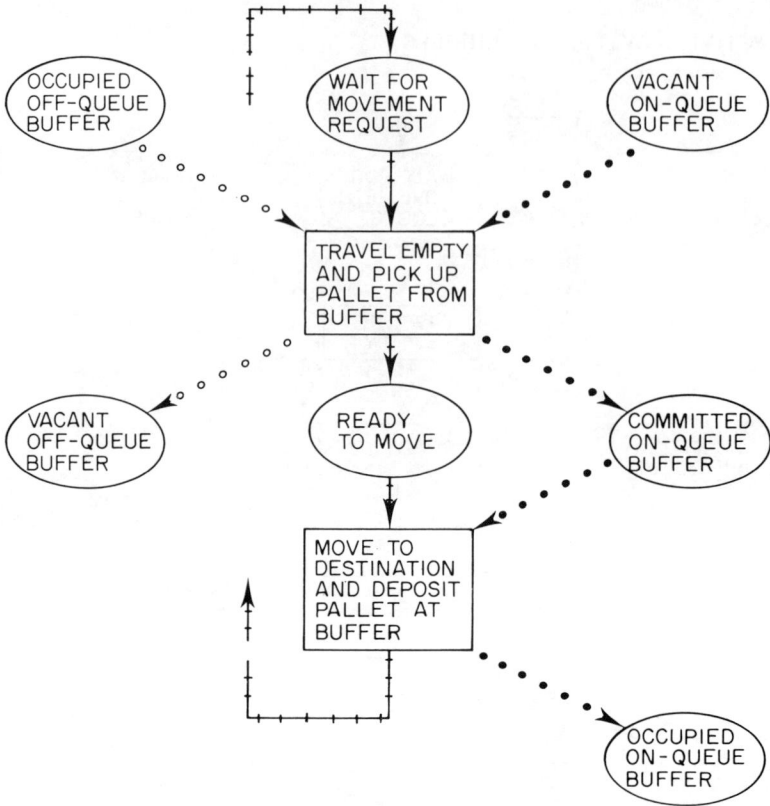

Figure 11.8   Simplification of figure 11.7 when dummy queues are omitted and bound activities are combined.

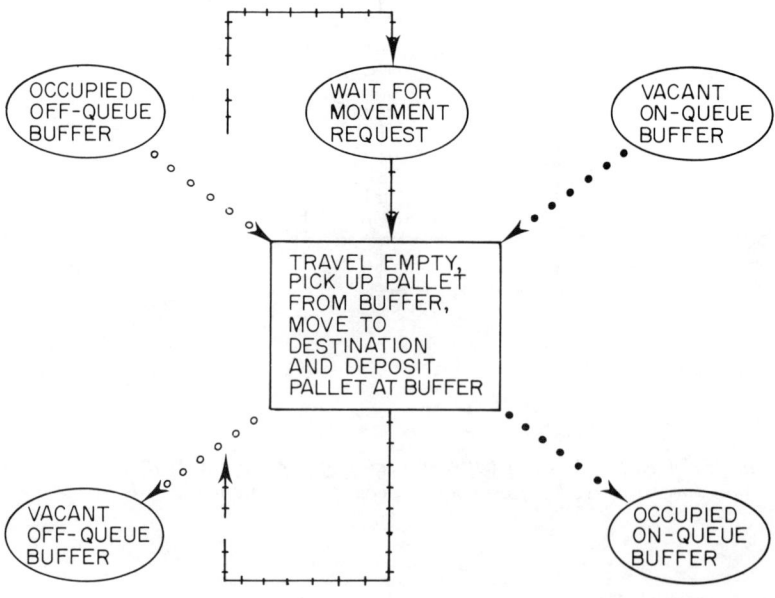

Figure 11.9   Further simplification of figure 11.7.

11.2   MOVEMENT REQUESTS

The conditions which are necessary for TRAVEL EMPTY to occur need consideration. So far the only queue leading into this activity is WAIT FOR MOVEMENT
REQUEST. Unless there were some other entity involved in TRAVEL EMPTY, the
vehicle would be able to proceed directly from DEPOSIT PALLET AT BUFFER to
TRAVEL EMPTY to the next pallet to be moved. Since it is clearly possible that
there could be a delay before the next pallet is ready to be moved, there must be
some other condition involved in the TRAVEL EMPTY activity which is not shown
in the diagram. To handle this situation a dummy entity, which we could call a movement request, can be defined. The activity cycle for movement requests would be as in
Figure 11.10. Activities which generate movement requests are all those which place
a pallet into a condition to be moved. For example, moving a pallet from the worktable of a machine to its off-queue would generate a move request for it to be moved to
some other workstation or to unloading. Similarly, activities such as TRAVEL EMPTY
in order to pick up a pallet, would convert the move request from active to inactive.
Different packages provide different methods for simplifying the process of setting up
movement requests. For example, the SIMAN package has a REQUEST block for this
purpose.

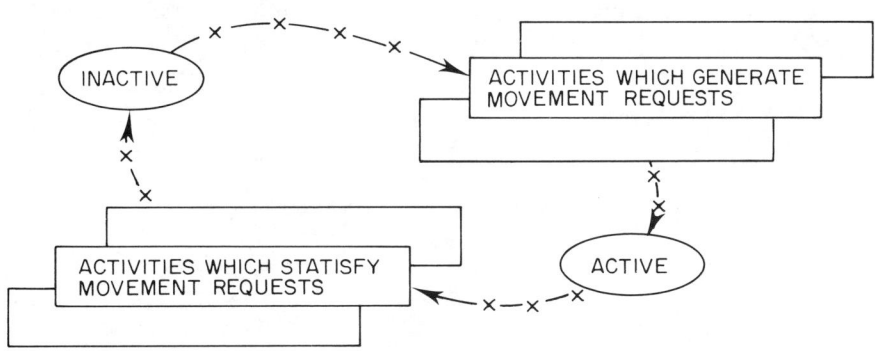

Figure 11.10   Activity cycle diagram for movement requests.

11.2.1   Effect on system activity cycle diagrams

Figure 11.11 shows how the extended vehicle cycle and the cycle for movement requests
would combine with the cycles for pallet, load stations, machine buffers, amended as
necessary to suit. Queues READY TO PICK UP PALLET, READY TO MOVE and
so on have been renamed by READY A, READY B, etc., to save space. As before,
when the cycles of two entities run in parallel, either or both can be compacted. In
figure 11.11 the pallet's cycle has been compacted. New queues have been introduced
to the pallet's cycle, ON VEHICLE GOING OUT and ON VEHICLE COMING
BACK. Alternatively, the pallet's cycle could be expanded to match that of the vehicle,

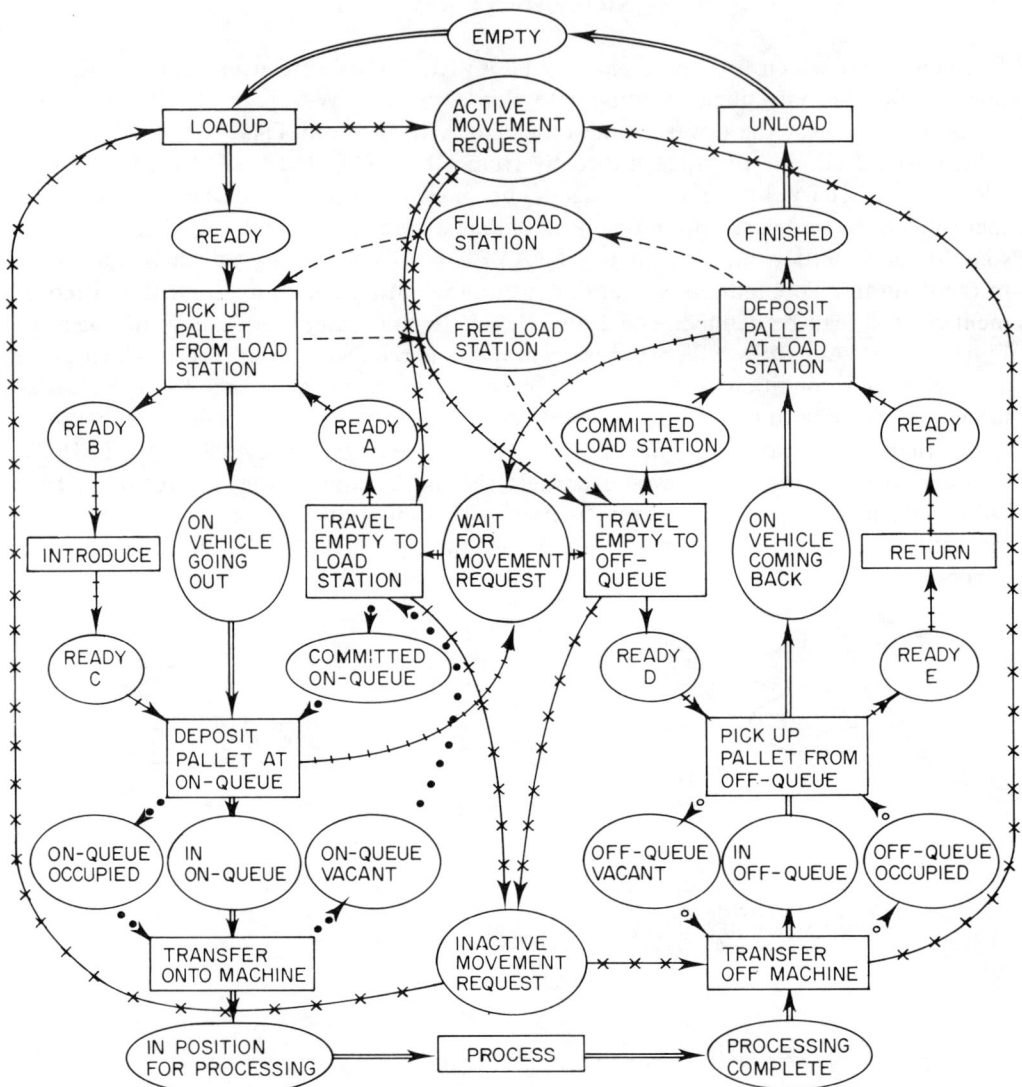

Figure 11.11   Extended activity cycles for vehicles and movement requests merged with pallet, buffer and load station cycles.

and the vehicle's cycle could be condensed with a single queue replacing READY A, INTRODUCE, READY B, and so on.

One point not to be overlooked is that there must be sufficient of them to handle the maximum number of simultaneous move requests. Typically this would be the maximum number of pallets which could be waiting to be moved at any time. This may be as large as the total number of pallets, or it might be slightly less depending on the configuration of the system. The dummy entity will have one or more attributes. At least one attribute will be needed to specify which pallet is to be moved. Another attribute will probably be used to record when the request arose. How the data is handled depends on whether it is more convenient to associate the data with the dummy entity, or whether the pallet, buffers or machines already have attributes holding the necessary data. For example, the dummy entity might only have one attribute indicating which pallet needs to be moved. The location of that pallet and the time the request arose might be attributes of the pallet. The destination might also be an attribute of the pallet, or perhaps, of the component on the pallet.

The reader may be thinking that this discussion is becoming over-elaborate, and that the models in chapter 8 seemed not to involve movement requests. In fact, all the earlier models have included movement requests implicitly. No dummy entity was needed in Model One of chapter 8, because one of the queues of the pallet entity implied that a movement was requested, namely the queue COMPLETE in figure 8.24. In that model the travelling empty and loaded were joined together, as in figure 11.8 above, in which it can be seen that, if there is a pallet in an off-queue and a vacant on-queue at its destination, then the movement can be initiated. We can usually avoid the use of dummy entities by placing the pallet itself in a set of objects needing to be moved. As mentioned several times, the set is a general concept and there is no restriction on the number of sets of which an entity can be a member simultaneously. Thus the pallet can be placed in a set of objects to be moved while remaining in a set representing a physical queue somewhere in the system. Indeed, this approach is generally preferable to the use of dummy entities.

11.2.2 Setting up movement requests in advance

One basic principle in simulation, explained in chapter 1, is that we do not try to jump ahead of the clock. We do so to avoid anticipating the situation which may exist at some point in the future, in case we create a situation which is inconsistent, or undesirable, or assumes a situation which does not in fact arise. Instead we examine the situation in the model at each point in time as the clock moves forward, and take the appropriate decision at that time. One of the consequences of that approach is that it is made more difficult to take a decision at the current clock time in the knowledge that a certain situation is about to arise.

This has practical significance in the field of material handling. Suppose that the vehicle were standing idle, because no pallets were waiting to be moved. With the logic outlined above the vehicle would remain in the queue WAIT FOR MOVEMENT REQUEST, and would only begin to TRAVEL EMPTY to the location of the pallet to be moved when the pallet was actually waiting for the vehicle. Ideally, we would like to examine the operations in progress and find out where the next pallet to be moved would appear, and then we could send the vehicle to that point so as to be ready when the pallet became available.

Another situation when it would be desirable to look ahead is when there are several pallets waiting to be moved. We might select the next pallet to move because its destination is close to the location of another pallet which is waiting to be picked up. This would reduce the amount of time spent travelling empty, and improve the overall efficiency of the vehicle. On the other hand, suppose that no such chain of movements can be set up. Conceivably, there may be a machine near the destination of one of the waiting pallets which has almost completed its operation. Then we could select that pallet to move next, because, by the time the vehicle would reach its destination another pallet would be ready for picking up close to the vehicle's then location.

We can in fact handle this problem quite easily without upsetting the simulation's list of forthcoming events, by providing an attribute of each pallet or location at which a pallet might wait to be picked up. For example, whenever we start an operation on a machine we could store the completion time as an attribute, perhaps of the machine's off-queue buffer. It would then be a simple matter to scan these attributes to identify the location to which the vehicle should be sent to wait for the next pallet, or for any related purpose. A new activity, SCAN OFF-QUEUE BUFFERS, could be invoked whenever

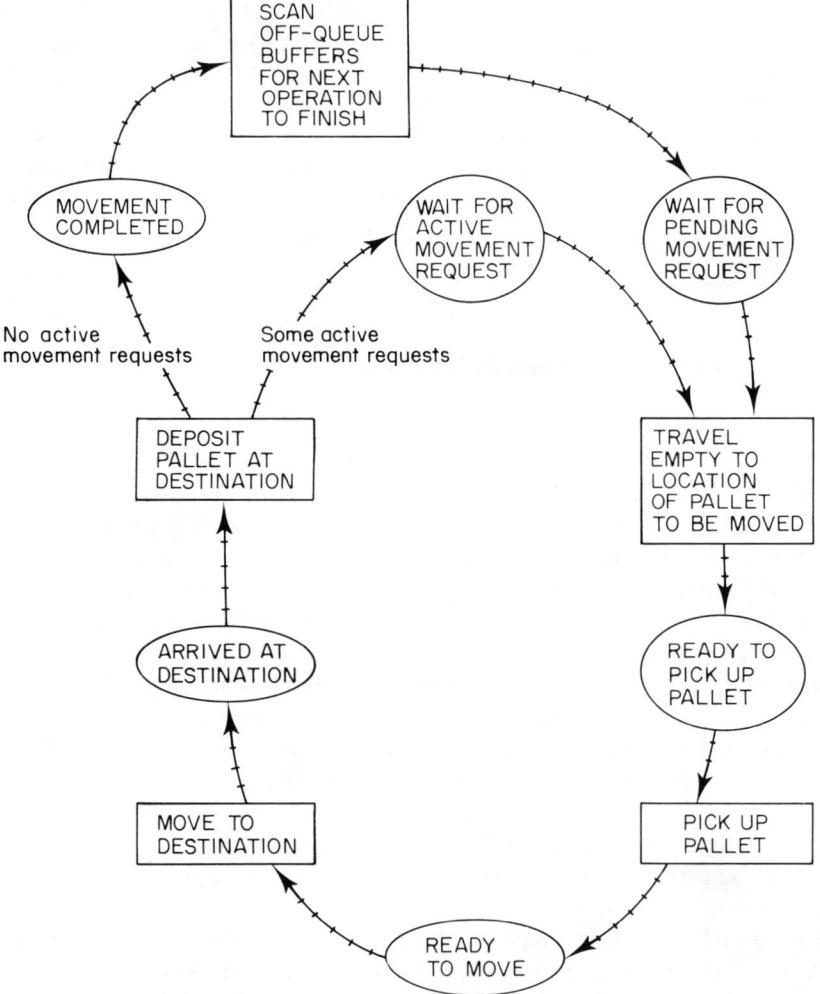

Figure 11.12   *Activity cycle diagram for vehicle, when scanning ahead to find next pallet to be moved is permissible.*

the vehicle would otherwise just wait for the next movement request, producing an activity cycle as in Figure 11.12.

Note that the WAIT FOR ACTIVE MOVEMENT REQUEST and WAIT FOR PENDING MOVEMENT REQUEST queues are now dummy queues, in the sense that the conditions for the entity to leave that block and go on to the subsequent activity will be satisfied even before the entity enters the queue. If an active movement request is waiting then the vehicle can immediately proceed to TRAVEL EMPTY, and if there is no active movement request then the vehicle identifies a "pending" one, travels empty and then waits in the queue READY TO PICK UP PALLET. There are some consequences to this approach. For example, the activity cycle for the movement request would no longer be the simple one shown in figure 11.10. In addition, the logic for selecting the next pallet to be moved might be quite complex. For example, suppose that while the vehicle was moving to where the next pallet was expected to appear, an urgent order entered the system. Would the anticipated movement be cancelled and the vehicle sent off to pick up the urgent order? For reasons of that type it is not desirable to extend the logic in the activity cycle diagram, but to incorporate the selection logic into the various activities, using whatever attributes and variables may be necessary. We will encounter a somewhat similar situation in section 11.5, dealing with vehicles which can handle more than one pallet at a time.

11.3 BATTERY CHARGING

Vehicles which are powered by on-board batteries require to stop every so often to have the batteries changed or re-charged. Usually there is a location at one end or on a loop of the track where this operation is done. If the system operates on three shifts, it will normally be necessary to include battery charging in the model, because the vehicle must be taken out of service for the required period. On the other hand, if the time taken to switch old and new batteries is small and the vehicles are not heavily utilised, then it can usually be omitted. Similarly, if the system operates on only two shifts, it can be omitted because battery charging can be done overnight.

If battery charging or changing is to be included, then we must arrange for the vehicle to be unavailable for the appropriate length of time, at the appropriate intervals. This is almost exactly the same problem, although at more predictable intervals, than that presented by breakdowns of machines, or of vehicles for that matter, and can be handled by the same logic.

11.4 MOVEMENT DURATIONS

The durations of movements have been discussed in various contexts. They can be computed quite simply from the distance to be travelled and the speed of the vehicle. Generally we need to know three locations:

1. the current location of the vehicle,
2. the location of the pallet to be moved,
3. the location of the workstation to which it is to be taken.

These will normally be held as attributes of the vehicle, pallet and workstation respectively. The speed of travel of the vehicle can also be one of its attributes. Alternatively, we may not have a location attribute for the pallet, but instead record which workstation it is at, and use the workstation's location attribute. Similarly, we may have an attribute which defines the destination of the pallet. Suppose MCAT and DEST are the attributes of the pallet, POSN is the current location of the vehicle, and LOCN is the location of the workstations. Then in ECSL we could calculate the duration of the movement of pallet number A as follows:

```
B = MCAT OF PALLET A
C = DEST OF PALLET A
EMPTYDIST = ABS ( POSN OF VEHICLE - LOCN OF MACHINE B )
FULLDIST = ABS ( LOCN OF MACHIN B - LOCN OF MACHIN C )
DURATION = ( EMPTYDIST + FULLDIST ) / SPEED OF VEHICLE
```

This coding assumes that the travelling empty and full are combined in a single activity, as in chapter 8 and earlier, and in figure 11.8. If they are separate activities then the appropriate statements will be placed in the respective activities.

If necessary, acceleration and deceleration times may be included. In the case of vehicles which travel in both horizontal and vertical directions, the times for the two directions can be added together. The speed of movement in the vertical direction is usually much slower than for horizontal movements. Some vehicles have a rotational capability to access pallets on either side of the track, such as that in Cummins' Makino line. This rotational movement is also usually quite slow. A fixed time can usually be applied.

At the start of a project the exact locations of the equipment may not be accurately known, and the distances given are only estimates. Fortunately, this does not normally cause a problem, because the speed of travel of the vehicle is usually so fast that a metre or so of error in distance will be negligible when converted to a time difference.

Even when distances are accurately known we may discover that the time difference between the longest movement and the shortest may be very small. Using a constant time value for all movements may be sufficiently accurate and can greatly simplify the model. This would normally be acceptable in systems in which the movement times are small in relation to the machine cycle times, and the vehicles are only lightly utilised. Of course, if the utilisation and efficiency of the vehicle is a critical factor in the model, or rather in the real system, then precise durations must be calculated. The travelling empty would normally be treated in exactly the same way as travelling with a load, as in the coding above. However, if it has been decided that a single value of travel time can be used with sufficient accuracy, then it may be permissible to omit the travel empty activity altogether. This would make no difference to movement times, but may enable the definitions of movement activities to be less precise, and the model simplified accordingly.

Another aspect of the movement durations arises with packages which use an integer clock. With these packages all the times must be expressed as integers and the most suitable size of the time unit has to be determined. Depending on the nature of the system one-minute units might be too coarse, while one-second units might be too small. The use of too small a time unit can lead to very large time values, which may overflow the word size of the computer, although this is not likely to be a problem with modern computers. For example, there are approximately 7,000,000 seconds in six months of two-shift operation. To overcome this problem, ECSL, for example, allows the run to be carried out over a number of equal time periods, with the clock being reset to zero at the start of each period. An artificial time unit, which gives a good compromise between accuracy and the size of the time step, may often be used efficiently. Units of half a minute, or of ten seconds, may be appropriate in some systems, while units of two or five minutes might be suitable in others. In general, the time unit should be the highest common factor of the actual times, rounded as necessary, so that the time unit is as large as possible without introducing a compromise on accuracy.

11.5 VEHICLES WITH TWO LOAD-CARRYING POSITIONS

In the earlier chapters it was assumed that the vehicle could carry only one pallet at a time. Frequently, especially in rail-car systems, the vehicles have two load-carrying positions, and we must consider this case. The main difference between these two conditions is that, whereas in the single pallet case the destination must be vacant before a pallet can be dropped off, this is not required when two pallets can be carried. The vehicle can bring a pallet to a workstation, or a workstation buffer, and if there is a pallet there waiting to be picked up, it can pick up that pallet and then drop off the pallet which it had brought. This capability is particularly appropriate when workstations are supplied directly by the vehicle or have only one buffer position, such as a pallet shuttle, rather than a pair of pallet stands. As with so many aspects of modelling there are short cut methods and elaborate ones. The unblocking routine used in chapter 9, in which a pallet was moved from a load station to the common storage, and another pallet was brought from common storage to a load station would be a simple movement for a vehicle with two load positions. In chapter 9 we had to place one of the pallets at some unspecified location while the second one was moved. This would be unnecessary if the vehicle had two load positions, since the second load position could be used. There would also be a saving of a movement since it would not be necessary to move a pallet to another place, and then to its destination; the pallet could be moved directly to its destination. Such simple routines can often be used, but we should examine the logic which is necessary if we wish to model the situation more comprehensively.

To start with, let us identify the movements which would be made by the vehicle in handling pallets. Since the two load-carrying positions will be a metre or so apart, the vehicle must make a small movement between picking up one pallet and dropping off the other at the same location. The sequence of activities would therefore be as follows:

1. Receive one or more movement requests,
2. Move empty to vicinity of location of pallet 1,
3. Move so that load position 1 is adjacent to pallet 1,
4. Pick up pallet 1 on load position 1,
5. Move loaded to vicinity of destination of pallet 1,
IF destination is vacant, go on to step 8:
6. Move so that load position 2 is adjacent to destination,
7. Pick up pallet 2 on load position 2,
8. Move so that load position 1 adjacent to destination,

9. Deposit pallet 1,

IF load position 2 is vacant, go to step 1 and wait for the next movement request, else:

10. Move loaded to vicinity of destination of pallet 2,

11. etc.

This sequence of activities would be repeated as long as a pallet was to be collected from the destination of the previous one. Notice that from step 11 onwards the occupied load position is position 2, while initially (step 4 onwards) it is position 1 which is occupied. Thus, each time a pallet is dropped off and one picked up, the occupied and vacant positions alternate. Step 11 could have been written:

11. Reset occupied and vacant load positions,

12. Go to step 5.

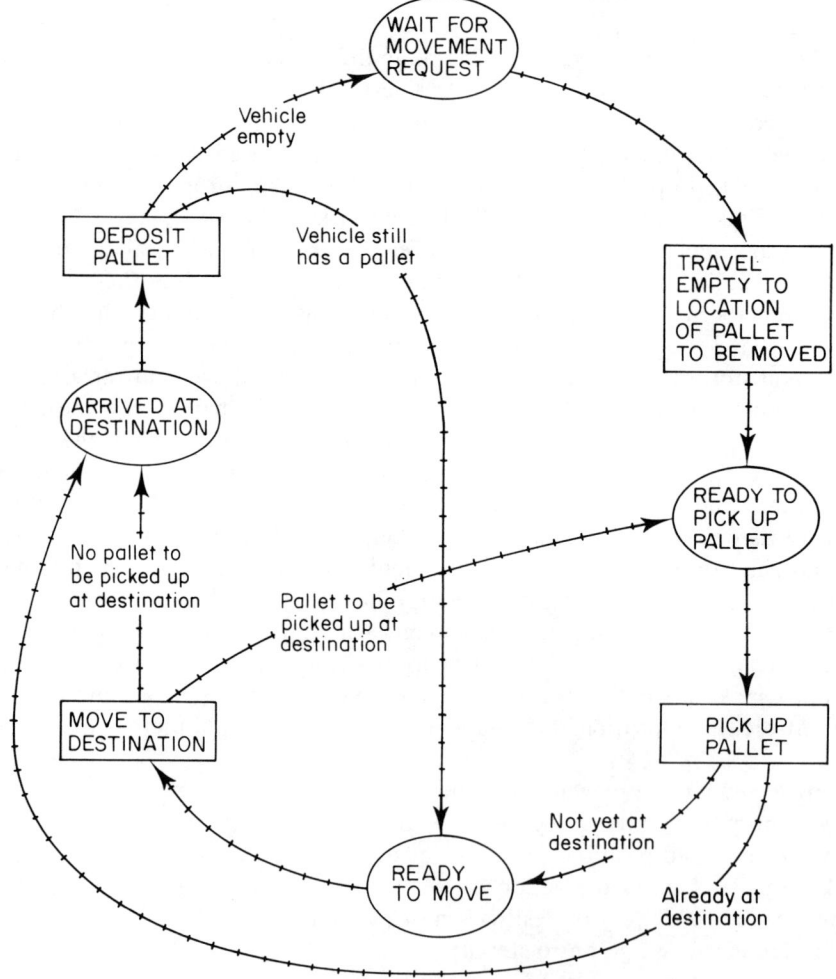

Figure 11.13   Activity cycle diagram for vehicle with two load-carrying positions.

with the additional amendment that the pallet numbers would now be general numbers or variables. The activity cycle for the vehicle given in figure 11.3 would become as in Figure 11.13.

The diagram contains exactly the same activities and queues as before but now has several branches which are necessary to express the more complex logic. To highlight the correspondence with the earlier diagram, we have omitted the adjustment of the vehicle's position to align the relevant load position with the pick up or deposit location at the workstation. That logic could either be included in the PICK UP and DEPOSIT activities or could be included as additional activities, on the branches from PICK UP PALLET to ARRIVED AT DESTINATION and from TRAVEL TO DESTINATION to READY TO PICK UP PALLET (with an extra queue in each case). There are several ways in which the tests could have been expressed. For example, testing

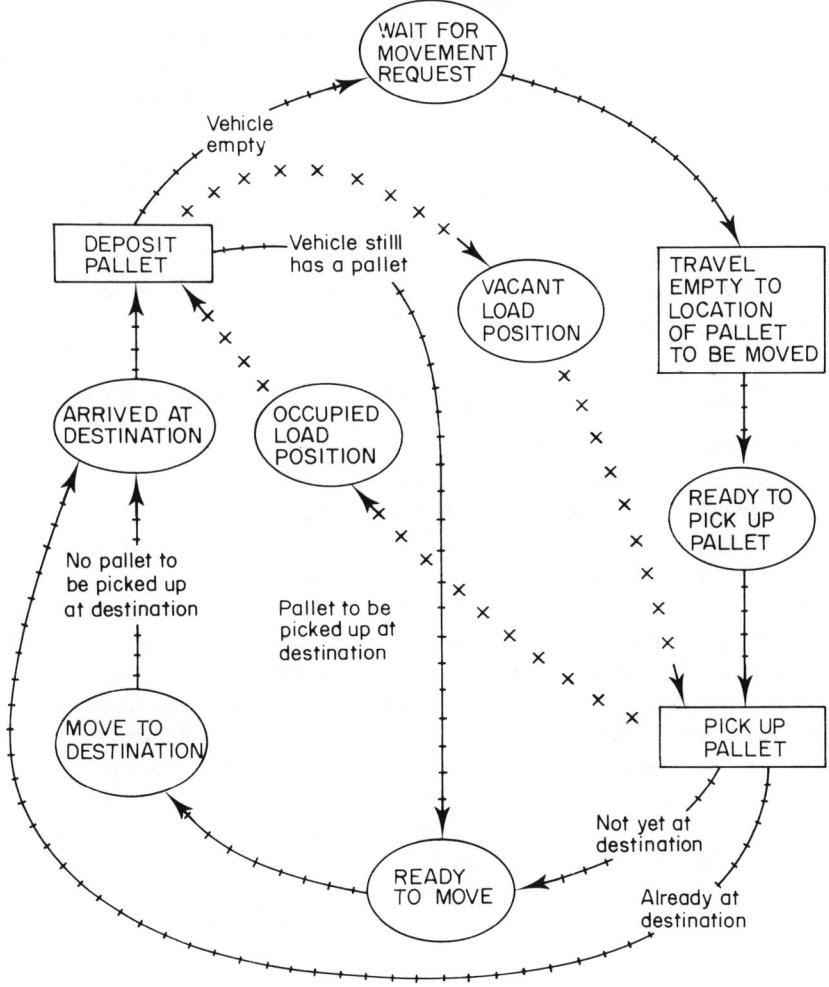

Figure 11.14 Activity cycle diagram for vehicle with two load-carrying positions and vehicle's load positions.

the contents of the two load-carrying positions would be a feasible method for some of the tests. We should express the conditions as succinctly as possible and allow for all possible conditions, whilst avoiding any ambiguity.

This logic describes, in fact, only a subset of the possible movements of the vehicle. The provision of a second load-carrying position allows the vehicle to serve two movement requests simultaneously in a quite general way. With the logic we have outlined, the main movement of pallets, activity TRAVEL TO DESTINATION, is done with only one pallet on board the vehicle. However, it would be quite possible for the vehicle to move to one location, pick up a pallet, move to a second location, pick up another pallet, and then travel to one of the destinations. In fact this would be very sensible if the pick up point for the second pallet lay on the route of the first pallet. The full list of possible movement sequences could be very extensive, depending on the sophistication of the control software, and how it decides which movement request to satisfy next. In fact moving a pallet to a location from which another pallet was to be collected might be the result of an algorithm designed to minimise travelling empty. Unfortunately, we cannot go into the possibilities further at present because they could be very numerous and depend greatly on the configuration and control logic of the particular system.

For completeness, it may be worthwhile to insert into this diagram the cycles for the load-carrying positions on the vehicle, as shown in Figure 11.14. This shows that for a pick-up activity to occur there must be a free load position, in addition to any conditions discussed in previous diagrams. To some extent this generalises the diagram for vehicles with any number of load positions. As indicated above the tests could be re-written to utilise the contents of the load-carrying positions.

11.6 SYSTEMS WITH MORE THAN ONE VEHICLE

Large FMSs normally have several vehicles, usually AGVs, and many routes over which they can travel, as illustrated by the Hattersley Newman Hender system. Also, many companies now use AGVs to transport materials around the plant even though they may not have an FMS as such. Activity cycle diagrams apply to a type of entity rather than to an individual entity, and therefore the diagrams presented above apply to any number of vehicles. For example, serving a movement request requires a free vehicle no matter how many are in the system. The interactions of vehicles with other types of entities have already been defined. However, when there are two or more vehicles various other factors have to be considered. We need to examine the interactions between vehicles, although not in the sense that they combine in activities. The principal factor is whether the vehicles can interfere with each other. To be more specific, although several vehicles can travel along a path in the same direction, none can travel along a path in the opposite direction to another. The inclusion of the logic to handle this problem can add considerably to the complexity of the model. There are two approaches to this problem:

1. Ignore it–assume that the vehicles can play leap-frog with each other,
2. Add the logic to model it correctly.

At first sight the first alternative seems ridiculous, and it certainly would be unacceptable as a method of controlling the vehicles in real life. However it may be acceptable in a model, and indeed quite a few models use it. If there are plenty of different routes to go from any point to any other point, then it is likely that any vehicle will be able to find an unobstructed path. Therefore there will be no great loss of accuracy in the results. The extra time and cost involved in building in proper logic might not be justified. However, can one be sure that the simplification is really valid? Unfortunately not. It is only possible to say that it is unlikely, and perhaps compute the probability of interference occurring. If there is any doubt at all then we must get down to the business of modelling vehicle movements in more detail.

The main principle in modelling the movement of a fleet of vehicles is that we have to record the path followed by every vehicle and ensure that no two vehicles try to use the same track in opposite directions. Strictly, travelling in opposite directions is not the only cause of problems. Another situation where interference could arise is when one vehicle has to stop on a section of track to deposit or pick up a pallet while another vehicle is required to move on through that point. We need to build into the model logic similar to that in the real control system. Perhaps we can get a statement of the rules used in the control software and incorporate it into the model. On the other hand, we may be using the simulation to help develop the rules for the real system, in which case there will be no short cut.

To record the path of each vehicle and prevent interference we need to ensure that only one vehicle has right of way at any time in any section of track. This is rather like the way in which the movement of trains is controlled in single track railway lines. The engine driver is given a token to enter a section of track. Without it he is not allowed to enter the section, and when he gets to the end of the track he hands back the token to the controller, who can then issue it to another driver. We divide the track up into sections, normally from one junction to the next. Long sections, in which we might permit more than one vehicle so long as they travel in the same direction and maintain a safe distance between each other, may be split up into two or more small sections. The size of the sections is normally the minimum distance between vehicles. However, if a long section is subdivided, we need to also ensure that the direction of travel is set the same for all of them. If the long section is empty, a vehicle can enter it from either end. However, as soon as a vehicle is in a long section we need to set the direction of travel for all the sub-sections to prevent a vehicle entering from the wrong end. In some systems the vehicles can only move along tracks in one direction, so the direction of flow along the track will never have to be altered, which simplifies the problem.

The logic associated with servicing a movement request can be summarised as follows:

1. Identify location of pallet to be moved,
2. Identify destination of pallet,
3. Identify any free vehicles and their locations,
4. Select a vehicle to move the pallet,
5. Find a path from the present location of the vehicle to the location of the pallet,
6. Reserve the sections of track along the path,
7. Initiate the movement of the vehicle,
8. Release the sections of track,
9. Pick up pallet,
10. Find a path to the destination of the pallet,
11. Reserve the sections of track,

12. Move the vehicle,
13. Release the sections of track,
14. Deposit the pallet.

Some of these steps could be done in a slightly different order. For example, the order of the first three is unimportant. The destination of the pallet might not be identified until the vehicle had picked up the pallet. Suppose that the vehicle moved to a pallet and picked it up, and then we discovered that there was no free path to the destination. Should the vehicle replace the pallet and go off in search of a pallet it could move? To avoid this it would be necessary to advance steps 11 and 12 to immediately follow steps 6 and 7 respectively. This would mean reserving quite a lot of track in advance, which might unnecessarily restrict the freedom of movement of other vehicles. Another approach would be to reserve only one section of track at a time. In this method we record where each vehicle is and its destination, and compute the next section of track it needs to enter, reserve it, move the vehicle on one section, release the previous section, and so on. Unfortunately, this might cause two vehicles to be moving towards the same section of track from opposite directions. The first one to arrive could enter the section but the second could not enter it until the first vehicle had vacated it. However the first one might not be able to vacate it because its exit was blocked by the second vehicle. In such circumstances a premium might be paid for vehicles which could play leap-frog! Normally, one vehicle has to reverse to let the other out. What if the vehicles can move in one direction only? Then we must prevent blockages occurring rather than try to cure them. This requires us to revert to reserving tracks well in advance. There is an interesting trade-off between the complexity of the control logic to avoid or handle such conditions, and the cost of providing additional paths, or some other hardware solution.

To reserve sections of track we can introduce another type of entity to represent a section of track. As with buffers, load positions and other entities we have met already, this entity is either reserved for use by a vehicle or is available. Its activity cycle is as in Figure 11.15.

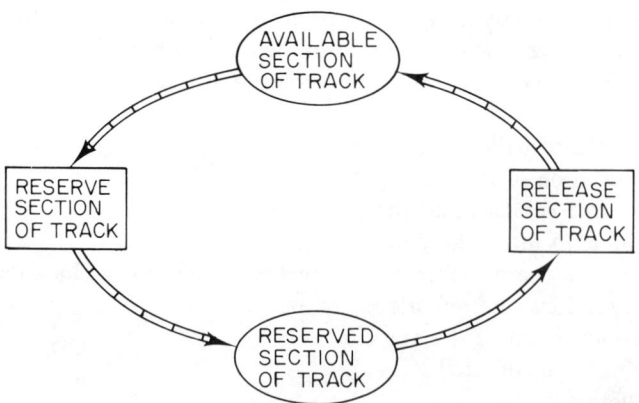

Figure 11.15   Activity cycle diagram for sections of track.

Combining this cycle with those of other entities concerned, involves adding a new activity and queue to the cycle of the vehicle, so that track can be reserved before the movement begins. Similarly, a new activity will have to be added for releasing the track. The activity cycle diagram for the vehicle would be as in Figure 11.16.

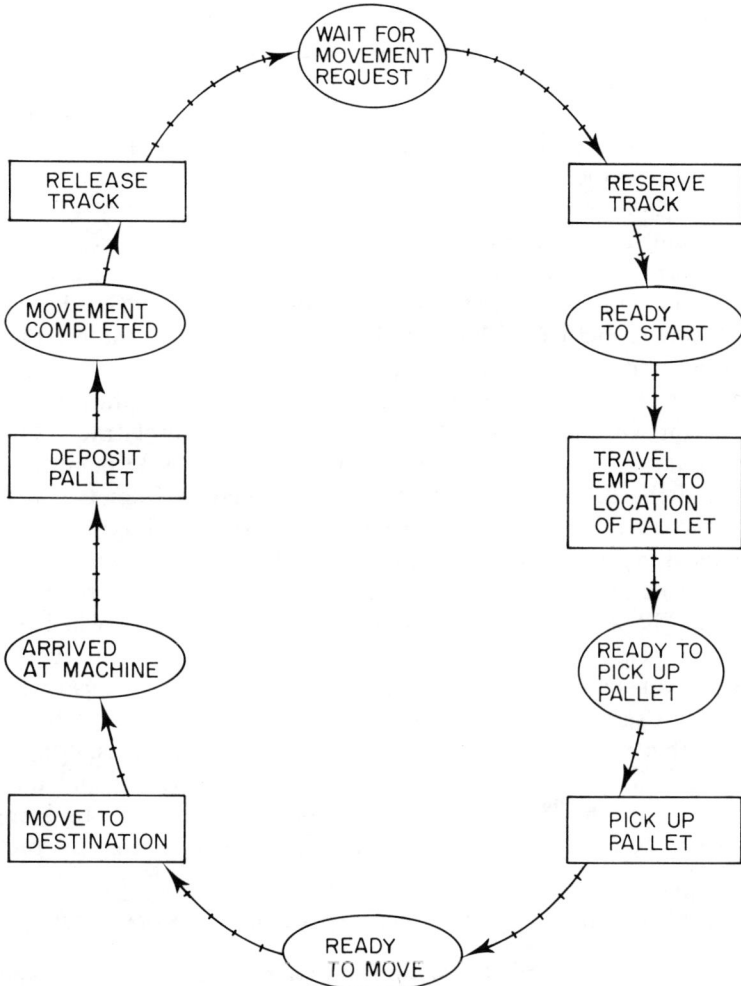

Figure 11.16   Activity cycle diagram for vehicle with reserving and releasing track sections included.

Strictly, the track should perhaps be released before depositing the pallet, as suggested above. However another factor enters the discussion. Where is the vehicle when it has finished the movement? Presumably it remains adjacent to the workstation to which it has delivered the pallet. It cannot release the section of track in which it remains. Releasing that section will have to wait until the vehicle moves off to collect its next pallet from elsewhere in the system. Consequently, the logic becomes rather more complex than can be shown easily in an ACD. As indicated above, one approach might be to reserve each section of track as the vehicle moves over it, and to release it immediately afterwards. This would involve incorporating the reserving and releasing activities within the move activities, although as was pointed out above this method might lead to traffic jams.

We must also consider how a free path from the vehicle's present location to its destination can be found. This is normally done using tables to specify the routes between locations. The data can be held in various ways, depending on the complexity of the algorithm used. The data could be held in the simplest possible form and analysed extensively each time a decision is needed. Alternatively, it could be held in a more complex form to reduce the computation needed to interpret it. This involves a trade-off between data storage and program execution speed.

The track network consists of sections of tracks. Each section of track connects two points. These points are either end points of a vehicle's journey, such as workstations, or they are points where two sections meet, such as junctions, at which the vehicle would not normally stop. These connecting points are often referred to as decision points, as in MAST. A simple method of holding the data would be in three tables. One table would give the number of the decision point nearest to each workstation, the second table would define which decision points are adjacent, and the third table would state the decision points (or workstations) bounding each section of track. To find a route from one workstation to another, we could identify the decision point next to the.first workstation and the decision point next to the destination from the first table, then examine the second table to find a series of decision points linking the first and last decision points. Finally, we would examine the third table to see which sections of track were involved and whether they were available or had been reserved already for another vehicle. The second stage involves quite complex logic, for which algorithms are available. It seems silly to have to go through this computation every time we needed to find a path. Instead, so long as the network was not too large for this to be practicable, we could set up a series of tables which held all the possible paths between any two points. Each path could be given an identifying number. The tables, or more accurately lists, would state the number of paths between any two points, and the list of path numbers. For each path we would have a list of the decision points along the route. Then there would be a table relating the decision points to sections of track.

11.7   MODEL EIGHT—COMPUTED MOVEMENT TIMES

The models in chapters 8, 9 and 10 all assumed that movement activities can be described as a single composite activity. Much of this chapter has been devoted to indicating how these should be broken down to provide a more accurate model, so far as movement durations are concerned, and to avoid involving entities for the complete activity, whereas in reality they are only involved for a small part. Let us therefore re-write Model Seven to separately identify the travelling empty and travelling loaded elements of the whole move, and calculate accurately the durations of the movements. To calculate the durations we need to know the distances involved. These can be read off the floor plan of the system. Figure 11.17 gives the layout plan of the system, which is as in Models Six and Seven, now showing the distances between locations.

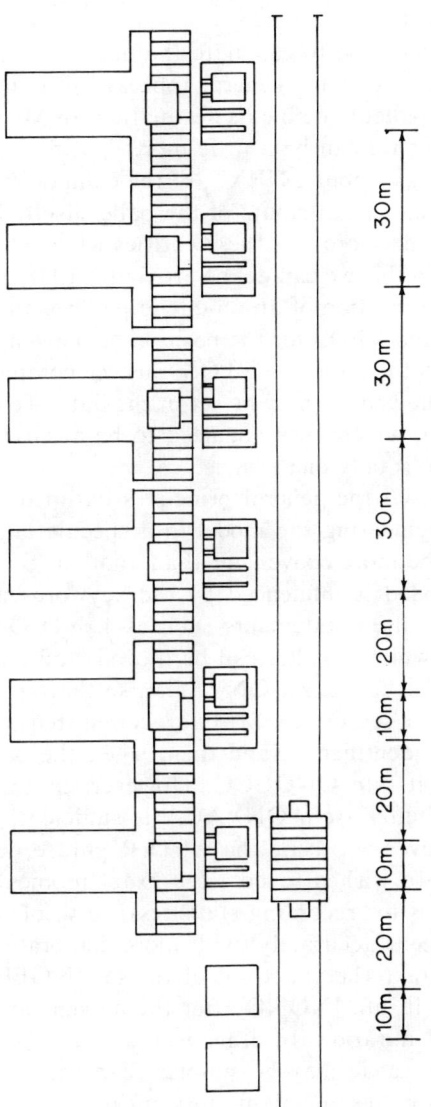

Figure 11.17 Layout of system for Model Eight

11.7.1 Amendments to model

The effect on the logic and code of the model will be in two areas. Firstly, the arithmetic of the duration calculation will be changed. Secondly, the tests for whether movements can be carried out may be altered, since the entities are not all involved for the total duration of the movement.

As explained in section 11.4, to calculate the durations it is necessary to define the distances involved. In that section a general approach was suggested in which we would have an attribute of the pallet to define its destination. In Model Eight we have a specific situation to simulate and we can be a little more specific in the way we model it. For example, we are using an attribute, NEXT, of the component to identify where a pallet is to be moved, rather than an attribute of the pallet itself. The pallet is moved to and from pallet stands, and therefore it is these entities which should have locations, rather than the machines. To do this we can define attributes LDPOS for the position of a load station, ONPOS for the position of an on-queue pallet stand, OFPOS for the position of an off-queue buffer, and VPOS for the position of the vehicle. LDPOS, ONPOS and OFPOS will be set up in the data, but VPOS must be computed after every movement. The initial position of the vehicle may be set in the data. To calculate the durations we need a variable SPEED, which does not need to be an attribute of the vehicle (as in section 11.4) since there is only one vehicle.

A further deviation from the general principles outlined above in sections 11.1 and 11.2, is that when implementing the model in a specific language, one method of expressing the logic may be more convenient than another. So it is with ECSL. The way Model Eight will be coded is influenced by the keywords and grammar of that language. Thus we have written statements such as FIND ONBUFF B IN ONVACA to find an on-buffer to which a pallet can be moved, and once the buffer has been selected we must take it out of the set ONVACA, so that no other pallet can be moved to it. We might define a set ONCOMM for committed on-buffers, place it in that set as soon as we have identified it, and then, when the pallet arrives, take it out of ONCOMM and place it into ONOCCU. However, in ECSL this is not necessary, since removing the on-buffer from ONVACA is sufficient, and it can be placed into ONOCCU after the movement using the AFTER phrase, just as we have been doing all along. However, if we wished to note the exact moment when the pallet goes on to the on-buffer (perhaps for recording statistics) we would have to have some means of identifying that moment accurately, with more elaborate statements. Similarly, we can use a statement to take the pallet out of the set INOFFQ at the start of a movement activity and place it into INONQ after the movement, even although it may not be involved in the total duration. In other words, the variant of figures 11.4 to 11.9 which is written into the code may be simpler than the accompanying calculation of movement durations. As with so many matters in simulation, we should use the simplest form consistent with (a) ensuring that the conditions used in the logical tests are correctly set, and (b) that the values of variables used for recording statistics are correctly maintained.

In addition to the travel times there is also the time to pick up and put down the pallet. We have already defined TRTIME as the time to move a pallet from a buffer to a machine, and the time to move a pallet from the vehicle to the buffer should be virtually the same, so we can use that variable for both purposes. Further, since the buffer does not become free until after the transfer from buffer to vehicle is complete, we can add the TRTIME to the travelling time.

The amendments to the INTROD, MOVE and RETURN activities are fairly straightforward. Figure 11.18 gives the additional or amended code for these three activities and the revised definitions and data sections. In each activity, we calculate the distance from the current position of the vehicle to the location of the pallet to be

```
THERE ARE 2 LOADSTN SET LDFULL LDFREE WITH LDPOS
THERE ARE 6 ONBUFF SET ONOCCU ONVACA ONPARK GROUP 7 WITH ONPOS
THERE ARE 6 OFFBUF SET OFFOCC OFFVAC OFPARK WITH OFPOS
THERE ARE 1 VEHICL WITH VPOS

BEGIN INTROD

. . . .
EDIST = ABS ( VPOS OF VEHICL     - LDPOS OF LOADSTN D )
FDIST = ABS ( ONPOS OF ONBUFF B - LDPOS OF LOADSTN D )
EDURN = 1 + EDIST/SPEED + TRTIME
DURATION = 1 + FDIST/SPEED +TRTIME + EDURN
LOADSTN D FROM LDFULL INTO LDFREE AFTER EDURN
PALLET A FROM READY INTO INONQ AFTER DURATION
ONBUFF B FROM ONVACA INTO ONOCCU AFTER DURATION
TIME OF VEHICL = DURATION
VPOS OF VEHICL = ONPOS OF ONBUFF B

. . . .

BEGIN MOVE

. . . .
E = MCAT OF PALLET A
EDIST = ABS ( VPOS OF VEHICL     - OFPOS OF OFFBUF E )
FDIST = ABS ( ONPOS OF ONBUFF D - OFPOS OF OFFBUF E )
EDURN = 1 + EDIST/SPEED + TRTIME
DURATION = 1 + FDIST/SPEED +TRTIME + EDURN
OFFBUF E FROM OFFOCC INTO OFFVAC AFTER EDURN
ONBUFF D FROM ONVACA INTO ONOCCU AFTER DURATION
PALLET A FROM INOFFQ INTO INONQ AFTER DURATION
TIME OF VEHICL = DURATION
VPOS OF VEHICL = ONPOS OF ONBUFF D

. . . .

BEGIN RETURN

. . . .
EDIST = ABS ( VPOS OF VEHICL     - OFPOS OF OFFBUF D )
FDIST = ABS ( LDPOS OF LOADSTN C - OFPOS OF OFFBUF D )
EDURN = 1 + EDIST/SPEED + TRTIME
DURATION = 1 + FDIST/SPEED +TRTIME + EDURN
OFFBUF D FROM OFFOCC INTO OFFVAC AFTER EDURN
LOADSTN C FROM LDFREE INTO LDFULL AFTER DURATION
PALLET A FROM INOFFQ INTO FINSHD AFTER DURATION
TIME OF VEHICL = DURATION
VPOS OF VEHICL = LDPOS OF LOADSTN C

. . . .

DATA
LDPOS 0 10
ONPOS 30 60 90 120 150 180
OFPOS 40 70 100 130 160 190
SPEED 30
```

Figure 11.18   Amended code for INTROD, MOVE and RETURN activities for Model Eight.

moved, and the distance from that location to where the pallet is to be taken. EDIST and FDIST have been used for these empty and full distances respectively. Then we compute the time involved in each part of the journey, and add the time to transfer the pallet to or from the vehicle. EDURN has been used for the duration of the empty movement and DURATION for the total duration of the complete activity. Finally, we update the position of the vehicle.

In the case of the activities which involve moving an empty pallet, ie BINTROD, BMOVE, BRETURN, BLOADUP, BTRFROF and BUNBLOCK, the amendments are more substantial. There are three main changes to the logic, as well as the duration calculations. Firstly, whereas in Models Five and Seven, we did not have to identify a vacant pallet stand, because we knew there must be one somewhere and we were only using it temporarily, now we must locate the vacant pallet stand, which could be an on-queue, an off-queue or a load station, so that we can compute the distances and times accurately. Secondly, whereas in these models we placed the empty pallet at the initial location of the loaded pallet which required to be moved, now we can leave it at the vacant pallet stand. Thirdly, whereas the unblocking movement activities included moving the loaded pallet to its desired destination, as well as moving the empty pallet, now it will not be necessary to include moving the loaded pallet. Once the empty pallet has been moved, the destination of the loaded pallet stand will be vacant, and the normal movement activity INTROD, MOVE etc, can occur normally. A spin-off from this change will be that the activity counts for the B--- activities will count the number of empty pallet movements, and the normal movements will include all the loaded pallet movements, which seems a sensible arrangement. The B--- activities will now also have precisely two movements, a travelling empty to the location of the empty pallet and then a loaded move to the vacant pallet stand. Figure 11.19 gives the complete code for the revised versions of the BINTROD, BMOVE, BRETURN, BLOADUP, BTRFROF and BUNBLOCK activities.

11.7.2   Results

Table 11.1 gives selected results when the Model Eight is run with 13 pallets and 16 fixtures, and the comparable figures for Model Seven, from table 10.5.

On the whole, there no great differences between the two sets of figures. This might suggest that a simpler, quicker to develop, model gives just as good results as a more elaborate one. Such comparisons depend on many factors, and we would not be justified in drawing this conclusion, although it is a useful question to bear in mind when developing any model.

The differences between Models Seven and Eight are in the durations of the movement activities, and in the fact that the buffer positions are vacated immediately the pallet is transferred to the vehicle and not when the pallet is finally transferred to the destination pallet stand. Consequently the differences in results between Models Seven and Eight will be not only in a more accurate estimate of the vehicle utilisation, but we might also obtain some improved output because machines can discharge pallets slightly sooner than in Model Seven. There should also be differences in queue lengths.

```
     BEGIN BINTROD
C MOVE EMPTY PALLET FROM AN ONBUFF REQUIRED BY A READY PALLET
     TIME OF VEHICL LE 0
     FIND FIRST PALLET A IN READY
        C = CMPNO OF PALLET A
        N = NEXT OF COMPON C
        FIND FIRST ONBUFF B IN GROUP N
           ONBUFF B IN ONPARK
     FIND PALLET E IN PARKON
        MCAT OF PALLET E EQ B
     EDIST = ABS ( VPOS OF VEHICL - ONPOS OF ONBUFF B )
     EDURN = 1 + EDIST/SPEED + TRTIME
     CHAIN
        FIND LOADSTN P IN LDFREE
        FDIST = ABS ( ONPOS OF ONBUFF B  - LDPOS OF LOADSTN P )
        DURATION = 1 + FDIST/SPEED + TRTIME + EDURN
        LOADSTN P FROM LDFREE INTO LDFULL AFTER DURATION
        PALLET E FROM PARKON INTO EMTY AFTER DURATION
        MCAT OF PALLET E = 0
        LDAT OF PALLET E = P
        VPOS OF VEHICL = LDPOS OF LOADSTN P
        OR FIND OFFBUF P IN OFFVAC
        FDIST = ABS ( ONPOS OF ONBUFF B  - OFPOS OF OFFBUF P )
        DURATION = 1 + FDIST/SPEED + TRTIME + EDURN
        OFFBUF P FROM OFFVAC INTO OFPARK AFTER DURATION
        PALLET E FROM PARKON INTO PARKOF AFTER DURATION
        MCAT OF PALLET E = P
        VPOS OF VEHICL = OFPOS OF OFFBUF P
        OR FIND ONBUFF P IN ONVACA
        FDIST = ABS ( ONPOS OF ONBUFF B  - ONPOS OF ONBUFF P )
        DURATION = 1 + FDIST/SPEED + TRTIME + EDURN
        ONBUFF P FROM ONVACA INTO ONPARK AFTER DURATION
        PALLET E FROM PARKON INTO PARKON AFTER DURATION
        MCAT OF PALLET E = P
        VPOS OF VEHICL = ONPOS OF ONBUFF P
     TIME OF VEHICL = DURATION
     ONBUFF B FROM ONPARK INTO ONVACA AFTER EDURN
     ADD 1 TO BINTROD

     BEGIN BMOVE
C MOVE EMPTY PALLET FROM ONBUFF REQUIRED BY PALLET IN ONFFQ
     TIME OF VEHICL LE 0
     FIND FIRST PALLET A IN INOFFQ
        B = CMPNO OF PALLET A
        N = NEXT OF COMPON B
        N GT 0
        FIND FIRST ONBUFF D IN GROUP N
           ONBUFF D IN ONPARK
     FIND PALLET C IN PARKON
        MCAT OF PALLET C EQ D
     EDIST = ABS ( VPOS OF VEHICL - UNPOS OF ONBUFF D )
     EDURN = 1 + EDIST/SPEED + TRTIME
```

Figure 11.19   Amended code for the routines for moving empty pallets.

```
    CHAIN
      FIND LOADSTN P IN LDFREE
      FDIST = ABS ( ONPOS OF ONBUFF D - LDPOS OF LOADSTN P )
      DURATION = 1 + FDIST/SPEED + TRTIME + EDURN
      LOADSTN P FROM LDFREE INTO LDFULL AFTER DURATION
      PALLET C FROM PARKON INTO EMTY AFTER DURATION
      MCAT OF PALLET C = 0
      LDAT OF PALLET C = P
      VPOS OF VEHICL = LDPOS OF LOADSTN P
    OR FIND OFFBUF P IN OFFVAC
      FDIST = ABS ( ONPOS OF ONBUFF D - OFPOS OF OFFBUF P )
      DURATION = 1 + FDIST/SPEED + TRTIME + EDURN
      OFFBUF P FROM OFFVAC INTO OFPARK AFTER DURATION
      PALLET C FROM PARKON INTO PARKOF AFTER DURATION
      MCAT OF PALLET C = P
      VPOS OF VEHICL = OFPOS OF OFFBUF P
    OR FIND ONBUFF P IN ONVACA
      FDIST = ABS ( ONPOS OF ONBUFF D - ONPOS OF ONBUFF P )
      DURATION = 1 + FDIST/SPEED + TRTIME + EDURN
      ONBUFF P FROM ONVACA INTO ONPARK AFTER DURATION
      PALLET C FROM PARKON INTO PARKON AFTER DURATION
      MCAT OF PALLET C = P
      VPOS OF VEHICL = ONPOS OF ONBUFF P
    ONBUFF D FROM ONPARK INTO ONVACA AFTER EDURN
    TIME OF VEHICL = DURATION
    ADD 1 TO BMOVE

    BEGIN BTRFROF
C MOVE AN EMPTY PALLET FROM OFFBUF OF MACHINE WAITING TO TRFROF A PALLET *
    TIME OF VEHICL LE 0
    FIND FIRST PALLET A IN COMPLE
      B = MCAT OF PALLET A
      OFFBUF B IN OFPARK
    FIND PALLET C IN PARKOF
      MCAT OF PALLET C EQ B
    EDIST = ABS ( VPOS OF VEHICL      - OFPOS OF OFFBUF B )
    EDURN = 1 + EDIST/SPEED + TRTIME
    CHAIN
      FIND FIRST LOADSTN P IN LDFREE                           *
      FDIST = ABS ( OFPOS OF OFFBUF B - LDPOS OF LOADSTN P )
      DURATION = 1 + FDIST/SPEED + TRTIME + EDURN
      LOADSTN P FROM LDFREE INTO LDFULL AFTER DURATION
      PALLET C FROM PARKOF INTO EMTY AFTER DURATION
      MCAT OF PALLET C = 0
      LDAT OF PALLET C = P
      VPOS OF VEHICL = LDPOS OF LOADSTN P
```

Figure 11.19   (continued).

```
          OR FIND OFFBUF D IN OFFVAC
          FDIST = ABS ( OFPOS OF OFFBUF D - OFPOS OF OFFBUF B )
          DURATION = 1 + FDIST/SPEED +TRTIME + EDURN
          PALLET C FROM PARKOF INTO PARKOF AFTER DURATION
          OFFBUF D FROM OFFVAC INTO OFPARK AFTER DURATION
          MCAT OF PALLET C = D
          VPOS OF VEHICL = OFPOS OF OFFBUF D
          OR FIND ONBUFF D IN ONVACA
          FDIST = ABS ( ONPOS OF ONBUFF D - OFPOS OF OFFBUF B )
          DURATION = 1 + FDIST/SPEED +TRTIME + EDURN
          PALLET C FROM PARKOF INTO PARKON AFTER DURATION
          ONBUFF D FROM ONVACA INTO ONPARK AFTER DURATION
          MCAT OF PALLET C = D
          VPOS OF VEHICL = ONPOS OF ONBUFF D
        TIME OF VEHICL = DURATION
        OFFBUF B FROM OFPARK INTO OFFVAC AFTER EDURN
        ADD 1 TO BTRFROF

        BEGIN BRETURN
C MOVE AN EMPTY PALLET FROM A LOADSTN NEEDED FOR A PALLET IN INOFFQ
        TIME OF VEHICL LE 0
        FIND FIRST PALLET A IN INOFFQ
          B = CMPNO OF PALLET A
          NEXT OF COMPON B EQ 0
        FIND PALLET E IN EMTY
        C = LDAT OF PALLET E
        EDIST = ABS ( VPOS OF VEHICL - LDPOS OF LOADSTN C )
        EDURN = 1 + EDIST/SPEED + TRTIME
        CHAIN
          FIND OFFBUF P IN OFFVAC
          FDIST = ABS ( LDPOS OF LOADSTN C - OFPOS OF OFFBUF P )
          DURATION = 1 + FDIST/SPEED + TRTIME + EDURN
          OFFBUF P FROM OFFVAC INTO OFPARK AFTER DURATION
          PALLET E FROM EMTY INTO PARKOF AFTER DURATION
          MCAT OF PALLET E = P
          LDAT OF PALLET E = 0
          VPOS OF VEHICL = OFPOS OF OFFBUF P
          OR FIND ONBUFF P IN ONVACA
          FDIST = ABS ( LDPOS OF LOADSTN C - ONPOS OF ONBUFF P )
          DURATION = 1 + FDIST/SPEED + TRTIME + EDURN
          ONBUFF P FROM ONVACA INTO ONPARK AFTER DURATION
          PALLET E FROM EMTY INTO PARKON AFTER DURATION
          MCAT OF PALLET E = P
          LDAT OF PALLET E = 0
          VPOS OF VEHICL = ONPOS OF ONBUFF P
        LOADSTN C FROM LDFULL INTO LDFREE AFTER EDURN
        TIME OF VEHICL = DURATION
        ADD 1 TO BRETURN
```

Figure 11.19   (continued).

```
BEGIN BLOADUP
TIME OF VEHICL LE O
TIME OF OPERAT LE O
FIND FIRST COMPON B IN LDWAIT
EMTY EMPTY
FIND LOADSTN C IN LDFREE
CHAIN
  FIND FIRST PALLET A IN PARKON
  E = MCAT OF PALLET A
  EDIST = ABS ( VPOS OF VEHICL - ONPOS OF ONBUFF E )
  EDURN = 1 + EDIST/SPEED + TRTIME
  ONBUFF E FROM ONPARK INTO ONVACA AFTER EDURN
  FDIST = ABS ( LDPOS OF LOADSTN C - ONPOS OF ONBUFF E )
  DURATION = 1 + FDIST/SPEED +TRTIME + EDURN
  PALLET A FROM PARKON INTO EMTY AFTER DURATION
  OR FIND PALLET A IN PARKOF
  E = MCAT OF PALLET A
  EDIST = ABS ( VPOS OF VEHICL - OFPOS OF OFFBUF E )
  EDURN = 1 + EDIST/SPEED + TRTIME
  OFFBUF E FROM OFPARK INTO OFFVAC AFTER EDURN
  FDIST = ABS ( LDPOS OF LOADSTN C - OFPOS OF OFFBUF E )
  DURATION = 1 + FDIST/SPEED +TRTIME + EDURN
  PALLET A FROM PARKOF INTO EMTY AFTER DURATION
MCAT OF PALLET A = O
LDAT OF PALLET A = C
LOADSTN C FROM LDFREE INTO LDFULL AFTER DURATION
TIME OF VEHICL = DURATION
VPOS OF VEHICL = LDPOS OF LOADSTN C
ADD 1 TO BLOADUP

BEGIN BUNBLOCK
TIME OF VEHICL LE O
LDFREE EMPTY
TIME OF OPERAT LE O
FIND FIRST OFFBUF A IN OFFVAC
FIND FIRST PALLET B IN INOFFQ
    M = MCAT OF PALLET B
    MACHIN M IN OCCUPI
    FIND FIRST PALLET D IN COMPLE WITH MCAT EQ M
    C = CMPNO OF PALLET B
    NEXT OF COMPON C EQ O
MCAT OF PALLET B = A
EDIST = ABS ( VPOS OF VEHICL     - OFPOS OF OFFBUF M )
FDIST = ABS ( OFPOS OF OFFBUF A - OFPOS OF OFFBUF M )
EDURN = 1 + EDIST/SPEED + TRTIME
DURATION = 1 + FDIST/SPEED +TRTIME + EDURN
OFFBUF A FROM OFFVAC INTO OFFOCC AFTER DURATION
OFFBUF M FROM OFFOCC INTO OFFVAC AFTER EDURN
VPOS OF VEHICL = OFPOS OF OFFBUF A
TIME OF VEHICL = DURATION
ADD 1 TO BUNBLOCK
```

Figure 11.19   (continued).

Table 11.1.   Results comparing Model Eight (calculated movement durations) compared with Model Seven (constant durations) for the case with 13 pallets and 16 fixture

Facility occupancy and utilisation (%)

Facility	Occupancy		Utilisation	
	Model Seven	Model Eight	Model Seven	Model Eight
Machine 1	55	54	48	46
Machine 2	81	78	74	66
Machine 3	60	70	56	61
Machine 4	60	48	55	41
Machine 5	34	41	32	36
Machine 6	56	71	49	59
All machines	58	60	53	52
Vehicle			54	75
Operator			43	41
Load Stations	69	90		

Average queue lengths

	Activity	Model Seven	Model Eight
Components:	LDWAIT	0.275	0.227
	FXWAIT	4.416	5.326
	ONPALL	5.199	6.136
	ONBENC	0.002	0.000
Pallets:	READY	0.090	0.124
	INONQ	1.122	1.304
	COMPLE	0.096	0.462
	INOFFQ	0.173	0.669
	FINSHD	0.030	0.012
	EMTY	0.903	1.211
	PARKON	1.626	1.231
	PARKOF	4.734	3.901
On-queue buffers:	ONOCCU	1.291	1.304
	ONVACA	2.943	3.095
Off-queue buffers:	OFFOCC	0.359	0.669
	OFFVAC	0.527	0.899
Machines:	OCCUPI	3.271	3.592
	IDLE	2.540	2.376

Component production

Component type	Number required	Number made		Average time in system	
Model		Seven	Eight	Seven	Eight
1	10	9	10	4422	5044
2	8	9	9	2040	2600
3	6	7	7	2050	2450
4	4	3	2	1600	1900
5	6	6	6	2075	2225
6	2	1	1	1200	1600
overall	36	35	35	2648	3067

Table 11.1.   (*continued*)

Activity counts

Activity	Model Seven	Model Eight
INTROD	88	118
BINTROD	33	23
RETURN	74	116
BRETURN	45	85
LOADUP	85	95
BLOADUP	13	5
TRFROF	12	306
BTRFROF	299	256
MOVE	129	190
BMOVE	65	29
BUNBLOCK	0	0
PUTON	97	92
TAKEOFF	96	93
RECLAMP	24	23
UNLOAD	96	93
TRFRON	313	307
ARRIVE	35	35

Fixture utilisation (%)

Fixture number	Fixture type	Utilisation	
		Model Seven	Model Eight
1	11	59	81
2	12	32	37
3	13	34	39
4	14	16	17
5	15	22	26
6	16	6	10
7	21	100	100
8	22	55	63
9	23	65	81
10	24	12	17
11	25	32	43
12	26	6	7
13	31	49	45
14	33	30	36
15	34	7	8
16	35	62	65
Overall		37	42

A closer look at the results shows that there are some interesting differences. The vehicle utilisation is 75% instead of 54%. This probably reflects an inconsistency between the average movement time used in Model Seven, with the values used to compute the movement times in Model Eight. If the estimate of the average had been more accurate this difference should not occur. However, when we interpret the other values, we may find longer queues and times in the system because the work load on the transporter has increased substantially from the previous value. The occupancy of the load stations has increased from 69% to 90%. This may be a result of the vehicle utilisation.

There are small differences in queue lengths, few of which are worth commenting upon. The average lengths of the PARKON and PARKOF queues are reduced, which may reflect the more accurate modelling of movement activities. However, the INOFFQ is longer, no doubt because the extra load on the transporter has outweighed the effects of more accurate modelling. The average number of components on a pallet has increased from 4.9 to 5.4, but is still small in relation to the number of pallets available. The number of components waiting for fixtures is larger, which may be due to the generally greater utilisation of the fixtures.

Output is the same as in Model Seven, but the average times in the system are longer than for Model Seven. This would follow directly from longer movement times, and from the extra utilisation of the vehicle. It would also cause the greater fixture utilisation.

The activity counts show little change. One difference is that, because of the new way of writing the B——— activities, the counts for the normal activities should be compared to the sum of the normal and B——— activities in the previous model. There is a noticeable difference in the proportion of BRETURN and BTRFROF movements. The number of BTRFROF activities has been reduced and the number of BRETURN moves has increased. Both of these results are predictable. We would expect a reduction in BTRFROF moves due to the more accurate modelling of the transfer of pallets to the vehicle. The increase in BRETURN moves would be expected from the increased occupancy of the load stations. The total number of B——— moves is approximately the same as before.

One of the most interesting results is that the work load on the machines was much more evenly spread than in Model Seven. For example, in Model Seven, machines 2 and 3 had utilisations of 0.74 and 0.56 respectively, whereas in Model Eight the figures became 0.66 and 0.61. There is no easy explanation for this result, which again illustrates the complex interactions between entities in complex systems.

Runs of Model Eight were performed with 16 and 20 fixtures and 9, 11 and 13 pallets, as for Model Seven. The results followed the pattern for Model Seven, table 10.6, with the differences discussed above consistently present. The detailed results are not quoted.

11.7.3 Discussion

The more astute reader may have formed the impression that the empty pallet movement routines (and the normal movement routines) are more elaborate than necessary. The logic described took the buffer position to which a pallet was to be moved out of the set of vacant locations at the start of the movement, so that no other pallet may be moved to it while the movement took place. Similarly, the pallet to be moved was taken out of the set of pallets waiting to be moved so that the model did not try to move it twice. Strictly, these cannot happen if there is only one vehicle, since the logic will not search for another pallet to move while one movement is taking place. However, if the system had two or more vehicles, then it would be possible for two vehicles to be looking simultaneously for pallets needing to be moved. In that case, it would be essential that we removed pallet from waiting to be moved set and pallet stands from the vacant set at the start of the move activity logic.

11.8   MODEL NINE—SYSTEM WITH A VEHICLE HAVING TWO LOAD POSITIONS

As already observed, in many systems with linear tracks and rail-guided vehicles, the vehicle has two load positions. It has already been observed that if the vehicle has two load positions the coding of the unblocking activities, BINTROD, BMOVE, etc in Model Seven could be simplified. It would not be necessary to move the empty pallet to a vacant pallet stand before picking up the loaded pallet. Instead of three movements, two with an empty pallet and one with a loaded pallet, only two would be necessary. These two movements would move the loaded pallet to its destination, pick up the empty pallet, drop off the loaded pallet and then move the empty pallet.

Rather than present a model incorporating this change to the logic, the reader is invited to amend the coding of the B––– activities in Model Seven to suit this case. Note that this is a very restricted exploitation of the vehicle's second load position, and excludes most of the considerations in section 11.5. Model Five, which is a simpler case, could be similarly amended.

The more ambitious reader might like to consider how Model Eight could be amended if the vehicle has two load positions.

11.9   MODEL TEN—A MAST MODEL OF A SYSTEM WITH SEVERAL VEHICLES

The complete activity cycle diagram and the coding of a model of a system with several vehicles and a complex track layout would be both extensive and difficult to follow without a detailed description. Instead of presenting a model of this type we will present the modelling of such a system using MAST, or more accurately using the SPAR-MAST-BEAM suite.[1] Since this software amounts to over 0.5 Mbytes, any model developed and presented here would have been rather trivial in comparison.

In chapter 7 a description was given of the Hattersley Newman Hender system, which had eight vehicles and a complex track layout. It is proposed to present a MAST model of a simplified version of that system. One of the simplifications we will make is that instead of having some 2000 component types which fall into three families we will just have three component types, and instead of real data on component processing times and production quantities we will use hypothetical data. We will have only four vehicles. We will also simplify the track layout and the number of buffer storage locations. At the end of this section we will discuss whether the real system could have been modelled by MAST in its full detail.

11.9.1   Defining the system to SPAR

When execution of SPAR begins, the initial screen give details of the licence holder and general information about the package. Thereafter the process is user-friendly and menu-driven. The main menu is shown in Figure 11.20. We select the new system evaluation option. Several subsequent screens invite the user to name the project, in this case ZHNH, give a time horizon for utilisation calculations, and then give details of components, workstations and transporters. Figure 11.21 shows the screen when the data for component 1 has been entered. Provided their name begins BUF..., SPAR will recognise buffer stations and automatically assign an operation time of zero.

System Planning for Aggregate Requirements

1) START NEW SYSTEM EVALUATION
2) EDIT EXISTING SYSTEM DESCRIPTION
3) PERFORM CAPACITY EVALUATION OF EXISTING SYSTEM
4) GENERATE DATA INPUT FOR MAST SIMULATION
5) GENERATE TOOLING CONFIGURATION
6) TERMINATE PROGRAM EXECUTION

SELECT OPTION

Figure 11.20   SPAR main menu.

System Planning for Aggregate Requirements

PART ID(10 CHAR) PART 1
PRODUCTION REQUIREMENT 30
PARTS PER PALLET 1

ENTER OPERATION DESCRIPTION(10 CHAR) AND OPERATION TIME
TYPE END TO TERMINATE OPERATION ROUTE

1	OPERATION ID:	LOAD	TIME:	2.00
2	OPERATION ID:	BUFF1	TIME:	0.00
3	OPERATION ID:	FM100	TIME:	30.00
4	OPERATION ID:	BUFF2	TIME:	0.00
5	OPERATION ID:	OUTFACE	TIME:	12.00
6	OPERATION ID:	WASH	TIME:	2.00
7	OPERATION ID:	LOAD	TIME:	2.00

STORAGE FACTOR 0.5

CORRECT PART ID(I), PRODUCTION(P), PARTS PER PALLET(A) OR FACTOR(F)

Figure 11.21   SPAR component definition screen.

STATION CAPACITY ANALYSIS (SPAR)

STATION TYPE	NUMBER AVAILABLE	TOTAL –TIME AVAILABLE	REQUIRED	STATION UTILIZATION	SLACK PER STATION
LOAD	2	960.0	540.0	56.3	210.0
BUFF1	1	480.0	90.0	18.8	390.0
FM100	5	2400.0	2640.0	110.0	−48.0
BUFF2	1	480.0	60.0	12.5	420.0
OUTFACE	2	960.0	600.0	62.5	180.0
WASH	1	480.0	360.0	75.0	120.0
BUFF3	1	480.0	30.0	6.3	450.0
ASSEMBLE	1	480.0	480.0	100.0	0.0

DO YOU WANT TO PRINT THIS REPORT ?

Figure 11.22   SPAR workstation capacity analysis printout.

MANUFACTURING SYSTEM CAPACITY ANALYSIS (SPAR)
ZHNH MANUFACTURING SYSTEM

INSUFFICIENT STATION CAPACITY DETECTED
FEASIBLE DESIGN OF THE FMS CAN BE ACHIEVED BY:

1.   EXTEND PLANNING HORIZON
2.   REQUIRED PRODUCTION REDUCTION
3.   ADD STATIONS
4.   ALTER PART ROUTES

FEASIBLE PRODUCTION QUANTITIES ARE CALCULATED BY A UNIFORM
PERCENT REDUCTION OF REQUIRED AMOUNT FOR EACH PART TYPE.
IF THESE FEASIBLE LEVELS ARE NOT ACCEPTABLE, THEN
THE NUMBER OF STATIONS NECESSARY TO MEET THE DESIRED
PRODUCTION IS FOUND.
IF THE ADDITIONAL STATIONS ARE NOT ACCEPTABLE, THEN THE
CAPACITY ANALYSIS IS TERMINATED.
ALTERATION OF THE PART ROUTES IS NOT PERFORMED
BY THIS PROGRAM AND CAN BE DONE BY CHANGING THE SYSTEM
DESCRIPTION THROUGH THE EDITING FEATURES IN SPAR.

TYPE ANY KEY TO CONTINUE

Figure 11.23   SPAR alternative decisions screen.

The storage factor is an estimate of the amount of time which a pallet holding the part will be held in queues or buffer storage as a proportion of the machining time. It is used by SPAR to calculate the number of pallets which will be needed. The user specifies the number available of each type of work station and transporter, the transportation time and the time for transferring pallets between workstation and transporter.

Work load calculations are then performed, load and capacity compared and an analysis presented. Figure 11.22 gives part of the accompanying printout. This shows that there is insufficient capacity at the FM100 workstation to produce the requirements within the time horizon. The utilisation would be over 100%, and there is negative slack. SPAR presents alternative courses of action to the user, Figure 11.23. If the second alternative is accepted SPAR reduces the quantities in proportion, down to a feasible level, and then re-calculates the station workloads and other statistics. Figure 11.24 shows final results.

```
                    MANUFACTURING SYSTEM CAPACITY ANALYSIS (SPAR)
                             ZHNH MANUFACTURING SYSTEM

              PRODUCTION REQUIREMENTS ADJUSTED FOR FEASIBILITY
                     ORIGINAL        FEASIBLE       PERCENT
                     PRODUCTION      PRODUCTION     REMAINING
                     ----------      ----------     ----------
        PART 1          30              27             90
        PART 2          30              27             90
        PART 3          30              27             90


        ARE THESE ADJUSTED PRODUCTION LEVELS ACCEPTABLE? Y
```

```
                        STATION CAPACITY ANALYSIS (SPAR)
                        ---------------------------------
```

STATION TYPE	NUMBER AVAILABLE	TOTAL-TIME AVAILABLE	REQUIRED	STATION UTILIZATION	SLACK PER STATION
LOAD	2	960.0	486.0	50.6	237.0
BUFF1	1	480.0	81.0	16.9	399.0
FM100	5	2400.0	2376.0	99.0	4.0
BUFF2	1	480.0	54.0	11.3	426.0
OUTFACE	2	960.0	540.0	56.3	210.0
WASH	1	480.0	324.0	67.5	156.0
BUFF3	1	480.0	27.0	5.6	453.0
ASSEMBLE	1	480.0	432.0	90.0	48.0

```
                    TRANSPORTER CAPACITY ANALYSIS (SPAR)
                    -------------------------------------
```

NUMBER TRANSPORT	TOTAL-TIME AVAILABLE	REQUIRED	TRANSPORT UTILIZED	SLACK PER TRANSPORT
4.0	1920.0	1539.0	80.2	95.3

```
                        STORAGE CAPACITY ANALYSIS (SPAR)
                        ---------------------------------
```

PART ID	PALLET CYCLE	PERCENT STATION	PERCENT STOR & TRAN	PRODUCTION PER PALLET	PALLETS REQUIRED
PART 1	90.0	53.0	47.0	5.3	5.1
PART 2	112.5	51.0	49.0	4.3	6.3
PART 3	61.5	54.0	46.0	7.8	3.5

```
        AVERAGE NUMBER OF PALLETS IN SYSTEM      14.85
```

Figure 11.24   SPAR final analysis reports.

```
SYSTEM        SPAR   C:ZHNH        1 08 11 1987
;
;
;
CONT,480*
;------------------------------------------------------------
;============================================================
;------------------------------------------------------------
;
;
;     PART #PART 1 AND ROUTE
;
PART,1,1,8,2,2,0,30,1*
ROUT,1,1,1,2,2,2*                                    LOAD
BUFF,1,2,3*                                          BUFF1
ROUT,1,3,4,30,5,30,6,30,7,30,8,30*                  FM100
BUFF,1,4,9*                                          BUFF2
ROUT,1,5,10,12,11,12*                               OUTFACE
ROUT,1,6,12,2*                                       WASH
ROUT,1,7,1,2,2,2*                                    LOAD
;------------------------------------------------------------
;
;     PART #PART 2 AND ROUTE
;
PART,2,2,8,2,2,0,30,1*
ROUT,2,1,1,2,2,2*                                    LOAD
BUFF,2,2,3*                                          BUFF1
ROUT,2,3,4,28,5,28,6,28,7,28,8,28*                  FM100
ROUT,2,4,12,2*                                       WASH
BUFF,2,5,13*                                         BUFF3
ROUT,2,6,14,15*                                      ASSEMBLE
BUFF,2,7,9*                                          BUFF2
ROUT,2,8,10,6,11,6*                                 OUTFACE
ROUT,2,9,12,2*                                       WASH
ROUT,2,10,1,2,2,2*                                   LOAD
;------------------------------------------------------------
;
;     PART #PART 3 AND ROUTE
;
PART,3,3,8,2,2,0,30,1*
ROUT,3,1,1,2,2,2*                                    LOAD
BUFF,3,2,3*                                          BUFF1
ROUT,3,3,4,27,5,27,6,27,7,27,8,27*                  FM100
ROUT,3,4,12,2*                                       WASH
ROUT,3,5,1,2,2,2*                                    LOAD
;------------------------------------------------------------
;============================================================
;------------------------------------------------------------
;
;
;     PALLET DESCRIPTION
;
PALL,1,0,8,1*
PALL,2,0,8,2*
PALL,3,0,8,3*
;------------------------------------------------------------
;============================================================
;------------------------------------------------------------
;
;
;     STATION  DESCRIPTION
;
STAT,1,0,0,0,1,1,1,29,29*                            LOAD
STAT,2,0,0,0,1,1,1,31,31*                            LOAD
STAT,3,0,0,6,1,0.5,0.01,3,3*                         BUFF1
STAT,4,0,0,1,5,0.5,0.5,7,7*                          FM100
STAT,5,0,0,1,5,0.5,0.5,9,9*                          FM100
STAT,6,0,0,1,5,0.5,0.5,11,11*                        FM100
STAT,7,0,0,1,5,0.5,0.5,13,13*                        FM100
STAT,8,0,0,1,5,0.5,0.5,15,15*                        FM100
STAT,9,0,0,13,1,0.5,0.01,20,20*                      BUFF2
STAT,10,0,0,1,5,0.5,0.5,17,17*                       OUTFACE
```

Figure 11.25   MAST data deck.

```
STAT,11,0,0,1,5,0.5,0.5,23,23*                    OUTFACE
STAT,12,3,3,0,1,0.5,0.5,24,25*                    WASH
STAT,13,0,0,3,1,0.5,0.01,26,26*                   BUFF3
STAT,14,0,0,0,1,0.5,0.5,27,27*                    ASSEMBLE
;-------------------------------------------------------------
;=============================================================
;-------------------------------------------------------------
;
;
;    CART DESCRIPTION
;
CART,1,4,1,60,2*
;-------------------------------------------------------------
;=============================================================
;-------------------------------------------------------------
;
;
;    TRACK LAYOUT
;
TRAC,1,0,0,2,10*
TRAC,2,0,0,3,10*
TRAC,3,3,0,4,10*
TRAC,4,0,0,5,10*
TRAC,5,0,0,6,10*
TRAC,6,0,0,7,10*
TRAC,7,4,0,8,10*
TRAC,8,0,0,9,10*
TRAC,9,5,0,10,10*
TRAC,10,0,0,11,10*
TRAC,11,6,0,12,10*
TRAC,12,0,0,13,10*
TRAC,13,7,0,14,10*
TRAC,14,0,0,15,10*
TRAC,15,8,0,16,10*
TRAC,16,0,0,17,10*
TRAC,17,10,0,18,10,40,10*
TRAC,18,0,0,19,10*
TRAC,19,0,0,20,10*
TRAC,20,9,0,21,10*
TRAC,21,0,0,22,10*
TRAC,22,0,0,23,10*
TRAC,23,11,0,24,10*
TRAC,24,12,0,25,10*
TRAC,25,12,0,26,10*
TRAC,26,13,0,27,10*
TRAC,27,14,0,28,10*
TRAC,28,0,0,29,10,42,10*
TRAC,29,1,0,30,10*
TRAC,30,0,0,31,10,43,10*
TRAC,31,2,0,32,10*
TRAC,32,0,0,1,10,33,10*
TRAC,33,0,0,34,10*
TRAC,34,0,0,6,10,35,10*
TRAC,35,0,0,8,10,36,10*
TRAC,36,0,0,10,10,37,10*
TRAC,37,0,0,12,10,38,10*
TRAC,38,0,0,14,10,39,10*
TRAC,39,0,0,16,10,40,10*
TRAC,40,0,0,41,10*
TRAC,41,0,0,23,10,42,10*
TRAC,42,0,0,43,10,44,10*
TRAC,43,0,0,33,10*
TRAC,44,0,0,36,10*
;-------------------------------------------------------------
;=============================================================
;-------------------------------------------------------------
;
;
;    SIMULATION PARAMETERS
;
PARM,24,3,2,2*
MAST*
```

Figure 11.25   (continued).

11.9.2   Running the MAST simulation model

One of the options on SPAR's main menu is to generate an input data file for the MAST simulation package from the data already supplied. To that data, various other details must be added, such as the layout of the track, the type of buffers at machines, the types of pallets and the parts which they can handle, the types of vehicles and their speed, and the decision rules to be used in running the model. The track layout is defined by means of decision points, which define junctions, locations of workstations and so on. For each decision point we give a list of the other points which are adjacent to it, and the distance to each of these points. MAST can then calculate the paths between any two points and the time which a transporter will take to make the journey. We also specify the location of each workstation by stating the decision point at which pallets are delivered to it and the point from which pallets can be picked up. The decision rules deal with such features as selecting the next operation on the part, selecting a workstation for the next operation, selecting a vehicle for a movement, scheduling vehicle movements, etc.

Once the input data has been prepared, the MAST program can be run. The input data can be printed out by MAST, Figure 11.25. During the MAST run no information is displayed on the screen other than the current clock value. Status reports can be produced at intervals, and details of each event can be written to the event trace file.

11.9.3   Graphical Animation using BEAM

At the end of the run the final MAST reports can be printed, or they can be displayed by BEAM, MAST's animation facility. BEAM involves two stages. First, the static background and track layout must be defined. The second stage is running the animation, in which BEAM reads the MAST event trace file and updates the display as the model progresses.

The layout of the track is defined by specifying pixel co-ordinates of each decision point. The sections of track connecting the decision points are then defined. Each workstation is located on the screen by specifying its location in relation to decision points, and selecting an icon from a menu. Figure 11.26 shows the screen for locating workstations, with the responses given for one of the load stations. BEAM provides two display modes, high resolution using enhanced graphics in two colours or medium resolution in four colours. In this case, the number of decision points involved in the model requires high resolution.

Figure 11.27 shows the screen display at a point in time while running the animation. Results statistics can be displayed at any time during the animation run or at the end of the run. Figure 11.28 shows a typical pie chart of station performance statistics, Figure 11.29 the equivalent for part statistics, and Figure 11.30 shows the output in relation to target for each type of part. Various other forms of chart are available. Depending on the number of parts and workstations these reports can extend over many screens.

Background and Enhanced Animation for MAST

Station 1 Description

1. Enter three character id   LOA   Group Name: LOAD
Select footprint (HMC,VMC,HTC,VTC,HS,BORE,PLANE,ROBOT) : HS

2. Enter Decision Point Number where parts
are DELIVERED to this station 29

3. Enter Decision Point Numberwhere parts
are PICKED UP from this station 29

4. Enter the number of storage positions
for IN—COMING parts   0

5. Enter the number of storage positions
for OUT—GOING parts   0

6. Is this station located ABOVE , BELOW , RIGHT or LEFT
of the delivery decision point   Enter A,B,R,or L: A

ITEM NUMBER TO CORRECT? N

Figure 11.26   BEAM screen for workstation definition.

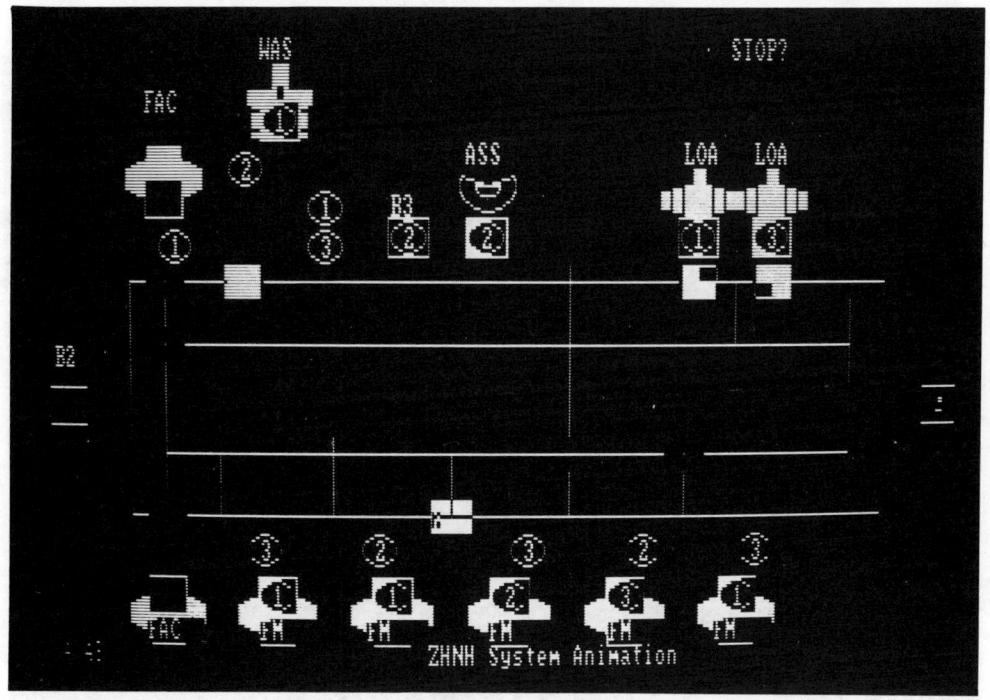

Figure 11.27   BEAM animation screen display.

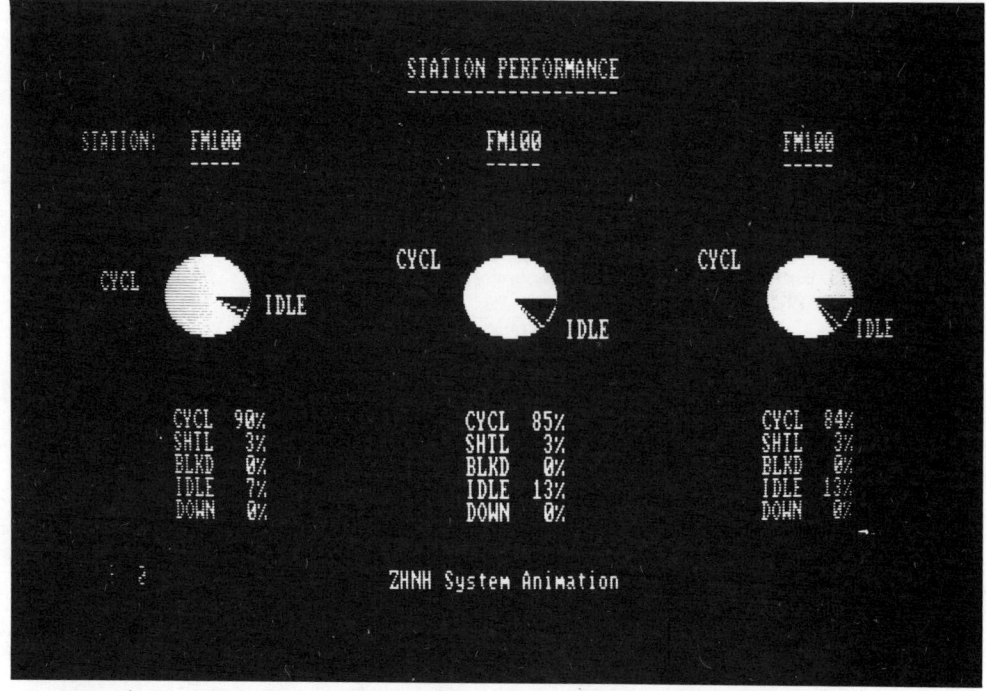

Figure 11.28   BEAM workstation performance chart.

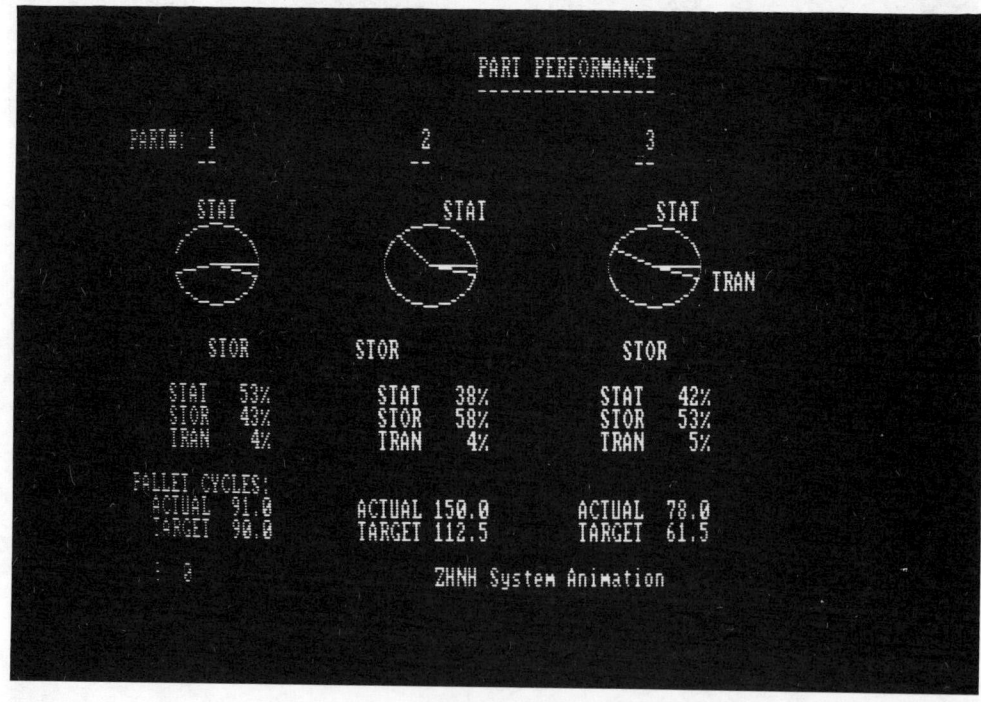

Figure 11.29   BEAM part performance chart.

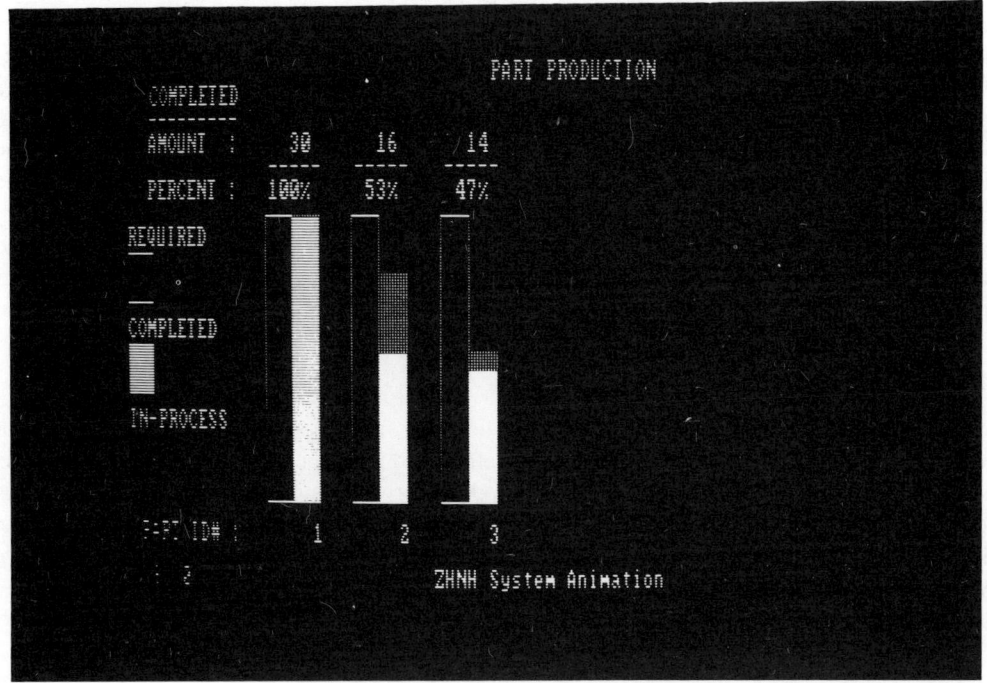

Figure 11.30   BEAM part production chart.

11.9.4   Comparison of MAST results with SPAR predictions

We can compare the SPAR predictions with the MAST results. From figure 11.28 the utilisation of three of the FM100 machines is found to be 90%, 85% and 84%, and for the other two the figures were 78% and 72% respectively. These are below the estimated value of 99% given in figure 11.24. Part of the explanation is that not all the parts have been produced. Figure 11.30 shows that although all the parts required of type 1 have been completed, only 53% of type 2 have been made and only 47% of type 3. This is a result of the part introduction rule specified, which attempted to introduce a type 2 part after a type 1, and a type 3 after a type 2. This rule enables MAST to produce parts in sets for assembly. It may also indicate that the model had not reached a steady state before statistics were collected or that the run was not long enough. We also notice that the first FM100 machine has the highest utilisation and the others have progressively less utilisation. This will be the result of the decision rule which schedules pallets to machines. When we asked the system to produce a MAST data file from the data supplied to SPAR, it automatically gave all the FM100 machines as suitable for each FM100 operation, but numbered them in sequence 1 through 5, with the result that the first one was selected if it was available, and the fifth only when all the other four were in use. We could smooth this out by a different rule, or by editing the file so that one type of part is sent to one machine as its first choice, and each other part type to a different machine. This might be very useful if the tool assignments had to be considered. Experimenting with alternative decision rules is advisable, to study the effects and to gain more insight into the workings of the package.

Figure 11.29 enables the time in the system for each type of part to be compared with SPAR's prediction. The pallet cycle times for type 1 agree very closely, 91.0 versus 90.0, but for the other part types the ratios are less good, 150.0 against 112.5 and 78.0 against 61.5. The actual values are affected by the decision rules used in the model and target values are computed using the storage factor defined for SPAR. These storage factor values were merely guesses, and may have been unrealistic. Another factor in the comparison is that in SPAR travelling times were assumed to be a constant 3 minutes, whereas in MAST the times are computed accurately. A possible effect of the increased cycle times is that more pallets might be needed. Some experimentation with different rules may help to determine what is achievable.

11.9.5 Discussion

Significant simplifications were made in defining this system. We should consider whether the full complexity of the real system could have been modelled using MAST, since a common failing of many generic models is that they cannot handle specific details of a real system. A definitive answer can only be given by someone who has an intimate knowledge of both the real system and the MAST package. Consequently, we will confine comments to a general discussion of the main simplifications and how the definiton of the model could be made more accurate.

1. Track layout

The layout defined here was a considerable simplification of the real layout. In principle, it is possible to define as extensive and complex a layout as necessary, merely by adding more decision points. However, in practice there is a limit to the number of decision points which MAST can handle. On the PC version used here, this limit is 80. The number of points was somewhat below this figure, but in running the model, it was found that the complexity of the track had to be reduced because the number of possible routes between workstations became too great for MAST's internal workings. Figure 11.31 shows the decision points on the layout which was specified. By comparing this with the layout of the system shown in chapter 7, it will be seen that it was not possible to define most of the links between inner and outer rectangles of track. These limits are therefore a result of array size definitions and of the total memory available in the computer. On more powerful computers there should be no such restriction.

2. Buffer stores

In the simplified model there were only three buffer stores, of varying capacity, in the system, whereas the real system has six groups of three or six pallet capacity. There would have been no problem in giving MAST the correct details, merely by additional input data. However, this would have compounded the problem mentioned above about the number of decision points and routes between them.

3. Part types and routings

The real system has some 2000 types of part, which fall into three basic families. This number greatly exceeds MAST's limit on the number of part types. The volume

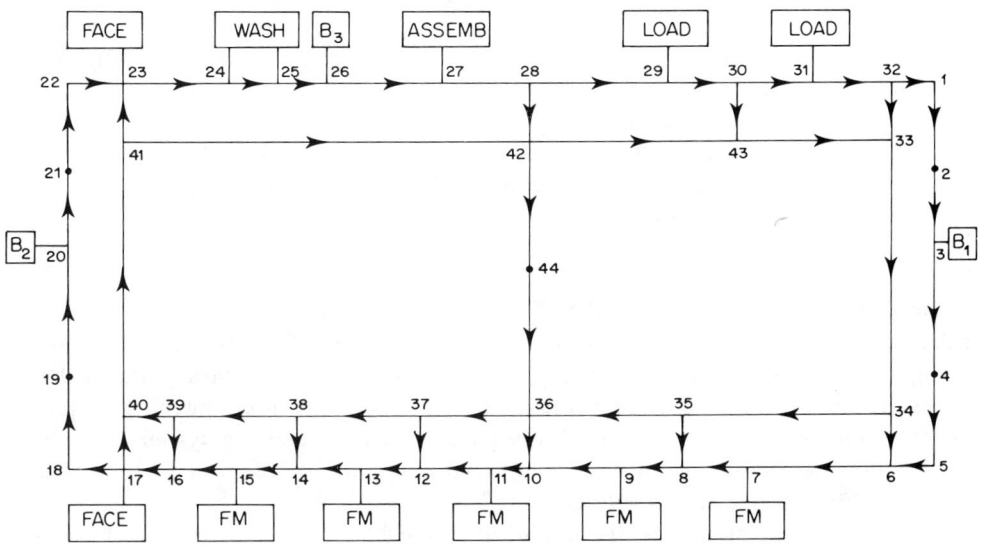

Figure 11.31   Track layout and location of decision points.

of data would exceed the computer's memory and entering all the data would be an extremely time-consuming and error-prone process. No doubt CMS Research could amend the software to hold data in disc file rather than arrays and increase the size of arrays, but this is not really the issue. The real issue concerns the most appropriate approach to simulating a system with so much data. An alternative approach would be necessary. As was done here, the data could be given for each part family. There would be another problem in doing so. Since MAST requires the operation times to be deterministic, it would not be possible to summarise statistically the variations between parts of the same family. A basic compromise on the realism of the model would be involved. Perhaps the most sensible solution would be to define as many part types as MAST can handle, selected from the highest volume parts or those representative of special cases, and adjusting their production quantities to indicate the overall level of production. A possible approach would be to define as many part types as there are pallets in the system.

4.   Defining certain physical facilities

It would be difficult to model certain facilities in the real system in MAST. One of these, which was omitted in the model above, is the fixture store. Because of the high variety of the parts, the demand pattern for them, and the extent to which certain fixtures are common to several parts, an automated   store is included in the real system. As yet

MAST has no facilities for modelling this. To do so we would have to consider whether MAST has a facility which might have a similar nature. In fact, if we define as many part types as there pallets, and assume that each fixture is permanently mounted on its pallet, then it should be possible to define the fixture store as some kind of buffer. Perhaps a dummy part type could be defined whose route would take it to the fixture store when no real part requiring the pallet was waiting at the load station. MAST provides a variety of rules for each type of decision. It also provides for the user to write his own functions, so that it might be possible to write a rule to load the dummy part at the correct time. The wash machine might also be difficult to model correctly, since it can process several pallets simultaneously. By adjusting the cycle time, a close approximation may be achieved.

The answer to the question "Can MAST model the real system in its full complexity?" is probably therefore "No, but ...", which is not the same as "No". As has been indicated throughout this text, simulation involves compromises between accuracy and realism of the results and the cost and time of the modelling process. A more meaningful question would be whether MAST, or indeed any package, can model the system in sufficient detail to provide usefully accurate results in within an acceptable time and cost budget. The answer to this question would be "Yes".

As a final postscript to this discussion, it should be pointed out that packages such as MAST, and indeed virtually all current packages, are being constantly developed, so that the types of facility and the range of decision rules available is steadily expanding.

11.10 MODEL ELEVEN—A SIMAN VERSION OF MODEL THREE

In chapter 6, some features of SIMAN[2] were mentioned, including the fact that it has functions specially designed for simulating manufacturing systems, especially some dealing with material handling. SIMAN will automatically compute the durations of movements using the distances and vehicle speed. It may be interesting to give a SIMAN version of one of the models already presented. For this purpose, Model Three has been selected, since it is not so voluminous as Model Eight would be.

Figure 11.32 gives the listing of the SIMAN model. It is not intended to give a comprehensive description of this model, but key features will be mentioned. The full details can be deduced with reference to the SIMAN manual.[2] The model is of modular construction. There is a sub-model for each major part, such as input buffers, machines, output buffers, etc, but splitting the model in this way requires the number of queues (or files) to be increased, and control to be transferred from one module to another in a more elaborate way than would otherwise be necessary. Transfer between the various modules is done by the TRANSPORT, ROUTE or NEXT statement in the final line of each module. These keywords direct the entity to a station number. Station numbers are assigned to the various components of the system. The model is general to an extent, in that it has been written for up to ten machines, although in this case there are only three. The SYNONYMS block allocates meaningful names to integer values to improve the readability of the model. The $A(n)$ values are attributes, $P(n,m)$ are parameters whose values are supplied in the experimental frame. $P(1,m)$ gives the number of machines and number of component types, $P(2,m)$ gives the mean time between arrivals, $P(3,m)$

```
BEGIN;
SYNONYMS:MACH=X(1):
         PART-MIX=X(2):
         M/C-MIN=1:
         M/C-MAX=10:
         INBUFF-MIN=11:
         INBUFF-MAX=20:
         OUTBUFF-MIN=21:
         OUTBUFF-MAX=30:
         LOAD STATION=31:
         JOB CONTROLLER=32:
         MANNING=33:
         HANDLING=34:
         QLOAD-NEW=41:
         QLOAD-FINISH=43:
         QLOAD-OUT=50:
         QWORLD=53:
         QTRANS-LI(N)=54:
         QTRANS-OL(C)=57:
         QWORKER-NEW=83:
         QWORKER-COMP=85:
         TPRIORITY-OL=1:
         TPRIORITY-LI=2:
         WPRIORITY-COMP=1:
         WPRIORITY-NEW=2:
         L/U-PRI-COMP=1:
         L/U-PRI-NEW=2;
  ;
  ;

         CREATE;
         ASSIGN:'MACH'=P(1,1);                NUMBER OF MACHINES
         ASSIGN:'PART-MIX'=P(1,2):DISPOSE;    NUMBER OF PART TYPES
  ;
  ;**********************
  ;*  ARRIVAL SUBMODEL  *
  ;**********************
  ;
         CREATE:EX(2,1),1000:MARK(1);         ARRIVAL TIME
         COUNT:1,1;                           NO. OF ARRIVAL
         ASSIGN:A(2)=DP(3,1);                 PART TYPE
         ASSIGN:A(3)=P(5,1);                  NO. OF OPERATION
         ASSIGN:A(4)=A(4)+1;                  INCREMENT OF SEQUENCE NO.
         ASSIGN:A(5)=P(7,A(4));               MACHINE NUMBER
         ASSIGN:A(6)=P(8,A(2));               PROCESSING TIME
  ;
         ASSIGN:A(9)=P(4,1);                  LOADING TIME
         ASSIGN:A(10)=P(4,2);                 UNLOADING TIME
  ;
         ROUTE:0,'JOB CONTROLLER';            TO JOB CONTROLLER
  ;
```

Figure 11.32   Listing of SIMAN model for Model Three.

```
;**********************************************************************
;*   CONTROLLER FOR NUMBER OF JOBS INSIDE THE SYSTEM SUBMODEL   *
;**********************************************************************
;
          STATION,'JOB CONTROLLER';                    ( 32 )
          QUEUE,'QWORLD';                          ( 53 )
          SEIZE:JOBLIM;
          ROUTE:0,'LOAD STATION';
;
;****************************
;*  INPUT BUFFER SUBMODEL  *
;****************************
;
          STATION, INBUFF-MIN'-'INBUFF-MAX';        ( 11-20 )
;
          FREE:CART;
          ROUTE:0,A(5);      ROUTE TO THE ASSOCIATED MACHINE
;
;******************************
;*  MACHINE STATION SUBMODEL  *
;******************************
;
          STATION,'M/C-MIN'-'M/C-MAX';     ( 1-10 )
;
          QUEUE,M;
          SEIZE:MACHINE(M);
          RELEASE:IBUFFER(M);
          DELAY:A(6);
          COUNT:M+1,1;
          ROUTE:0,M+'INBUFF-MAX';      ROUTE TO THE OUTPUT BUFFER(M+20)
;
;****************************
;*  OUTPUT BUFFER SUBMODEL  *
;****************************
;
          STATION,'OUTBUFF-MIN'-'OUTBUFF-MAX';      ( 21-30 )
;
          ASSIGN:A(8)=M;
          QUEUE,M;
          SEIZE:OBUFFER(M-'INBUFF-MAX');            ( M-20 )
          RELEASE:MACHINE(M-'INBUFF-MAX');          ( M-20 )
          QUEUE,M+'M/C-MAX';                        ( M+10 )
          ASSIGN:A(4)=A(4)+1;
          QUEUE,'QLOAD-FINISH';                     ( 43 )
          SEIZE,'L/U-PRI-COMP':LOAD;
;
          ASSIGN:J=1;
          ROUTE:0,'HANDLING';
;
```

Figure 11.32   (continued).

```
;********************************
;*   MATERIAL HANDLING SUBMODEL   *
;********************************
;
          STATION,'HANDLING';              ( 34 )
          BRANCH,1:
            IF,J.EQ.1,MOVE1:
            ELSE,MOVE2;
;
MOVE1     ASSIGN:M=A(8);
          QUEUE,'QTRANS-OL(C)';                      ( 57 )
          REQUEST,'TPRIORITY-OL':CART;
          RELEASE:OBUFFER(M-'INBUFF-MAX');
          TRANSPORT:CART,'LOAD STATION';
;
MOVE2     ASSIGN:M='LOAD STATION';
          QUEUE,'QTRANS-LI(N)';                      ( 54 )
          REQUEST,'TPRIORITY-LI':CART;
          RELEASE:LOAD;
          TRANSPORT:CART,A(7);             TO NEXT INPUT BUFFER
;
;******************
;*   EXIT SUBMODEL   *
;******************
;
EXIT      TALLY:A(2),INT(1);
          TALLY:13,INT(1):DISPOSE;
END;

                COUNTER:1,NO. OF ARRIVAL:
                        2,JOBS FROM M1:
                        3,JOBS FROM M2:
                        4,JOBS FROM M3;
                DSTAT:1,NR(1),MACHINE 1 UTIL.:
                        2,NR(2),MACHINE 2 UTIL.:
                        3,NR(3),MACHINE 3 UTIL.:
                        4,NT(1),CART UTIL.:
                        5,NR(31),LOAD STATION UT.:
                        6,NR(33),WORKER UT.:
                        7,NR(32),CONTROLLER UT.:
                        8,NQ(41),LOAD QUEUE:
                        9,NQ(53),ARRIVAL QUEUE;
                REPLICATE,1,0,8640.,,,1440.;
                END;


;**************************
;*   LOAD/UNLOAD SUBMODEL   *
;**************************
;
          STATION,'LOAD STATION';          ( 31 )
;
          BRANCH,1:
            IF,A(4).EQ.1,FIRST:
            ELSE,CARRY;              CHECK FOR NEW JOBS
FIRST     QUEUE,'QLOAD-NEW';     FILE FOR LOAD STATION ( 41 )
          SEIZE,'L/U-PRI-NEW':LOAD;
```

Figure 11.32   (continued).

```
            ASSIGN:A(1)=TNOW;
            ASSIGN:J=1;
            ROUTE:0,'MANNING';
CRT         ASSIGN:J2='M/C-MAX'+'MACH'
            FINDJ,'INBUFF-MIN',J2:MIN(NR(J));
            ASSIGN:A(5)=J-'M/C-MAX';
            ASSIGN:A(7)=J;
            ASSIGN:M=A(7);
            QUEUE,M+'QLOAD-OUT';                              ( 61-70 )
            SEIZE:IBUFFER(M-'M/C-MAX');        ( M-10 )
            ASSIGN:J=2;
            ROUTE:0,'HANDLING';
CARRY       FREE:CART;
            ASSIGN:J=2;
            ROUTE:0,'MANNING';
;
;*********************
;*   MANNING SUBMODEL   *
;*********************
;
            STATION,'MANNING';                        ( 33 )
;
            BRANCH,1:
              IF,J.EQ.1,MAN1:
              ELSE,MAN2;
;
MAN1        QUEUE,'QWORKER-NEW';                        ( 83 )
            SEIZE,'WPRIORITY-NEW':WORKER;
            DELAY:A(9);
            RELEASE:WORKER:NEXT(CRT);
;
MAN2        QUEUE,'QWORKER-COMP';                       ( 85 )
            SEIZE,'WPRIORITY-COMP':WORKER;
            DELAY:A(10);
            RELEASE:WORKER;
            RELEASE:LOAD;
            RELEASE:JOBLIM:NEXT(EXIT);
;
```

Figure 11.32   (continued).

gives the cumulative frequency of the part type distribution, P(4,m) gives the loading and unloading times, P(5,m) gives the number of operations to be performed, P(6,m) gives the number of alternative machines for the operation, P(7,m) gives the list of alternative machines for the operation and P(8,m) gives the machining time for each type of part.

One or two simplifications have been made relative to the ECSL model. The load stations have been defined as a single resource with two servers, rather than two separate resources. This slightly simplifies the coding of the model, and illustrates one of SIMAN's facilities. Consequently, they have the same location so far as calculating movement durations is concerned. The time to transfer pallets between machine buffers and machines or vehicle has been assumed to be zero. Times could have been inserted by including DELAY statements at appropriate points. Another simplification of this model relative to Model Three is that the number of jobs (components) which can be in the system has been limited to the number of pallets, but the pallets are assumed to be lifted off the load station when the component is unloaded. This means that any other pallet can be brought to the load station, whereas in Model Three it was assumed that the pallet would remain at the load station and no other pallet could be returned from a machine for unloading until a new component had arrived, been loaded on the pallet and been taken to a machine.

```
BEGIN;
PROJECT,MODEL 3,FELIX,8/21/87;
DISCRETE,1000,22,100,34;
RESOURCES:1-10,MACHINE:
          11-20,IBUFFER:
          21-30,OBUFFER:
          31,LOAD,2:
          32,JOBLIM,3:
          33,WORKER,1;
TRANSPORTERS:1,CART,1,1,10.0,31-A;
DISTANCES:1,11-31,1,1,1,1,1,1,1,1,1,1,1,1,1,1,1,1,1,1,1,1,1,11/
               1,1,1,1,1,1,1,1,1,1,1,1,1,1,1,1,1,1,1,1,1/
               1,1,1,1,1,1,1,1,1,1,1,1,1,1,1,1,1,1,1,9/
               1,1,1,1,1,1,1,1,1,1,1,1,1,1,1,1,1,1/
               1,1,1,1,1,1,1,1,1,1,1,1,1,1,1,1,1/
               1,1,1,1,1,1,1,1,1,1,1,1,1,1,1,1/
               1,1,1,1,1,1,1,1,1,1,1,1,1,1,1/
               1,1,1,1,1,1,1,1,1,1,1,1,1,1/
               1,1,1,1,1,1,1,1,1,1,1,1,1/
               1,1,1,1,1,1,1,1,1,1,1,1/
               1,1,1,1,1,1,1,1,1,9/
               1,1,1,1,1,1,1,1,1/
               1,1,1,1,1,1,1,11/
               1,1,1,1,1,1,1/
               1,1,1,1,1,1/
               1,1,1,1,1/
               1,1,1,1/
               1,1,1/
               1,1/
               1;
PARAMETERS:1,3,12:
           2,17.:
           3,0.1889,1,0.3483,2,0.5018,3,0.6081,4,
             0.6907,5,0.7615,6,0.8205,7,0.8677,8,
             0.9102,9,0.948,10,0.9763,11,1.,12:
           4,3.,3.:
           5,1:
           6,3:
           7,1,2,3:
           8,18.,60.,36.,40.,36.,24.,32.,28.,20.,20.,30.,32.;
TALLIES:1,T. IN SYS. P1:
        2,T. IN SYS. P2:
        3,T. IN SYS. P3:
        4,T. IN SYS. P4:
        5,T. IN SYS. P5:
        6,T. IN SYS. P6:
        7,T. IN SYS. P7:
        8,T. IN SYS. P8:
        9,T. IN SYS. P9:
        10,T. IN SYS. P10:
        11,T. IN SYS. P11:
        12,T. IN SYS. P12:
        13,T. IN SYS.ALL;
```

Figure 11.33   Experimental frame for SIMAN version of Model Three.

Figure 11.33 gives the experimental frame. The RESOURCES element defines the various workstations. Each is given a number which is used to identify it. The SYN-ONYMS block in the model assigns appropriate names to key station numbers. In the TRANSPORTER element, the vehicle is named and its speed set to 10 metres per minute—this is rather slow, but will give movement durations roughly equivalent to the constant value used in chapter 9. The durations are calculated using the distances between stations. These are defined in the DISTANCES element. The machines are assumed to be at 10-metre spacing and the pallet stands at 2-metre spacing, with the load stations opposite the centre machine. By coincidence this greatly simplifies the distances table. Default values of one have been inserted, and real values inserted where necessary, eg 11 metres from the load stations to the input buffer of machine 1. The next set of entries in the experimental frame are the values of the PARAMETERS. Then come the captions for the statistics which are being collected. Finally, the REPLICATE element causes the simulation to run for 8640 minutes with the statistics being cleared after 1440 minutes, giving a recording period of 7200 as in the ECSL model.

11.10.1 Results

Table 11.2 gives the SIMAN results printout for 5 pallets. The values can be compared with those obtained in the ECSL model given in table 9.5. The output of the system was 428 components, with 430 arrivals, compared with 441 and 441 for the ECSL model. In both models, the expected number of arrivals was 423 (7200/17). Some slight difference is to be expected, since the two packages use their own random number generator. The overall average time in the system is 60.0 compared with 64.9 for the ECSL model. The utilisations of machines 1,2 and 3 were 94%, 77% and 34% respectively. These average 68%, the same value as in the ECSL model, however the SIMAN value is strictly more comparable to the occupancy statistic from ECSL which was 74%. The average time in the system for the different types of part can be compared, as can the mix of parts which was produced. These values are also influenced by the differences in random number generators. Overall the results from the two models compare well, with the exception of the vehicle utilisation which is affected by the change in the method determining the duration of movements. The SIMAN model gave a utilisation of 11%, versus 24% for the ECSL model.

The individual machine utilisations were not recorded in the ECSL model, although they could have been, as was done in Models Seven and Eight. Another interesting statistic, also not collected in the ECSL model, is the number of components produced by each machine. These figures were 196, 160 and 72, and correlate with the machine utilisation figures. These figures result from machine 1 being always selected if available, and machine 2 only being selected if machine 1 is busy and machine 3 only selected if the other two are both busy. The model was run with between 3 and 8 pallets. The results showed that when there were 3 or 4 pallets output was reduced to 378 and 409 respectively, and machine 3 was never used. The maximum output was obtained with 5 pallets. Increasing the number above 5 increased the time in the system but did not increase output.

The same machine load distribution would have been found in the ECSL model, if the data had been recorded, because the same machine selection decision rule was used. (The same effect was observed in the MAST model, Model Ten.) The statement which

Table 11.2   Results from SIMAN version of Model Three, with 5 pallets

SIMAN RUN PROCESSOR RELEASE 3.0
COPYRIGHT 1985 BY SYSTEMS MODELING CORP.

SIMAN SUMMARY REPORT

RUN NUMBER 1 OF 1

PROJECT: MODEL 3
ANALYST: FELIX
DATE : 8/21/1987

RUN ENDED AT TIME : 0.8640E+04

TALLY VARIABLES

Number	Identifier	Average	Standard deviation	Minimum value	Maximum value	Number of Obs.
1	T. IN SYS. P1	43.55985	15.18832	25.39844	78.79980	65
2	T. IN SYS. P2	87.50317	17.06206	66.29993	129.29248	65
3	T. IN SYS. P3	59.72095	16.15012	43.39844	98.49365	70
4	T. IN SYS. P4	65.94979	16.60657	46.30029	103.71094	48
5	T. IN SYS. P5	59.57687	18.83319	42.30029	101.91846	38
6	T. IN SYS. P6	54.23240	16.47853	31.92383	88.28613	40
7	T. IN SYS. P7	59.47323	15.81070	39.34863	92.99219	32
8	T. IN SYS. P8	52.82848	16.62168	36.10010	88.69971	18
9	T. IN SYS. P9	44.20538	21.02032	27.51709	85.50049	12
10	T. IN SYS. P10	44.60384	14.26929	27.40039	68.67065	16
11	T. IN SYS. P11	53.53304	13.76741	36.30029	70.30127	13
12	T. IN SYS. P12	51.99845	13.31425	38.36719	72.00000	11
13	T. IN SYS.ALL	59.96385	21.16144	25.39844	129.29248	428

DISCRETE CHANGE VARIABLES

Number	Identifier	Average	Standard deviation	Minimum value	Maximum value	Time period
1	MACHINE 1 UTIL.	0.93691	0.24312	0.00000	1.00000	7200.00
2	MACHINE 2 UTIL.	0.76658	0.42301	0.00000	1.00000	7200.00
3	MACHINE 3 UTIL.	0.33685	0.47263	0.00000	1.00000	7200.00
4	CART UTIL.	0.11417	0.31802	0.00000	1.00000	7200.00
5	LOAD STATION UT.	0.48778	0.63463	0.00000	2.00000	7200.00
6	WORKER UT.	0.35724	0.47919	0.00000	1.00000	7200.00
7	CONTROLLER UT.	3.57947	1.43946	0.00000	5.00000	7200.00
8	LOAD QUEUE	0.00441	0.07374	0.00000	2.00000	7200.00
9	ARRIVAL QUEUE	0.64662	1.39936	0.00000	9.00000	7200.00

COUNTERS

Number	Identifier	Count	Limit
1	NO. OF ARRIVAL	430	INFINITE
2	JOBS FROM M1	196	INFINITE
3	JOBS FROM M2	160	INFINITE
4	JOBS FROM M3	72	INFINITE

selects the machine is in the load/unload submodel, immediately following that labelled CRT, which reads

FINDJ,'INBUFF-MIN',J2:MIN(NR(J));

This searches for the machine input buffer with the minimum number of pallets at it. Since these have either zero or one pallet it causes the model to assign the next job to the machine input buffer with the lowest station number.

It would be interesting to see whether the performance of the system would be different if the work load were allocated more evenly among the machines. For example, we might base a rule on the level of congestion at each machine, by looking not merely at whether the machine and its input buffer were occupied, and also the number of jobs already assigned to that machine but still waiting tobe moved to it. We could send the pallet to the least congested machine. The model can be amended, and in doing so we can illustrate one of SIMAN's features, that of linking the standard logic blocks with user-written routines. Figure 11.34 gives the amendments. In the arrival sub-model some additional setting of attributes is required, because when we come to select a machine we need to know on which machines the operation can be carried out. In the load–unload sub-model the statement at label CRT and the following two lines are replaced by a call

```
;**********************
;*   ARRIVAL SUBMODEL   *
;**********************
;
          CREATE:EX(2,1),1000:MARK(1);          ARRIVAL TIME
          COUNT:1,1;                            NO. OF ARRIVAL
          ASSIGN:A(2)=DP(3,1);                  PART TYPE
          ASSIGN:A(3)=P(5,1);                   NO. OF OPERATION
          ASSIGN:A(4)=A(4)+1;                   INCREMENT OF SEQUENCE NO.
          ASSIGN:A(5)=P(7,A(4));                MACHINE NUMBER
          ASSIGN:A(6)=P(8,A(2));                PROCESSING TIME
;
          ASSIGN:A(15)=A(5);
          ASSIGN:A(18)=A(6);
          ASSIGN:A(14)=P(6,1);        NO. OF ALT. ROUTES IN THIS OP.
          ASSIGN:A(21)=A(14);
          ASSIGN:X(21)=A(14);
          ASSIGN:X(21)=X(21)-1;
          BRANCH,1:
            IF,X(21).EQ.0,CM3:
            ELSE,CM1;
CM1       ASSIGN:X(21)=X(21)-1;
          ASSIGN:A(16)=P(7,2);
          ASSIGN:A(19)=A(6);
          BRANCH,1:
            IF,X(21).EQ.0,CM3:
            ELSE,CM2;
CM2       ASSIGN:A(17)=P(7,3);
          ASSIGN:A(20)=A(6);
CM3       ASSIGN:A(9)=P(4,1);                   LOADING TIME
          ASSIGN:A(10)=P(4,2);                  UNLOADING TIME
;
          ROUTE:0,'JOB CONTROLLER';       TO JOB CONTROLLER
```

Figure 11.34   Amendments to SIMAN model for revised rule for allocating pallets to machines.

```
;
;***************************
;*  LOAD/UNLOAD SUBMODEL   *
;***************************
;
          STATION,'LOAD STATION';                ( 31 )
;
          BRANCH,1:
            IF,A(4).EQ.1,FIRST:
            ELSE,CARRY;              CHECK FOR NEW JOBS
FIRST     QUEUE,'QLOAD-NEW';    FILE FOR LOAD STATION ( 41 )
          SEIZE,'L/U-PRI-NEW':LOAD;
          ASSIGN:A(1)=TNOW;
          ASSIGN:J=1;
          ROUTE:0,'MANNING';
CRT       EVENT:1;
          ASSIGN:A(7)=A(5)+'M/C-MAX';                      A(5)+10
          ASSIGN:M=A(7);
          QUEUE,M+'QLOAD-OUT';                         ( 61-70 )
          SEIZE:IBUFFER(M-'M/C-MAX');        ( M-10 )
          ASSIGN:J=2;
          ROUTE:0,'HANDLING';
CARRY     FREE:CART;
          ASSIGN:J=2;
          ROUTE:0,'MANNING';


          SUBROUTINE EVENT(JOB,I)
C
          GO TO (1) I
1         CALL ALTOPT(JOB)
          RETURN
          END
C
C
C
          SUBROUTINE ALTOPT(JOB)
          DIMENSION J1(3),J2(3),NITEM(3)
          COMMON/SIM/D(50),DL(50),S(50),SL(50),X(50),DTNOW,TNOW,TFIN,J,NRUN
C
C   NOP= NO. OF ALTERNATIVE ROUTES AVAILABLE IN THE PRESENT OP.
C     ( MAX.= 3 )
C
          NOP=A(JOB,14)
          IF(NOP.EQ.1) RETURN
          DO 10 IM=1,NOP
          N=IM+14
          JJ=A(JOB,N)
          J1(IM)=JJ+10
          J2(IM)=JJ+60
C         NUMBER QUEUING FOR MACHINE + AT MACHINE + IN INPUT BUFFER
          NITEM(IM)=+NQ(J2(IM))++NR(J1(IM))+NR(JJ)
10        CONTINUE
C
          MP=1
          DO 20 JM=2,NOP
          IF(NITEM(MP).LE.NITEM(JM)) GO TO 20
          MP=JM
20        CONTINUE
C
          KM=MP+14
          KT=MP+17
          CALL SETA(JOB,5,A(JOB,KM))
          CALL SETA(JOB,6,A(JOB,KT))
          RETURN
          END
```

Figure 11.34   (continued).

to the event routine with the event type set to 1. A file of FORTRAN segments is linked with the model file. This file comprises the event routine and the user-written routine, ALTOPT. The user-written routine contains a loop which scans the machines which can do the next operation on the job, and counts the number of pallets assigned to the machine, at the machine, and in its input buffer. The machine with the fewest pallets is selected. When the results were analysed it was found that the maximum output could be achieved with 3 pallets, compared to 5 previously, and that the average time in the system was considerably reduced. There was still a slightly heavier load on machine 1 than 2, and 2 than 3, because when they were similarly congested the lower-numbered machine would still be selected. The rule could have been based on the number of jobs processed so far by each machine, but that rule would not consider the current state of congestion at each machine. This illustrates that the simple matter of the way of writing a few statements in a computer program may affect the results in a way which would not apply in real life if a human being were making the decision. The remedy is, of course, to write a better decision rule!

SUMMARY

In this chapter, various aspects of modelling vehicles and their movements have been considered in greater detail. The subdivision of movements into their elements so that their duration could be determined accurately was shown. It was also shown that these aspects are closely tied up with setting the correct logic for the state of the buffers to or from which pallets are moved. We have also considered the case when the vehicle has two load-carrying positions and when there are several vehicles in a system. Various models were presented.

Model Eight, in which the durations of the movements were accurately calculated, was presented. It was noted that, as a result of the more accurate method of modelling movements, the unblocking routines of Model Seven were re-written as empty pallet movements in Model Eight, with some simplifying of the logic.

Model Nine, in which the vehicle had two load positions, was proposed but not presented. Readers were invited to consider how the code of the earlier models would be amended to cater for this situation.

Model Ten was a MAST model of a simplified version of a real complex system. It illustrated the processes involved in using a generic flexible manufacturing system model. Some discussion was included concerning the choice of decision rules and their effects.

Model Eleven was a SIMAN model roughly equivalent to Model Three. This gave an opportunity to compare the results of the SIMAN and ECSL models, and to gain additional insight into the behaviour of the system being modelled. It also provided an illustration of some of SIMAN's facilities. It was found that by using an improved work allocation rule, the same output could be achieved with 3 pallets as previously achieved with 5. This illustrates the value of simulation.

EXERCISES

1. Which of the tests in figure 11.14 can be expressed in terms of the number of occupied or free load positions on the vehicle? In which activities would this number be altered? Do the tests require any attributes? If so, of what entity are they attributes?

2. Consider how Model Five might be revised to cater for the situation in which the vehicle has two load positions.

3. Consider how Model Seven might be revised to cater for the situation in which the vehicle has two load positions.

4. Consider how your revised version of Model Seven might be further revised if movement durations were to be accurately computed. How would it then differ from Model Eight?

5. In general terms, consider how the code of Models Five, Seven and Eight would be revised if there were two (or more) vehicles. Omit consideration of the physical layout of the track.

6. What rule do you consider would be best in selecting a machine from the alternatives for an operation? What factors would you take into account? What data would you need?

REFERENCES

1. *SPAR*, *MAST* and *BEAM User Manuals*, CMS Research Inc, Oshkosh, Wisconsin, USA.

2. Pegden, C.D., *Introduction to SIMAN*, Version 3.0, Systems Modeling Corporation, Calder Square, State College, Pennsylvania, 1985.

CHAPTER 12

Building FMS models—
5: Robots, conveyors and AS/RS systems

In this chapter we will introduce more aspects of materials handling systems. As before, the notation which will be used in activity cycle diagrams will be given at the start, Figure 12.1.

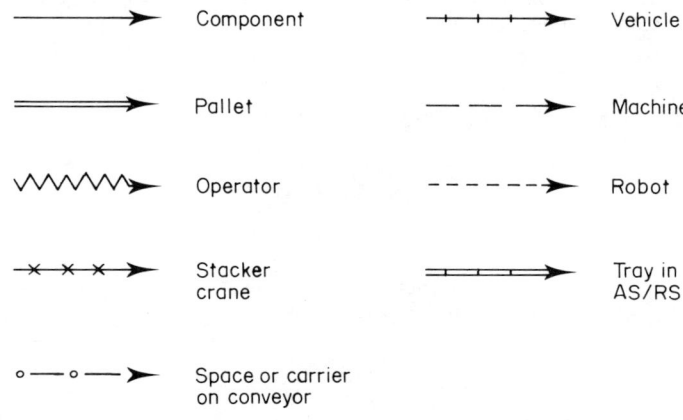

Figure 12.1   Notation for activity cycle diagrams in chapter 12.

12.1   ROBOTS

We must consider the modelling of robots, since they are becoming widely used in manufacturing systems. However they do not actually add much to what we have already covered when considering workstations. In general, robots are used for one of two purposes:

1. as a productive work station, for example for welding or assembly,
2. as a means of handling components, for example for loading and unloading machines.

Robots which are used as productive tools will have activity cycles similar to those described for machines, or workstations in general. Because of their nature, robot workstations do not normally have buffer positions like pallet stands, but may well have buffers in the form of conveyor queues. Whereas prismatic components are normally palletised for supply to machining centres, components which are to be processed by robots will often not be palletised, because robots can grip a component directly. Where a pallet is used it may be a simple tray type so that several components may be transported together. Consequently, some of the remarks made when describing workstations will not apply.

In assembly operations, ancillary equipment, such as fixtures, bowl feeders, and so on will probably be present. In a simulation model it will not normally be necessary to model these features in detail, unless the model is specifically to focus upon them. More often the model will be to assess how the productivity of the cell as a whole depends on the capacity of buffers, reliability of equipment, and so on.

Robots used for handling components will have an activity cycle similar to many we have already encountered. They are either working or idle, and have a very simple activity cycle diagram, as shown in Figure 12.2.

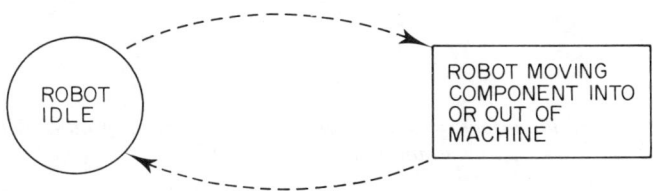

Figure 12.2 Activity cycle for robot used for machine loading and unloading.

However, a robot is frequently used to service a group of machines, through which the component must be processed in turn. For example, Figure 12.3 shows the layout of one of the cells in Cummins' Con-rod line. It consists of several machines served by one robot, with input and output conveyors linking it to preceding and succeeding cells. The components are moved between the input conveyor, the machines and the output conveyor by the robot. How should the activity cycle diagram in figure 12.2 be amended? Can a single activity describe all the working states of the robot? Should each one be defined as a different type of activity? If so, would Figure 12.4 give an accurate ACD?

This diagram indicates that there is no specific sequence between the movements of the robot. If we wish to ensure that it moves a component from one machine to another in strict sequence we would have to delete the central ROBOT IDLE queue and insert new ones between each pair of activities. However this would imply that the robot could not pick up a part until the previous one had gone right through the system, which would mean that only one machine could be active at one time. This would obviously not be very productive, and the diagram given seems more sensible. This is another example of a situation where it is important to know the rules used by the real system controller. There is no point in writing a different set of rules into the model, unless by doing so the model can be greatly simplified or speeded up without significant loss of accuracy in the results.

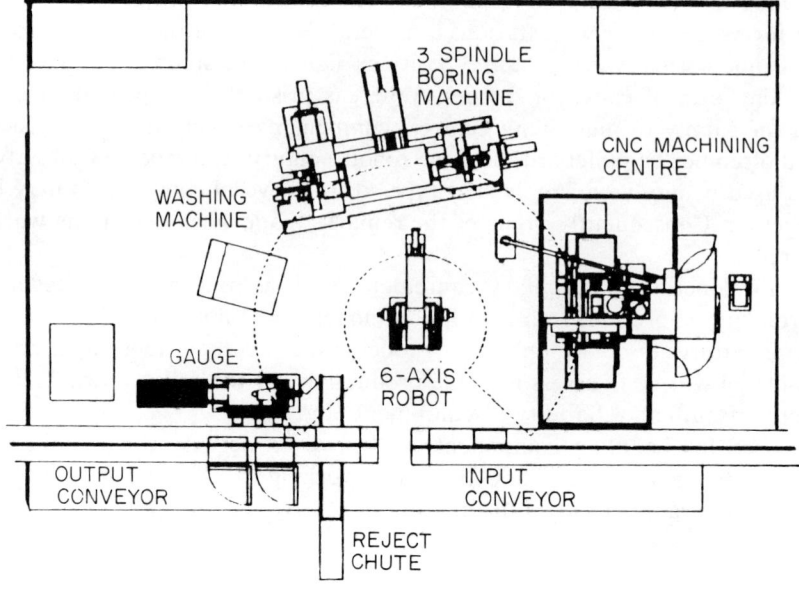

Figure 12.3   Layout of manufacturing cell served by a robot (reproduced by permission of Cummins Engines Co Ltd).

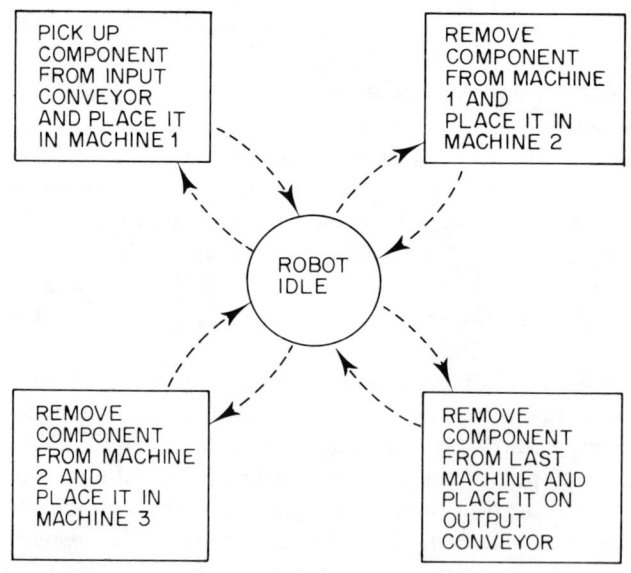

Figure 12.4   Expanded activity cycle for robot used for machine loading and unloading.

12.2 CONVEYORS

Another major type of material handling equipment which we must consider is the conveyor. There are many varieties of conveyor, including some very specialised ones. Since the more recent FMSs tend to use vehicles rather than conveyors, we will only consider the general characteristics of the principal types and their use in manufacturing systems.

Conveyors are mainly used in systems producing relatively small, often cylindrical, components. These components can be placed into tray type pallets which are moved from workstation to workstation along the conveyor. The conveyors may be of either powered or free roller type. Most systems using conveyors for the primary movement of material will consist of both powered and free sections. The powered sections provide the motive force, while the free sections allow pallets or components to stop and perhaps queue up at workstations. Lifting the components to machines will normally be done by a robot.

As with so many other aspects of manufacturing systems, the approach we adopt to modelling conveyors depends on the part they play in the system as a whole, and on the objective of the model. On the one hand we can model them in great detail, observing closely the movement of materials along them. On the other hand, we might only record the presence of a component on the conveyor and their entry to and exit from it, but not its location on the conveyor. To illustrate these possibilities let us consider an example.

Figure 12.5 shows a section of conveyor connecting two machines. Components are discharged on to the conveyor from the first machine and carried on the conveyor to the second machine. They must wait at the end of the conveyor until the second machine is idle, before leaving the conveyor. Let us assume for the present that there is no restriction on the movement of components along the conveyor, so that they can accumulate at its end. This is sometimes known, as in WITNESS, as a queueing conveyor.

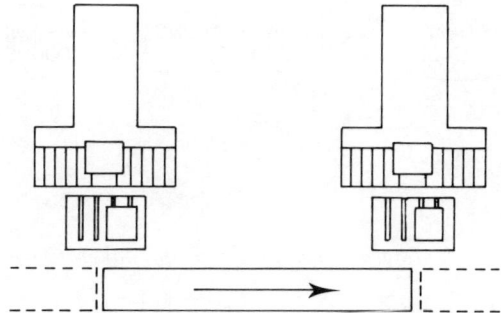

Figure 12.5 Diagram of conveyor linking two machines.

We can develop an activity cycle diagram for the entities. A simple version would be as in Figure 12.6. It should be noted that the component's cycle is incomplete since it omits the world outside the two machines. A queue depicting the rest of the world could be shown, but in a real system there would normally be various other activities and queues. This diagram shows that OP 2 depends on Machine 2 being idle (MC 2 IDLE) and there being a component in ON CONVEYOR.

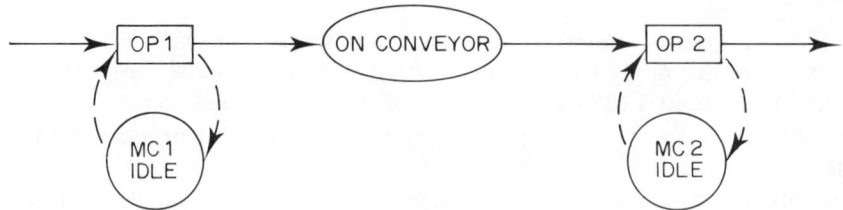

Figure 12.6   Elementary activity cycle diagram for system in which components travel along a conveyor between two machines.

Figure 12.6 does not depict the time taken to move along the conveyor. To introduce this we would need another activity and to redefine the queues, as shown in Figure 12.7. The conditions for OP 2 are now that there is a component at the end of the conveyor and Machine 2 is idle. The duration of the TRAVEL ALONG CONVEYOR activity is the time to travel the full length of the conveyor if no other components interfere. The value will normally be a constant, calculated from the length of the conveyor and the speed at which it moves components. This value would be used even if there are other components on the conveyor ahead of the component concerned. It represents the minimum delay between entering and leaving the conveyor. If there are other components already on the conveyor then the component will take longer than this time to reach the end of the conveyor. However, the model remains accurate because the machine will not be ready to start on this component until it has dealt with all the other components ahead of this one. In other words, this logic correctly models the operation of Machine 2 but does not model in detail the movement of jobs along the conveyor or the accumulation of jobs at its end.

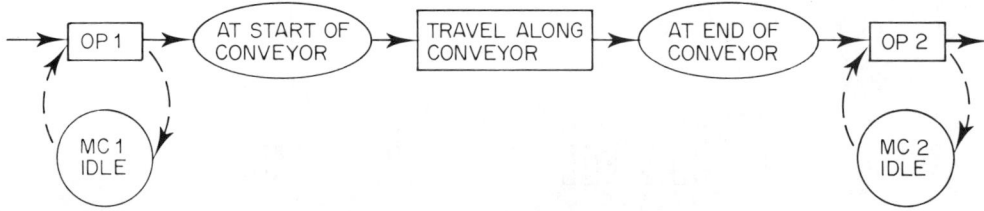

Figure 12.7   Activity cycle diagram including the time taken to travel along the conveyor.

If the conveyor is of limited length then a further complication may arise. If the conveyor is full then there will be no space on it for Machine 1 to discharge its component. Machine 1 would be prevented from starting to work on another component until Machine 2 has completed work on its current component and taken the next one from the conveyor, thereby releasing space for Machine 1 to discharge its component. This would require a limit to be set on the number of components which can be held on the conveyor at one time. The easiest way to model this constraint is to add a variable to the model which gives the number of free spaces on the conveyor. This variable is incremented every time Machine 2 takes a component from the conveyor and decremented each time Machine 1 puts one on it. Provided the value of the variable is greater than zero, Machine 1 may discharge its component. We can represent this variable in an ACD by a dummy entity, of which there is a number equal to the number of components which can be held on the conveyor. We must also subdivide OP 1 and OP 2 into performing the operation itself and discharging a component to, or taking one from, the conveyor. Figure 12.8 shows the resulting diagram. Notice that although the queues following OP 1 are labelled COMPONENT HELD UP and MACHINE ONE BLOCKED, normally both the component and entities will pass through these queues instantaneously. They will only be held up or blocked when there is no free space on the conveyor.

Although we have a queue of components at the end of the conveyor, it may seem that we have not kept track of the sequence in which the components are arranged on the conveyor, and which one is actually at the end. As described in chapters 4 and 5, in addition to the logic of the model we must specify queue disciplines and other details. The queue discipline of the set AT END OF CONVEYOR will keep the components in the correct order. Software packages frequently arrange the items in a set in the order in which they entered the set, ie in first-in-first-out sequence (FIFO), and will assume FIFO unless an alternative is specified. In this case FIFO is exactly what we need. It will ensure that the components are removed from AT END OF CONVEYOR in the order in which they entered AT START OF CONVEYOR. Thus, even although the duration of TRAVEL ALONG CONVEYOR may underestimate the time to reach the true end of the conveyor, this logic correctly models the sequence and timing of components for the TAKE COMPONENT FROM CONVEYOR activity.

So far our model correctly models components entering and leaving the conveyor, and the time delay in travelling along it, on the assumption that they can accumulate at the end of the conveyor. It does not model the position of the component as it moves along the track. It would be important to incorporate this aspect if we wish to display the movement of components on a display screen, or if the conveyor is of a type in which the spacing of components on the track is fixed. This would occur in carousel conveyors or pallet chain types, where carriers are fixed to the conveyor mechanism, known in WITNESS as a Fixed Conveyor. The carriers are either occupied by a component or are empty. Suppose one component enters a carrier, but the next component is not ready to enter when the next carrier passes the entry point, but is ready for the third carrier. There will be a gap on the conveyor between the successive components. This gap will travel along the conveyor with the components and will eventually reach the end. When it gets to the end Machine 2 would not be able to take a component from the conveyor because the carrier is empty. Machine 2 will have to wait until the next loaded carrier reaches it. To handle this we need further refinements of the model. As it stands the simulation software would not know that there are gaps on the conveyor and will only recognise the components. This would have the effect of components entering and arriving at the end of the conveyor too soon, because the spacing of components on the conveyor would not be controlled.

To overcome this problem we could enlarge the responsibilities of our dummy entity. Instead of merely using it to control the total number of components on the conveyor we could use it to ensure they occupy the correct spaces. Every time a carrier comes round to the entry point either a gap or a component will be placed on it. Similarly, every time a carrier comes round to the exit point we remove either a gap or component.

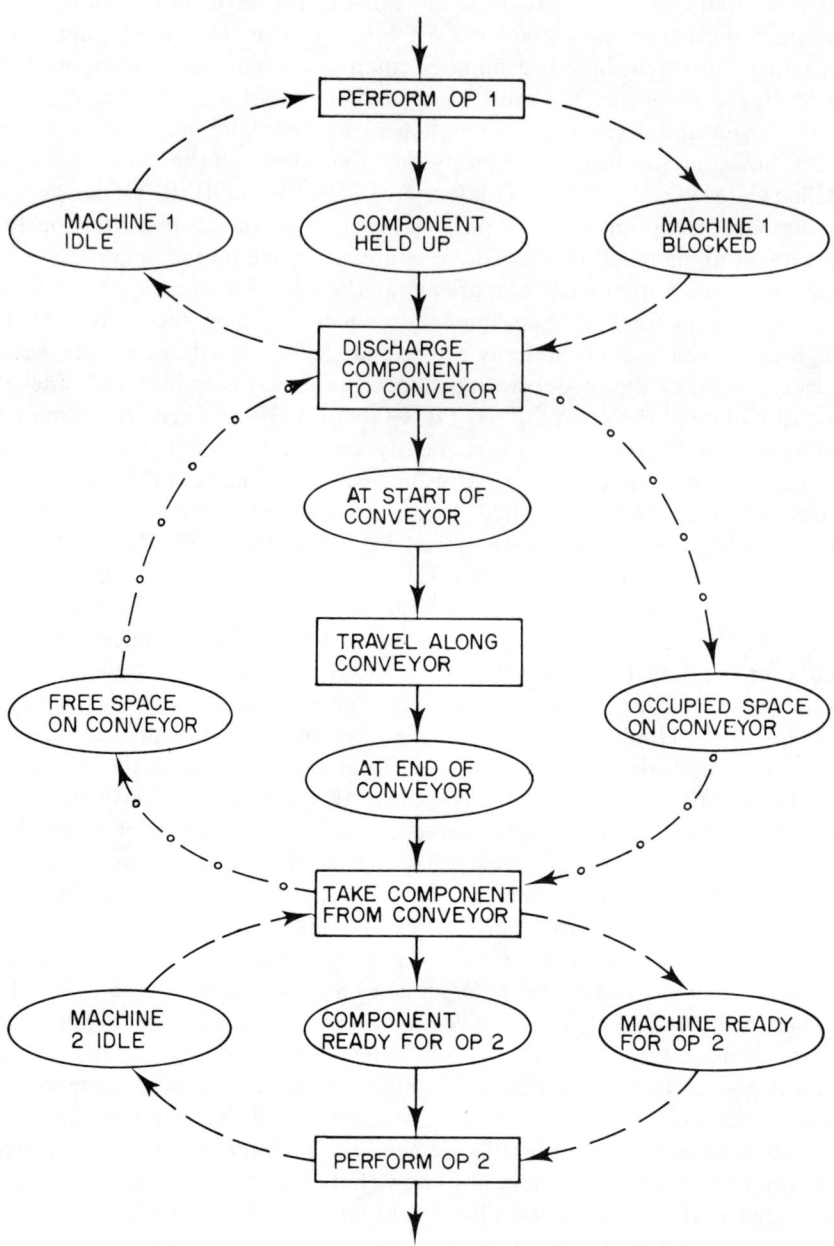

Figure 12.8   Activity cycle diagram when conveyor has limited capacity.

In fact with this type of conveyor, we can replace the dummy entity, space, by a real entity, the carrier, which will have an attribute to indicate what is on it, either a gap or a component. The number of carrier entities will be equal to the number of carriers between the pick-up and deposit positions, inclusive. Figure 12.9 describes the necessary logic for the component and carrier entities, but omits the machine entities which will be as they were in figure 12.8.

Several points in this diagram may require a few words of explanation. The conditions necessary for PUT COMPONENT ON CONVEYOR are that there must be a COMPONENT READY TO GO ON CONVEYOR and a CARRIER AT ENTRY POINT, whereas the only condition for PUT GAP ON CONVEYOR is that there must be a CARRIER AT ENTRY POINT. Since there will always be a carrier at the entry point we can always put a gap on the conveyor. However, there will only be one carrier at the entry point so that only one of these activities can be done each time the conveyor moves forward. If a component is available we would want it to be placed on the carrier, not a gap put on it. Therefore we must ensure that the simulation processor always checks whether it can begin PUT COMPONENT ON CONVEYOR before it tries to start a PUT GAP ON CONVEYOR activity, and only puts a gap on the conveyor if no component is ready. Different packages have different methods of achieving this. Chapter 4 showed the dialogue used by CAPS to ensure that ECSL handled this type of requirement correctly, which asked the order in which the activities should be attempted on each activity scan.

We no longer need to have an activity TRAVEL ALONG CONVEYOR, because the carrier entity possesses the necessary information. Nor do we need an elaborate mechanism for identifying where each carrier is on the conveyor. The carriers in CARRIER SOMEWHERE ON CONVEYOR will be kept in correct order in the simulation package by the ranking used on the set, ie the FIFO rule. The duration of the INDEX CONVEYOR activity is the time for the conveyor to move a carrier from one position to the next, less the time to put a component on and to take a component from the conveyor, so that the total time of the three activities on either branch of the complete cycle for the carrier is the time to move on one position. This assumes that the conveyor will pause even if no component is placed on or taken from the conveyor. If this is not the case, then the duration can be adjusted according to whether a gap or a component is on the carrier. On completion of an INDEX CONVEYOR activity the conditions will exist for either TAKE COMPONENT FROM CONVEYOR or TAKE GAP FROM CONVEYOR, depending on the value of the carrier's attribute. As soon as that activity has been completed, the carrier entity returns to CARRIER AT ENTRY POINT and we can put either a component or a gap on the conveyor. The carrier taken out of CARRIER SOMEWHERE ON CONVEYOR was the one at the head of the set, and it will be added at the tail of the set and all the carriers in between will have been moved up one place automatically by the simulation software.

If we wish to display the contents of the conveyor on the screen it might be thought that we must compute the exact location of each carrier. However, once again we can use the capabilities of the software package. Most packages which provide a display require the user to specify where on the screen the first member of a set is to be displayed, the direction of the line along which the subsequent members should be displayed and the spacing between them. The package will then do the necessary computation for us, so long as we add entities to the set, and remove them from the set correctly. To distinguish

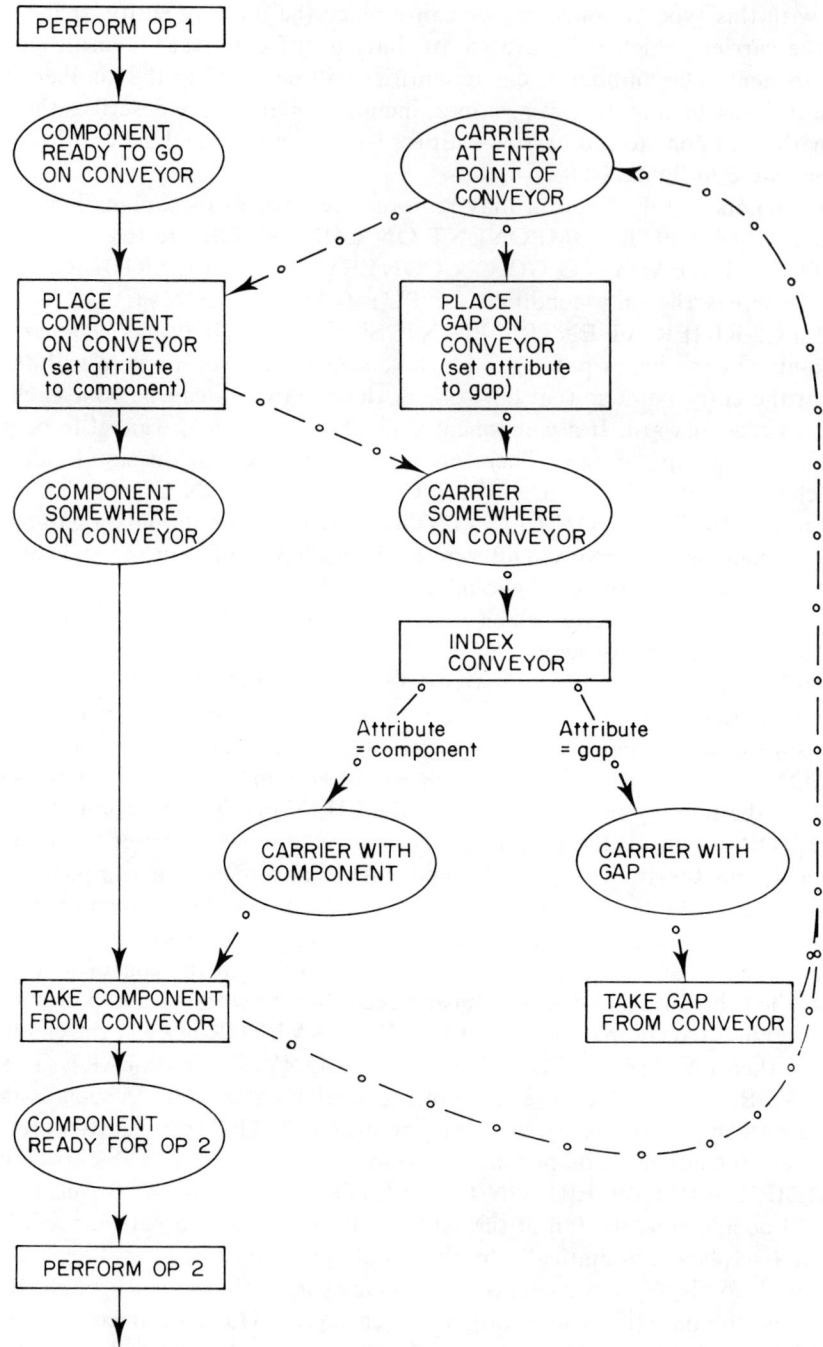

Figure 12.9   Activity cycle diagram for conveyor which has carriers at fixed spacing, or when position of components on conveyor is to be displayed.

between gaps and components we can use different colours or characters, depending on the capabilities of the software package. For example, gaps could be displayed in the background colour and components in a different colour, or we might display gaps with a space, and components with a non-space character.

12.3   AUTOMATIC STORAGE AND RETRIEVAL SYSTEMS

Automatic storage and retrieval systems (AS/RS) are included in several FMSs for such purposes as fixture stores, central buffer storage, or as a store for raw material and finished components. When building AS/RS into a model we have to consider the logic for the AS/RS itself and also the effect on the activity cycles of the other entities.

12.3.1   Basic logic for modelling AS/RS

AS/RS generally consist of one or more stacker cranes and a large number of storage locations for trays or pallets, which hold the material. The stacker crane delivers trays or pallets to operators' stations, where material can be placed in them or taken out of them, and replaces them in the store. The operators are responsible for placing material into trays and for taking them out of the trays, and perhaps other administrative duties.

The logic required to model AS/RS is very similar to some of what we have already encountered. The stacker is either idle or moving a tray to or from the operator. The operator is either idle or placing material into a tray, or taking out of a tray material which is required elsewhere, or perhaps performing some administrative task. The storage locations are either occupied or vacant. Since there are usually an equal number of storage locations and trays it is not normally necessary to model both of these. Since it is the tray which is moved around it would be better to model it, rather than the storage location. We will assume that the trays are either occupied or empty. Strictly, this is an over-simplification, since we might store several components (either all the same or all different) in one tray if they are small enough. A further qualification of these remarks is that if the components are large, or are required in large quantities, they may be supplied (from outside suppliers) already palletised so that manual handling is not required. In this case the number of pallets and locations would not necessarily be equal, and the number of locations would be an upper limit on the number of pallets which could be introduced to the system. We will overlook this case for the present.

The basic activity cycle diagram for modelling AS/RS would be as in Figure 12.10. Although this diagram gives the basic logic for an AS/RS it is not a full model. We have omitted everything associated with triggering the storage or retrieval of material, such as the arrival of material at the store or a demand for material to be retrieved, and the delivery of material away from the store. In an FMS these would be handled by other parts of the model. However, comprehensive modelling of AS/RS is a large subject and specialist simulation packages are available (eg Reference 2). We will confine our present treatment to the FMS context. The figure does not show the detailed breakdown of the movements of the stacker crane. As with vehicles in FMSs, each move consists of several sub-activities, such as travel empty, pick up tray, move loaded, deposit tray, and wait for movement request. We could apply the thinking which was described for vehicles, but, to avoid to repetition, we will adopt the simpler format shown in this figure.

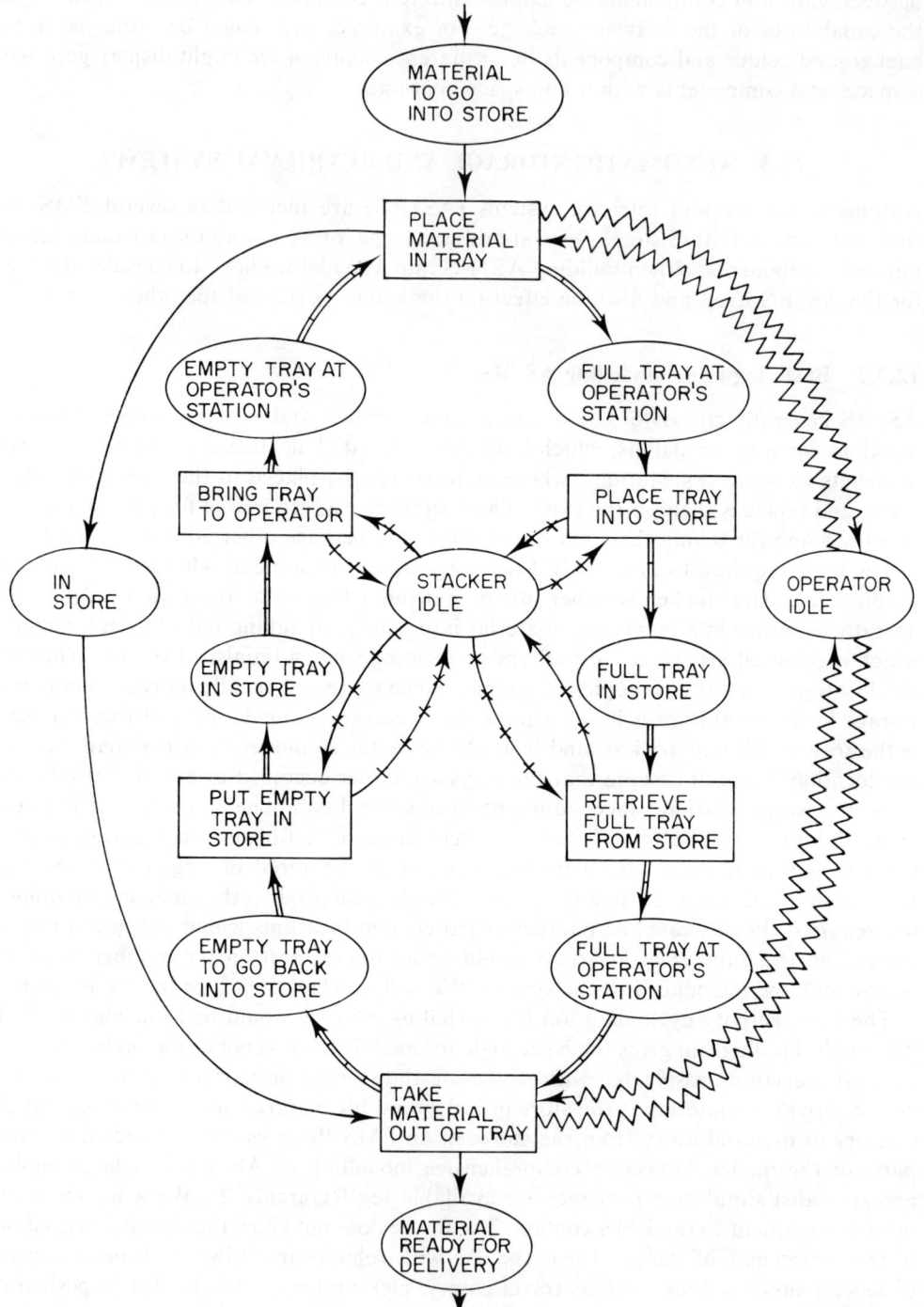

Figure 12.10   Activity cycle diagram for AS/RS.

12.3.2 Effect of AS/RS on component and pallet activity cycles

We have already discussed modelling fixture stores and central buffers, and the logic we developed in the earlier sections would need little modification to incorporate an AS/RS. On the other hand, if an AS/RS is used for storing raw and finished components, we will have to expand our discussion of the UNLOAD and LEAVE activities, since we generally avoided detailed consideration of what happened to the component after it was unloaded.

Let us consider the case where an AS/RS is used for storing raw components prior to machining and also for storing finished components after machining, prior to delivery to the assembly shop, and that material is handled by operators as described above. If we add the pallet and component activity cycles to figure 12.10 we would obtain Figure 12.11.

An attribute, type, has been introduced to distinguish between new components arriving in the store from outside and old components returning to the store after completion of machining operations. It will be observed that there is no interaction between the cycle of the pallet and of the AS/RS entities. In fact, we might define the arrival of components as being from the AS/RS, and in this way we could omit the AS/RS from the model altogether. However, if an AS/RS is being used there will almost certainly be pressure to include it in the model, so that, for example, the number of storage locations which would be required, could be assessed. This suggests that two separate models may serve as well as one large model, and would be simpler to create.

Another aspect of this diagram which bears comment is that the material must be handled three times before it is sent to the FMS. These are, once to place it in a tray to go into the AS/RS, once when it is retrieved and once to place it on a pallet for sending to the FMS. After machining, further handling operations are necessary. It would be preferable to palletise the components for handling within the FMS before they go into the AS/RS, so that when they come out for manufacture no further handling is necessary. This would be difficult to organise if the AS/RS serves other manufacturing areas or functions as well as the FMS, because it would be virtually essential to have one method of containing components in the AS/RS, and FMS-type pallets would probably not be compatible with the other functions. Another problem which would arise is that holding components on FMS pallets while in the AS/RS would greatly increase the number and cost of pallets and fixtures.

If the AS/RS is to serve only the FMS then we could store the components ready fixtured for the FMS. The AS/RS would be placed between the load–unload operations and the machining centres, and would truly be part of the FMS. In this arrangement the AS/RS serves as a central buffer and store for pallets and fixtures. The Citroen Meudon plant has a store of this type, but it is an unusual arrangement due to the quantities of raw parts requiring to be stored. Its principal function would be to hold the work to be processed during a shift or day, thereby providing a buffer of raw and in-progress components. Figure 12.12, based on figure 7.12, shows a possible layout for this arrangement.

This arrangement allows components to be delivered directly to the FMS by the stacker transferring pallets from the load–unload area across to the pallet stands, where they can be collected by a vehicle. This adds several activities and queues to the prev-

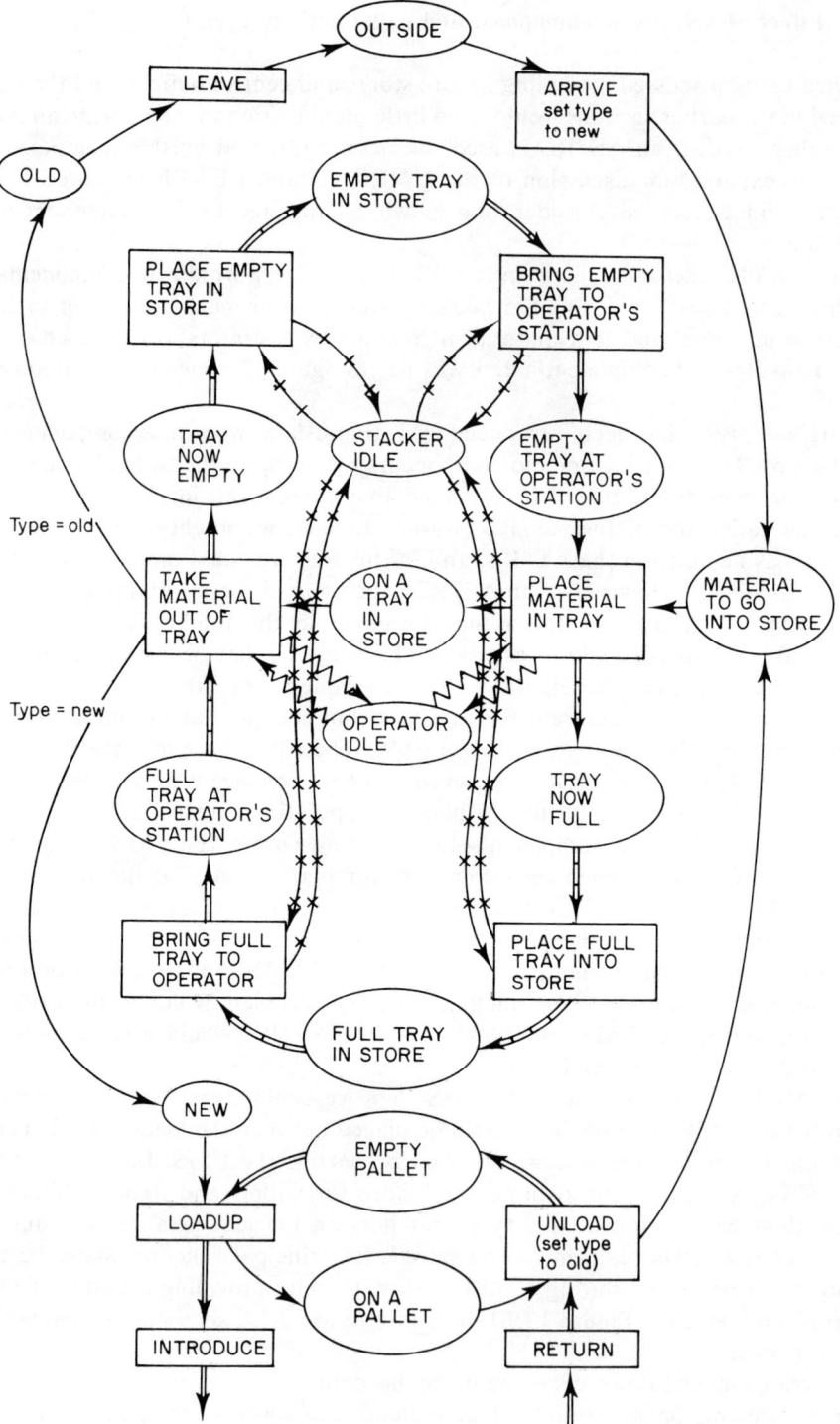

Figure 12.11   *Activity cycle diagram when AS/RS is used for storing material and finished components.*

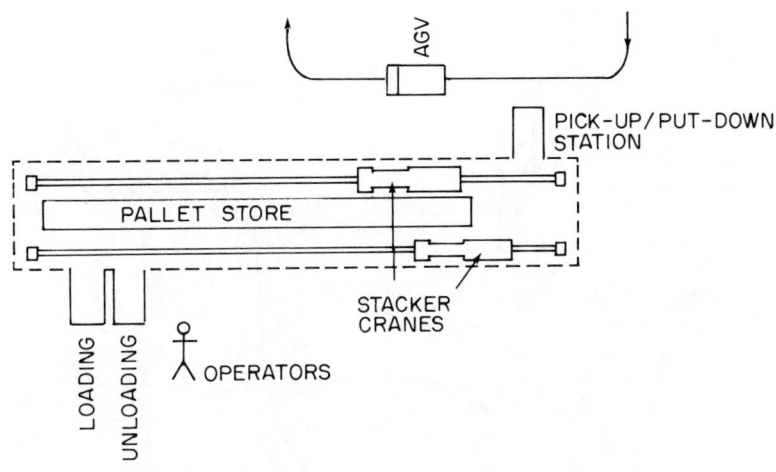

Figure 12.12   Schematic of AS/RS in FMS.

ious diagram. The activity cycle diagram would be as in Figure 12.13. The different types of movement have all been shown explicitly, rather than merge them and introduce one or more attributes, to distinguish between circumstances, as was done in figure 12.11. The movements between load stations and pallet stands, and if necessary into store, are equivalent to the six types of movement identified in the section on central buffers. Indeed the AS/RS system now serves largely as a central buffer store. In principle, although in figure 12.13 the pallets are only stored immediately after loading and before unloading, they could be brought to the store between successive operations if necessary. Of course storing components between operations inflates work in progress inventories and should be discouraged.

With this use of an AS/RS, where the store contains pallets used in the FMS, with fixtures on them, we have to ensure that the correct pallet is brought to the load station to place a new component on it. That is we have to match the type of component in the queue NEW with the type of pallet in EMPTY PALLET IN STORE. Type attributes can be used to handle this, but since the component is not involved in BRING EMPTY PALLET TO LOAD STATION this would have to be done indirectly, at least in some software packages. A convenient method is to create a movement request when a component arrives, allocate an attribute to it which defines the type of component, and make a necessary condition for BRING EMPTY PALLET TO LOAD STATION that there are active movement requests and empty pallets with matching type attributes.

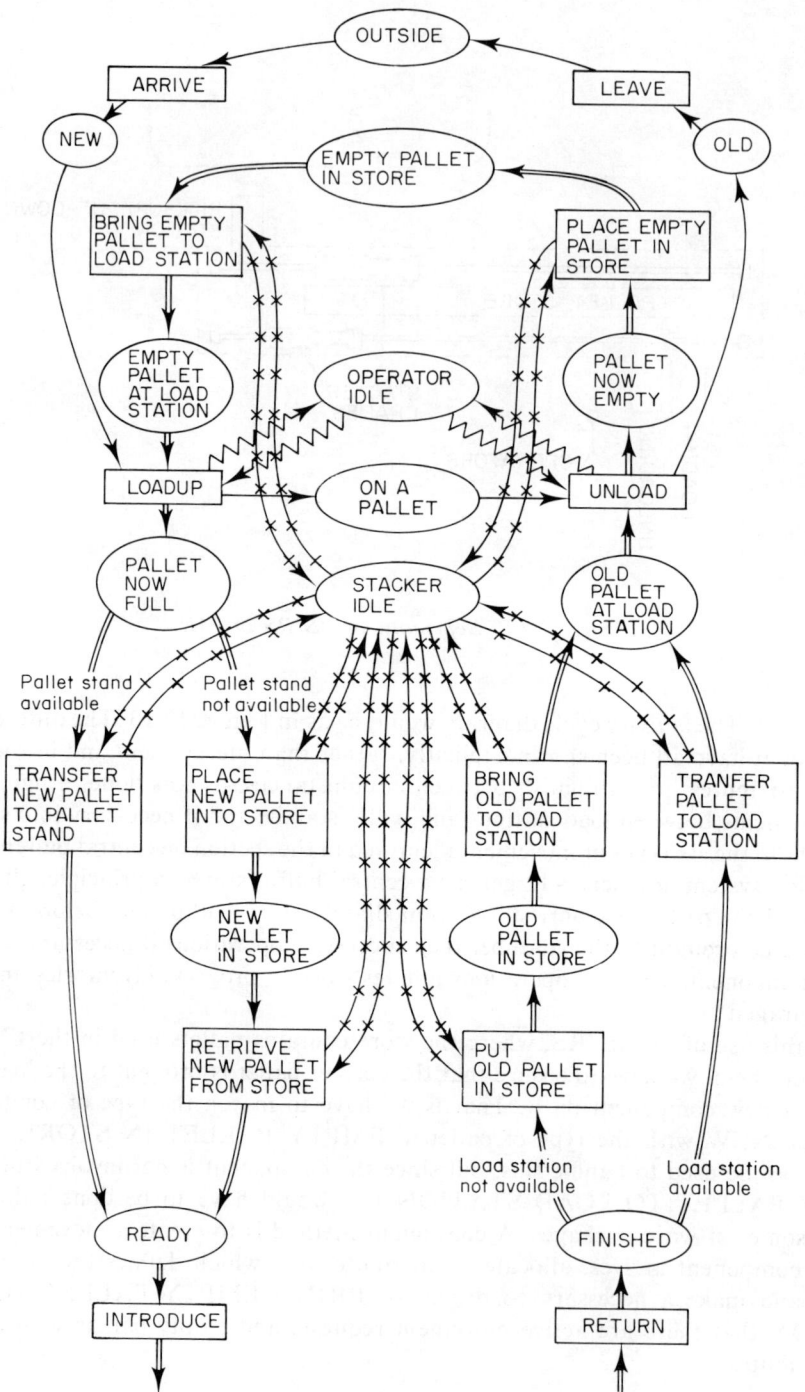

Figure 12.13   Activity cycle diagram when components are held in the AS/RS already palletised for the FMS.

12.4 MODEL TWELVE—A FLEXIBLE ASSEMBLY SYSTEM

Some of these concepts can be illustrated by considering the flexible assembly system, discussed by Ekere and Hannam,[3] whose layout is depicted in Figure 12.14. The system consists of a loop conveyor serving 25 workstations, at each of which a machine performs some assembly task. Each machine has a buffer position at which an assembly may wait for the machine to complete the operation on its previous assembly. There are six types of assembly, each of which requires a different sequence of assembly operations, and therefore visits a different set of workstations. The conveyor is of the indexing type with 30 carriers on it.

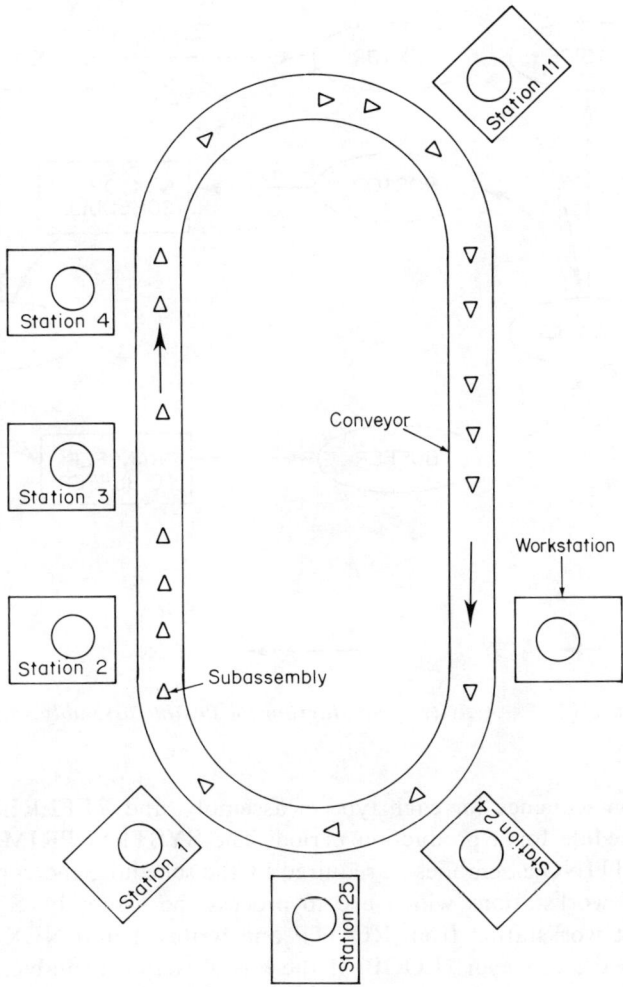

Figure 12.14 Layout of flexible assembly system, from Ekere and Hannam[3] (reproduced by permission of Dr Hannam).

Figure 12.15 is an activity cycle diagram based on that presented by Ekere and Hannam. Several interesting concepts are used, in particular in the uses of sets, to simplify the logic of the model and the activity cycle diagram. OPSEQ is a group of sets defin-

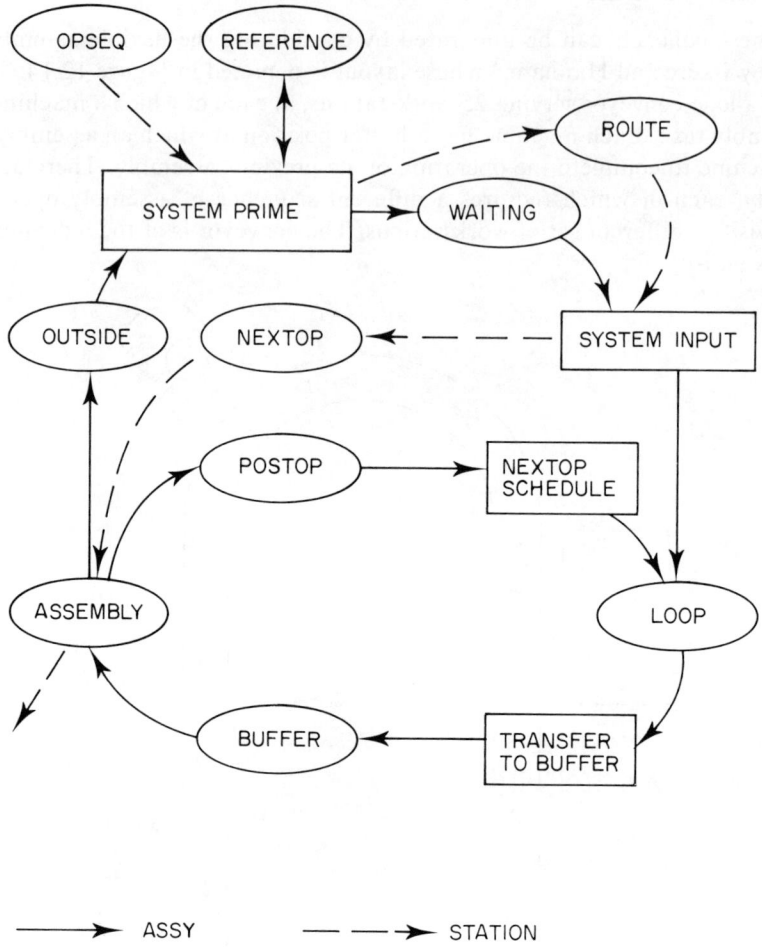

Figure 12.15   Activity cycle diagram for flexible assembly system.

ing the operation sequence for each type of assembly, and **REFERENCE** defines the requirement schedule for a production period. The **SYSTEM PRIME** activity places into the set **WAITING** assemblies as required by the schedule, and copies into **ROUTE** the sequence of workstations which are to process the assembly. **SYSTEM INPUT** extracts the first workstation from **ROUTE** and writes it into **NEXTOP**, and places the assembly on the conveyor (**LOOP**) if there is a carrier available. The assembly is taken from the conveyor and placed into the buffer of the workstation for its next operation by **TRANSFER TO BUFFER**. **ASSEMBLY** is the actual assembly operation at the workstation. It takes the assembly from the buffer, processes it and places it into a dummy queue **POSTOP** or, if this is the final workstation in the system, out of the system, ie into the queue **OUTSIDE**. **POSTOP** is a dummy queue in the sense that the following activity is a logical one rather than a physical one. **NEXTOP SCHEDULE** places the assembly back on the conveyor (in queue LOOP) and extracts the name of the workstation for the next operation from ROUTE and inserts it into NEX-TOP. Thus by the stage that an assembly has been completed the set ROUTE will be empty. This is an interesting alternative method to those suggested in chapter 10. It simplifies the logic, although the data has to be held twice, ie in both OPSEQ and ROUTE.

ECSL coding for this logic, which is based on that developed but not published by Ekere and Hannam, is given in Figure 12.16. Although only two types of entities are defined explicitly, two more are implied. INHAND is an attribute of station used to count whether the buffer at each station is occupied. An entity could have been used for the buffer space as in earlier chapters. Similarly, the variable CONVEYOR is used to count the number of assemblies on the conveyor. A type of entity representing a vacant carrier could have been used to depict this condition. INHAND is increased by one in activity TRANSFER TO BUFFER and decremented in ASSEMBLY. CONVEYOR is increased by one in SYSTEM INPUT and in ASSEMBLY if the assembly is not complete. It is decreased in TRANSFER TO BUFFER. NEXTOP defines the station for the next operation on an assembly, and the station buffer in which an assembly is located. Since the station buffers can only hold one assembly at a time, NEXTOP uniquely defines the buffer contents. Therefore there is no need to define separate buffer queues for each station, and so only one set BUFFER has been defined to hold all assemblies in station buffers.

The data used is given in figure 12.16. The operation sequences are somewhat arbitrary. The operation times were drawn from table 3.1, the random number table in chapter 3. The use of the set ROUTE avoids the need to define an array of the number of operations of each type of assembly, although the volume of data would be small.

The results obtained when the model is run for a two-shift period are given in Table 12.1. Rather few statistics have been collected; other aspects of the system could have been observed. The utilisations of the stations vary widely, from less than 20% to over 80%, reflecting the arbitrary routings.

Some observations on this model can be made. One potentially restricting feature is that, in order for an assembly to leave the conveyor, it must have an assembly operation at the final station on the line, so that the test S EQ 25 can become true. An apparent omission from the model is the time for the conveyor to circulate, so that the delay between entering the conveyor and reaching the station for the first operation is assumed to be zero. Similarly, the time to travel from one station to another is assumed to be zero. This would only be acceptable if the conveyor circulates quickly relative to the operation times. Although the operation sequences are arbitrary, the requirement that all types of assembly use station 25 is reflected in the utilisation of that station.

This model could be used for various studies in flexible assembly systems or in group technology flow lines. In fact, although the diagram is of a carousel conveyor, the logic could also be applied to a linear conveyor. Many studies have been carried out over the years on the layout of multi-product lines. The reader is invited to consider whether this model could be used to evaluate the line designs developed by, and with the data quoted by, Singleton,[6] Hollier[4] and Carrie.[1]

```
THERE ARE 180 ASSY SET REFERENCE OUTSIDE WAITING LOOP BUFFER
+POSTOP WITH ATYPE NOP
THERE ARE 25 STATION SET OPSEQ 6 ROUTE.ASSY NEXTOP.ASSY
+WITH INHAND TIMIDLE
ARRAY OPTIME(15,6) NOOPS (6)
FUNCTION PICTURE

RECYCL
RUNINZ = 2 * SHIFT
PREVCLOCK = RUNINZ
SWITCH ADD ON AFTER RUNINZ
ACTIVITIES    3 * SHIFT
DURATION = CLOCK - PREVCLOCK
ADD CONVEYOR TO HIST CONHIST(31 0 1) DURATION
FOR STATION I WITH TIME OF STATION I LE 0
   ADD DURATION TO TIMIDLE OF STATION I
PREVCLOCK = CLOCK

BEGIN SYSTEM PRIME
TIME OF SYSTEM LE 0
FOR ASSY I IN REFERENCE
     ASSY I FROM OUTSIDE INTO WAITING AND NOP OF ASSY I = 0
     ROUTE OF ASSY I GAINS OPSEQ (ATYPE)
TIME OF SYSTEM = SHIFT

BEGIN SYSTEM INPUT
CONVEYOR LE 29
FIND ASSY A IN WAITING ANY SA
FIND FIRST STATION B IN ROUTE OF ASSY A
STATION B FROM ROUTE OF ASSY A INTO NEXTOP OF ASSY A
ASSY A FROM WAITING INTO LOOP
CONVEYOR + 1
REPEAT 30

BEGIN TRANSFER TO BUFFER
FIND FIRST ASSY A IN LOOP
   FIND FIRST STATION S IN NEXTOP OF ASSY A
      INHAND OF STATION S LE 1
ASSY A FROM LOOP INTO BUFFER
CONVEYOR - 1
INHAND OF STATION S + 1
REPEAT
```

Figure 12.16   ECSL code for Model Twelve, a flexible assembly system.

```
    BEGIN ASSEMBLY
    FIND FIRST ASSY A IN BUFFER
      FIND FIRST STATION S IN NEXTOP OF ASSY A
        TIME OF STATION S LE 0
    NOP OF ASSY A + 1
    DURATION = OPTIME (NOP OF ASSY A, ATYPE OF ASSY A)
    TIME OF STATION S = DURATION
    CHAIN
      S EQ 25
      ASSY A FROM BUFFER INTO OUTSIDE AFTER DURATION
      ADD 1 TO PRODUCED
      OR ASSY A FROM BUFFER INTO POSTOP AFTER DURATION
      CONVEYOR + 1 AFTER DURATION
    INHAND OF STATION S - 1
    STATION S FROM NEXTOP OF ASSY A
    REPEAT

    BEGIN NEXTOP SCHEDULE
    FIND FIRST ASSY A IN POSTOP
    FIND FIRST STATION B IN ROUTE OF ASSY A
    STATION B FROM ROUTE OF ASSY A INTO NEXTOP OF ASSY A
    ASSY A FROM POSTOP INTO LOOP
    REPEAT

    FINALISATION
    TYPE **"Final report from simulation MODEL TWELVE"/
    TYPE 'NUMBER PRODUCED 'PRODUCED
    FOR STATION I
      TYPE 'Utilization of STATION'+4,I,(1-TIMIDLE/( 1. *(CLOCK -RUNINZ)))
    TYPE 'Histogram of number on conveyor'/PICTURE(CONHIST)

    DATA
SHIFT 480
OUTSID 1 TO *
ATYPE 30*1 30*2 30*3 30*4 30*5 30*6
REFERENCE 1 TO 20 31 TO 50 61 TO 80 91 TO 110 121 TO 140 151 TO 170
NOOPS   15 12 12 13 13 13
OPSEQ 1 1 2 3 7 8 9 10 12 13 15 16 17 18 21 25
OPSEQ 2 1 4 5 6 11 14 19 20 22 23 24 25
OPSEQ 3 2 7 8 9 10 14 19 20 21 23 24 25
OPSEQ 4 1 5 10 11 12 13 17 19 21 22 23 24 25
OPSEQ 5 2 4 6 8 10 12 14 16 18 20 22 24 25
OPSEQ 6 1 3 5 7 9 11 13 15 17 19 21 23 25
OPTIME 8 1 7 8 8 9 5 9 4 10 3 3 6 10 6
        7 9 4 5 1 3 7 1 9 3 1 8 0 0 0
        9 3 4 6 2 2 1 7 5 2 8 1 0 0 0
        8 2 10 1 10 7 6 4 3 3 5 8 8 0 0
        9 10 8 8 5 3 10 10 8 10 7 7 6 0 0
        10 9 4 4 9 7 5 6 9 2 6 9 6 0 0
SA 11111
    END
```

Figure 12.16   (continued).

Table 12.1   Results from Model Twelve, a flexible assembly system

Final report from simulation MODEL TWELVE

NUMBER PRODUCED   84

Utilization of STATION	1	.8395
Utilization of STATION	2	.4979
Utilization of STATION	3	.4645
Utilization of STATION	4	.4562
Utilization of STATION	5	.1958
Utilization of STATION	6	.3041
Utilization of STATION	7	.3875
Utilization of STATION	8	.5062
Utilization of STATION	9	.6770
Utilization of STATION	10	.4958
Utilization of STATION	10	.2104
Utilization of STATION	12	.5500
Utilization of STATION	13	.3979
Utilization of STATION	14	.3604
Utilization of STATION	15	.4187
Utilization of STATION	16	.3520
Utilization of STATION	17	.4729
Utilization of STATION	18	.3708
Utilization of STATION	19	.2812
Utilization of STATION	20	.5062
Utilization of STATION	21	.6083
Utilization of STATION	22	.4895
Utilization of STATION	23	.4583
Utilization of STATION	24	.6791
Utilization of STATION	25	.8166

Histogram of number on conveyor

CELL	FREQUENCY
23	12****
24	16*****
25	25********
26	26********
27	49***************
28	77************************
29	113**************************************
30	162**

12.5 MODEL THIRTEEN—A FLEXIBLE TRANSFER LINE

Another approach to modelling transfer lines can now be shown. An increasing number of advanced manufacturing systems are based on linked cells. Examples were given in chapter 7. The Cummins Con-rod line, for example, consists of several cells, which are connected by conveyors and which have robots for picking up the components from the input conveyor, feeding them to the machines of the cell, and then putting them on to the output conveyor.

Model Thirteen is a simple model of such a system. A plan of the system, which could be notionally based on the Cummins system, is shown in Figure 12.17. The system has six cells and seven conveyors. With the exception of the first and last, the conveyors connect a pair of cells. There is a limit on the number of components which may be present within each cell. This limit may be due to the number of working positions in the cell or the number of pallets or other type of carrier for the components. Similarly, there is a limit on the number of components which can be accommodated on each conveyor, due to its limited length. If the conveyor fills up, the components will be held up in the cell feeding the conveyor, and in an extreme case the cell and whole line could fill up.

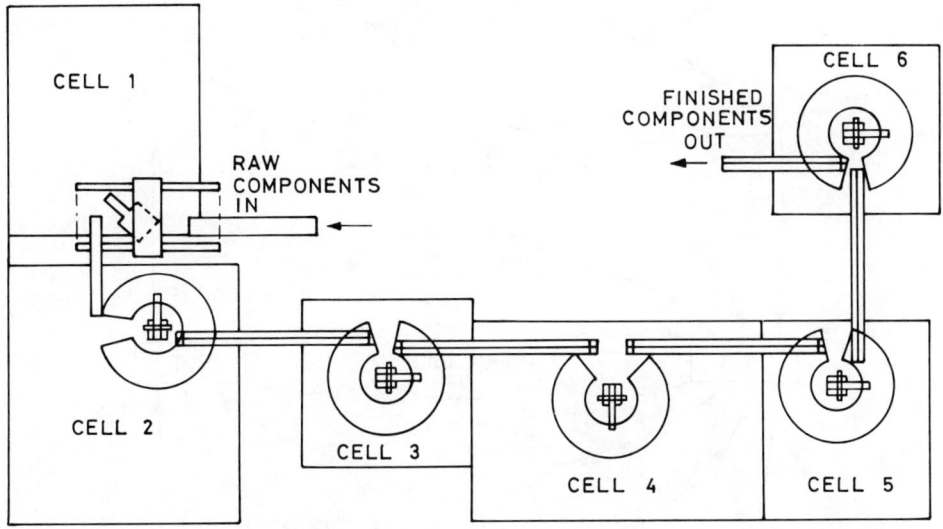

Figure 12.17   Layout of flexible transfer line for Model Thirteen.

The logic of a model of such a system can be expressed in various ways. Figure 12.18 gives an activity cycle diagram for one approach. The conveyor linking each pair of cells is represented by two queues, the output queue for the cell feeding the conveyor and the input queue of the cell fed by the conveyor. A time delay to travel along the conveyor is represented by a TRAVEL activity which moves components from the OUTQ to the INQ. A LEAVE activity has been included to represent the disappearance of components from the OUTQ of the last cell, perhaps into some container or package, and to enable the time which the item took to pass through the system to be recorded.

A queue DONE has been introduced, to represent components which have had all the work in a cell done on them and are waiting for the robot to place them on the output conveyor. To avoid unnecessarily complicating the logic of the arrival mechanism, we will assume that the first conveyor has a large capacity.

The main assumption made by the model, which would not apply in a real system, is that the robot in a cell is used only to supply the component to the cell, and to remove it from the cell after it has been processed. In a real system, the robot would probably also move the component between the processing stations within the cell, as in figure 12.4. These operations could have been added to the model, but have been excluded since we are here interested in the effects of the capacity of the cells and conveyors on system performance, and to keep the model simple.

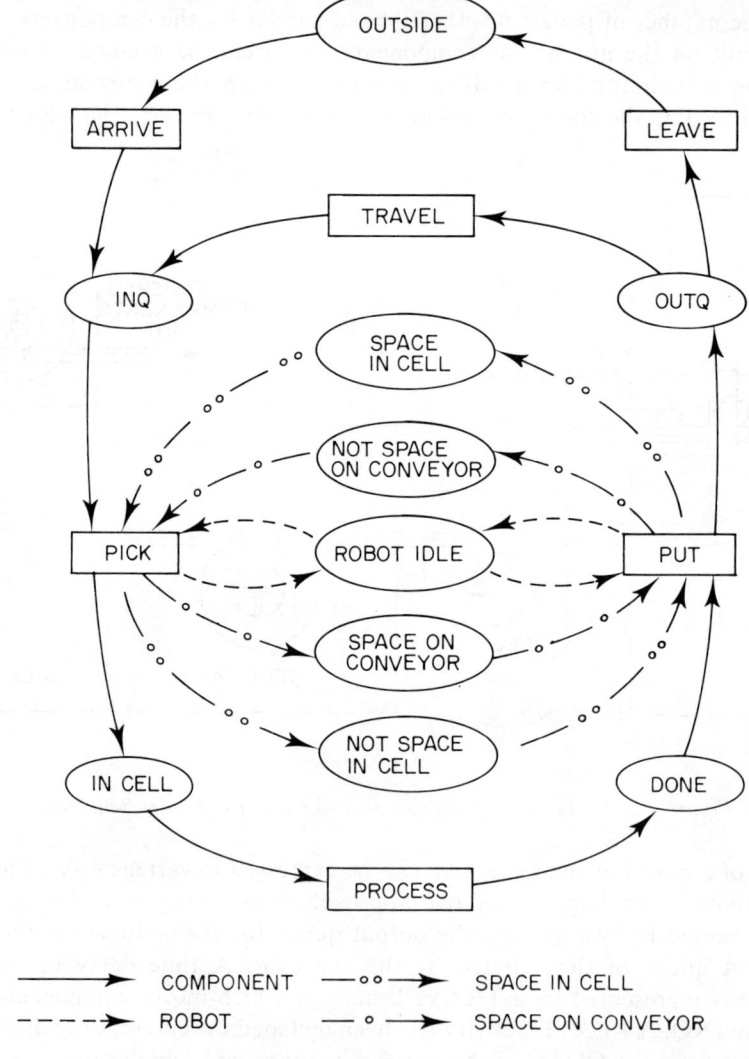

Figure 12.18   Activity cycle diagram for flexible transfer line model.

The ECSL coding for this model is given in Figure 12.19. The capacity of each cell and conveyor has been defined by attributes CLSIZE and CNSIZE, of cell and conveyor respectively, whose values are set in the data section. Another pair of attributes, INCELL and ONCONV indicate the number of components currently in each cell or on each conveyor. These attributes are equivalent to the "space on conveyor" and "space in cell" entities which were shown in the activity cycle diagram, but are slightly more convenient for expressing the statements for recording statistics.

The data section includes statements defining the mean and standard deviation of the processing time for components in each cell, TPM and TPS, the average inter-arrival time, ART, and the times for the PICK, PUT and TRavelling along the conveyor. The TRAVEL and LEAVE activities include a test, so that if the conveyor can hold only one component the time taken to travel along it will be zero. These times are not based on the times in a real system, but are arbitrary ones which can be used for experimentation.

```
THERE ARE 7 CONVEYOR WITH CNSIZE ONCONV HIST CNHIST(20 0 1)
THERE ARE 6 CELLS WITH CLSIZE INCELL HIST CLHIST(20 0 1)
+HIST DONHIST(20 0 1) HIST INQHIST(20 0 1) HIST OUTHIST(20 0 1)
THERE ARE 6 ROBOT
THERE ARE 100 COMPONENT SET INQ.CELLS DONE.CELLS OUTQ.CELLS OUTSIDE
+WITH TIMIN
THERE ARE 1 ARRIVAL
ARRAY TRTIME(7) ZIROBOT(6) TPM(6) TPS(6)
HIST TIMHIST(20 20 20)
FUNCTION PICTURE MEAN NORMAL NEGEXP
RECYCL
RUNINZ=480
SWITCH ADD ON AFTER RUNINZ
PREVCLOCK= RUNINZ
ACTIVITIES 2880
DURATION = CLOCK - PREVCLOCK
FOR ARRIVAL WITH TIME OF ARRIVAL LT 0 ADD DURATION TO ZARRI
FOR   CONVEYOR I = 1 TO 7
  ADD ONCONV OF CONVEYOR I TO HIST CNHIST OF CONVEYOR I DURATION
FOR CELLS I = 1 TO 6
  ADD OUTQ OF CELLS I TO HIST OUTHIST OF CELLS I DURATION
  ADD DONE OF CELLS I TO HIST DONHIST OF CELLS I DURATION
  ADD INQ OF CELLS I TO HIST INQHIST OF CELLS I DURATION
  ADD INCELL OF CELLS I TO HIST CLHIST OF CELLS I DURATION
FOR ROBOT I = 1 TO 6
  TIME OF ROBOT I LT 0
  ADD DURATION TO ZIROBOT (I)
PREVCLOCK = CLOCK
```

Figure 12.19   ECSL code for Model Thirteen, a flexible transfer line.

```
BEGIN ARRIVE
TIME OF ARRIVAL LE 0
FIND COMPONENT A IN OUTSIDE
COMPONENT A FROM OUTSIDE INTO INQ OF CELLS 1
DURATION = 0 + NEGEXP(ART,SA)
TIMIN OF COMPONENT A = CLOCK
TIME OF ARRIVAL = DURATION
ONCONV OF CONVEYOR 1 + 1
ADD 1 TO ARRIVE
REPEAT

BEGIN PICK
FOR CELLS I
  TIME OF ROBOT I LE 0
  INCELL OF CELLS I LT CLSIZE OF CELLS I
  FIND FIRST COMPONENT A IN INQ OF CELLS I
  DURATION = PICKTIME
  TIME OF ROBOT I = DURATION
  OPTIME = NORMAL(TPM(I) TPS(I) SB)
  DURATION = DURATION + OPTIME
  COMPONENT A FROM INQ OF CELLS I INTO DONE OF CELLS I AFTER DURATION
  ADD 1 TO PICK
  INCELL OF CELLS I + 1
  ONCONV OF CONVEYOR I - 1

BEGIN PUT
FOR CELLS I
  TIME OF ROBOT I LE 0
  J = I+1
  ONCONV OF CONVEYOR J LT CNSIZE OF CONVEYOR J
  FIND FIRST COMPONENT A IN DONE OF CELLS I
  DURATION = PUTTIME
  COMPONENT A FROM DONE OF CELLS I INTO OUTQ OF CELLS I AFTER DURATION
  ONCONV OF CONVEYOR J + 1
  INCELL OF CELLS I - 1
  TIME OF ROBOT I = DURATION
  ADD 1 TO PUT

BEGIN TRAVEL
FOR CONVEYOR J = 2 TO 6
  I = J - 1
  FIND FIRST COMPONENT A IN OUTQ OF CELLS I
  CHAIN
    CNSIZE OF CONVEYOR J GT 1
    DURATION = TRTIME(J)
    OR DURATION = 0
  COMPONENT A FROM OUTQ OF CELLS I INTO INQ OF CELLS J AFTER DURATION
  ADD 1 TO TRAVEL

BEGIN LEAVE
FIND FIRST COMPONENT A IN OUTQ OF CELLS 6
CHAIN
  CNSIZE OF CONVEYOR 7 GT 1
  DURATION = TRTIME(7)
  OR DURATION = 0
COMPONENT A FROM OUTQ OF CELLS 6 INTO OUTSIDE AFTER DURATION
TIMIN OF COMPONENT A = CLOCK + DURATION - TIMIN OF COMPONENT A
ADD TIMIN OF COMPONENT A TO TIMHIST
ONCONV OF CONVEYOR 7 - 1
ADD 1 TO LEAVE
REPEAT
```

Figure 12.19   (continued).

```
FINALISATION
TYPE 'ARRIVE was started ' ARRIVE ' times'
TYPE 'PUT    was started ' PUT ' times'
TYPE 'PICK   was started ' PICK ' times'
TYPE 'LEAVE  was started ' LEAVE ' times'
TYPE 'TRAVEL was started ' TRAVEL ' times'
TYPE 'Utilisation of ARRIVAL'+4, (1-ZARRI/(1.*(CLOCK-RUNINZ)))
FOR ROBOT I = 1 TO 6
   TYPE 'Utilisation of robot'+4,I,(1-ZIROBOT(I)/(1.*(CLOCK-RUNINZ)
TYPE'Histogram of time in system'/PICTURE(TIMHIST)
FOR CELLS I
   TYPE 'Histogram of INQ'I,/PICTURE(INQHIST)
   TYPE 'Histogram of DONE'I,/PICTURE(DONHIST)
   TYPE 'Histogram of OUTQ'I,/PICTURE(OUTHIST)
   TYPE 'Histogram of number in Cell'I/PICTURE(CLHIST)
FOR CONVEYOR I
   TYPE'Histogram of number on Conveyor'I/PICTURE(CNHIST)
TYPE'       CELL    AV. NO.     AV. NO.    AV. NO.    AV. NO.'
TYPE'       NUMBER  IN INQ      IN DONE    IN OUTQ    IN CELL'
FOR CELLS I
   TYPE+4,I,MEAN(INQHIST),MEAN(DONHIST),MEAN(OUTHIST),MEAN(CLHIST)
FOR CONVEYOR I
   TYPE'Average number on conveyor'+4,I,MEAN(CNHIST)

 DATA
OUTSIDE 1 TO *
CNSIZE 100 1 1 1 1 1
CLSIZE 1 1 1 1 1 1
TPM 4 4 4 4 4 4
TPS 1 1 1 1 1 1
ART 6
SA 11111
SB 12221
PICKTIME 1
PUTTIME 1
TRTIME 1 1 1 1 1 1 1
   END
```

Figure 12.19   (continued).

12.5.1   Results

The model can be used to examine the sensitivity of the system to the capacity of the conveyors and cells. There is a substantial literature on the effects of varying the capacity of inter-operation buffers in production lines with variable operation times and workstations which are subject to breakdown. Some papers on the subject are more than 30 years old, eg Koenigsberg.[5] These studies show that increasing buffer capacity yields increased system output until a limit is reached. In this case the machines are not subject to breakdown, but the operation times are stochastic, so we would expect the system output to increase with increasing conveyor length.

The model was run with various values of cell and conveyor capacity. Some of the results obtained when this model was run (for one shift) with cell and conveyor capacities of only one component are given in Table 12.2. The utilisation of the dummy entity "arrival" was less than 1.0, indicating that the system became starved. We also observe that the number of LEAVE activities, 389, was well below the number of ARRIVES, 461, we can conclude that the system cannot provide the intended output. This is also indicated by the histogram of the time taken by components to pass through the system which shows values overflowing the upper cell value. Remember that in ECSL the last cell collects all instances greater than the upper limit of the second last cell. Therefore many components took well over 400 minutes to pass through the system, compared to an average work content of just 36 minutes, including pick and put operations.

Table 12.2   Some results from Model Thirteen, a flexible transfer line, with cell and conveyor capacities of one component

ARRIVE	was started	461 times
PUT	was started	2328 times
PICK	was started	2328 times
LEAVE	was started	389 times
TRAVEL	was started	1939 times
Utilisation of ARRIVAL		.9916
Utilisation of robot 1		.3237
Utilisation of robot 2		.3233
Utilisation of robot 3		.3225
Utilisation of robot 4		.3225
Utilisation of robot 5		.3233
Utilisation of robot 6		.3237

Histogram of time in system

CELL	FREQUENCY
100	10***
120	48****************
140	35***********
160	19******
180	13****
200	14****
220	4*
240	8**
260	7**
280	8**
300	20******
320	28*********
340	8**
360	18******
380	12****
400	137***

Table 12.3 gives some results when the conveyors have capacities of 2, 3 and 10 components respectively. These show only a marginal improvement in output over that obtained when the conveyor capacity is one component. The system output does not rise above 394 components, and the required output still cannot be met.

Table 12.4 gives results when the number of components which can be in a cell is increased to 2, 3 and 5, for a conveyor capacity of just one component. There is a dramatic improvement, and the system appears capable of producing the required output. The model was also run with both conveyor and cell having a capacity of two components, but the results showed no improvement over the case where the conveyor had a capacity of one.

Table 12.3. Results from Model Thirteen, with increased conveyor capacity, and cell capacity = 1

Conveyor capacity:		1	2	3	10
ARRIVE	was started	461	467	467	467 times
PUT	was started	2328	2362	2364	2381 times
PICK	was started	2328	2363	2367	2383 times
LEAVE	was started	309	393	394	394 times
TRAVEL	was started	1939	1969	1972	1989 times
Utilisation of ARRIVAL	0.9916	1.0000	1.0000	1.0000	

Table 12.4 Results from Model Thirteen, with increased cell capacity, and conveyor capacity = 1

Cell capacity:		1	2	3	12
ARRIVE	was started	461	467	467	467 times
PUT	was started	2328	2786	2784	2784 times
PICK	was started	2328	2785	2786	2786 times
LEAVE	was started	309	464	464	464 times
TRAVEL	was started	1939	2319	2319	2319 times
Utilisation of ARRIVAL	0.9916	1.0000	1.0000	1.0000	

Table 12.5 gives additional results, such as average queue lengths, for the case when the cells have capacity of 2 components. Although they are not quoted the results for the cases with larger cell capacity showed that the additional capacity was fully used on a few occasions. There was no effect on output, and the only major difference was that the number of components waiting on conveyor one to enter the system was slightly reduced. In other words, when the cells could absorb more components there was less need for a large capacity input conveyor at the start of the system.

These results, which show that the conveyor capacity had no effect on the output of the system, are perhaps surprising. In the real Cummins Con-rod line the conveyors have a capacity of about twelve components. This may be more than necessary, but before any such suggestion can be seriously considered, a more detailed model would have to be developed.

Table 12.5   Additional results from Model Thirteen, for cell capacity = 2 and conveyor capacity = 1

Utilisation of robot	1	.3879
Utilisation of robot	2	.3879
Utilisation of robot	3	.3870
Utilisation of robot	4	.3858
Utilisation of robot	5	.3850
Utilisation of robot	6	.3862

Histogram of time in system
```
CELL       FREQUENCY
 40        225**************************************************************
 50        229*************************************************************
 60         10**
```

Histogram of number in Cell   1
```
CELL       FREQUENCY
  0        731********************************************
  1        881**********************************************************
  2        788*****************************************************
```

Histogram of number in Cell   2
```
CELL       FREQUENCY
  0        703******************************************
  1        988*****************************************************************
  2        709*******************************************
```

Histogram of number in Cell   3
```
CELL       FREQUENCY
  0        692************************************
  1       1007*******************************************************
  2        701*************************************
```

Histogram of number in Cell   4
```
CELL       FREQUENCY
  0        674**********************************
  1       1050**************************************************************
  2        676*********************************
```

Histogram of number in Cell   5
```
CELL       FREQUENCY
  0        674**********************************
  1       1070*********************************************************
  2        656********************************
```

Histogram of number in Cell   6
```
CELL       FREQUENCY
  0        656********************************
  1       1069*********************************************************
  2        675**********************************
```

Table 12.5   (*continued*)

Histogram of number on Conveyor   1
CELL   FREQUENCY
  0   1789**
  1   358************
  2   170*****
  3   60**
  4   19
  5   4

Histogram of number on Conveyor   2
CELL   FREQUENCY
  0   1371***
  1   1029***

Histogram of number on Conveyor   3
CELL   FREQUENCY
  0   1365***
  1   1035**

Histogram of number on Conveyor   4
CELL   FREQUENCY
  0   1402**
  1   998**

Histogram of number on Conveyor   5
CELL   FREQUENCY
  0   1414**
  1   986***

Histogram of number on Conveyor   6
CELL   FREQUENCY
  0   1408**
  1   992**

Histogram of number on Conveyor   7
CELL   FREQUENCY
  0   1936**
  1   464**************

CELL NUMBER	AV. NO. IN INQ	AV. NO. IN DONE	AV. NO. IN OUTQ	AV. NO. IN CELL
1	.4058	.0483	.0000	1.0237
2	.2350	.0370	.0000	1.0025
3	.2375	.0350	.0000	1.0037
4	.2225	.0312	.0000	1.0008
5	.2179	.0383	.0000	.9925
6	.2208	.0320	.0000	1.0079

Average number on conveyor	1	.4058
Average number on conveyor	2	.4287
Average number on conveyor	3	.4312
Average number on conveyor	4	.4158
Average number on conveyor	5	.4108
Average number on conveyor	6	.4133
Average number on conveyor	7	.1933

SUMMARY

In this chapter, we have discussed the logic associated with modelling robots, conveyors and automated storage and retrieval systems.

Models of a flexible assembly system and a flexible transfer line have been presented. Despite the fact that arbitrary data was used in both cases to run the model, the models could be used for serious experimentation. However, various aspects, such as the operations done within the cells in the transfer line case, would have to be modelled in more detail. The results obtained suggest that the capacity of the conveyors linking the cells in a flexible transfer line is of less significance than the capacity of the cells themselves.

EXERCISES

1. In figure 12.9, if, when a component reaches the end of the conveyor, machine 2 is busy and the component has to re-circulate, how should the figure be amended?

2. Add to figure 12.13 the cycle for a movement request to ensure matching of pallet and component types prior to loading a component on a pallet.

3. How could the time to travel around the loop conveyor be added to Model Twelve?

4. What other results statistics might it be worthwhile obtaining from Models Twelve and Thirteen?

5. What further experiments would you suggest carrying out to investigate more fully the performance of the systems in Models Twelve and Thirteen?

6. Why do you think that increasing the capacity of the conveyors in Model Thirteen had so little effect?

REFERENCES

1. Carrie, A.S., Layout of multi-product lines, *Int. J. Prod. Res.*, 13, 6, 541, 1975.

2. Dangelmaier, W. and Bachers, R., "SIMULAP - a simulation system for material flow and warehouse design", *Proc. SIM-1, First International Conference on Simulation in Manufacturing*, Stratford-upon-Avon, March 1985, IFS (Publications) Ltd, Bedford, England, 1985.

3. Ekere, N.N. and Hannam, R.G., A comparison of two approaches to the simulation of manufacturing systems, *Proc. 26th MTDR Conference*, Manchester, 1986, pp185–190, UMIST and Macmillan Press, Basingstoke, England, 1986.

4. Hollier R.H., The layout of multi-product lines, *Int. J. Prod. Res.*, 2, 47, 1963.

5. Koenigsberg, E., Production lines and internal storage—a review, *Management Science*, 5, 410, 1959.

6. Singleton, W.T., Optimum sequencing of operations for batch production, *Work Study and Industrial Engineering*, 6, 3, 1962.

CHAPTER 13

On simulation projects

This book has had two parts so far. Chapters 1 to 6 were about simulation as a technique. Chapters 7 to 12 dealt with its application to manufacturing systems. They concentrated on the "How" of simulation. We will now conclude with a more general discussion of the role of simulation, and the conduct of simulation projects. We will also speculate on some of the developments likely to become available over the next few years.

13.1 BUILDING AN FMS MODEL

We can begin with a reminder of the steps involved in a simulation project.[1,2] In general, the steps in modelling an FMS, or any manufacturing system, are:

— define the system to be modelled,
— build the model,
— collect data,
— validate the model,
— run experiments using the model,
— revise the model and repeat experiments.

As has been observed in chapter 6, the second and fourth steps may be largely eliminated if a generic package is used. Some comments on these may be helpful.

13.1.1 Define the system

Unless we are fully familiar with the system and its operation we cannot expect to model it effectively. As we have discussed in the preceding chapters, questions which we need to answer include: What is the basic configuration of the FMS? How many machines are there to be? Are they all of the same type? Are they grouped in some way? Is transport by conveyor, or by AGV? How many AGVs are there to be? Do they follow a single track, or are there junctions? Are parts palletised? How are loading and unloading done? What work in process storage facilities are included? Are there buffer positions for each machine? What about fixtures and tools? How many part types are to be produced? Do they have similar processing requirements? How are operations allocated to machines? How is the production programme determined? This information defines the entities and activities involved in the model, and the logic which determines the conditions which are required for each activity to occur.

Of just as much importance is the nature of the results which we want the model to provide. We will certainly wish to obtain facility utilisations, ouput of the system, throughput times and so on. But we will also need to investigate the more subtle factors which limit the performance of the system. We may not be aware of what these are until we have done some experimentation. Then there is the question of how the results should be presented? Is tabular or graphical output required?

Similarly, the data the model will need must be considered. What are the operation times? How are transport times determined? What information is available concerning reliability, breakdowns and maintenance? The availability and accuracy of data may well constrain the model and the reliability of its results.

13.1.2 Build the model

Our description of the real system must be turned into a computer model. In chapter 6 we indicated that this may involve writing a computer program, or it may merely require us to specify the data for a generic model via menus or structured data file. Defining the model involves various compromises and decisions. The level of detail which the model involves must be defined. In the model building chapters, 8 to 12, we discussed such questions as can the movement of a pallet from an AGV to a machine be considered as instantaneous? Can the movement from one machine to another be considered as one composite activity? If not, we must include additional activities in the model and specify the logic involved. Simulation involves an abstraction of the real world. No model is ever identical to the real system. Some detail is lost, and the results obtained may reflect this.

Most systems are subject to random influences, and simulation models have to allow for these effects, usually by drawing random samples from statistical distributions. This makes their results probabilistic and places some uncertainty on their accuracy. Consequently, simulation is more appropriate for comparing alternatives than providing precise predictions of performance. In modelling FMSs, however, we have a major advantage over modelling other types of system. Since FMSs are computer controlled, many of the sources of random disturbances have been eliminated, and the rules governing the behaviour of the system are precisely defined. There can therefore be much greater confidence in simulation results.

Building the model is not complete until the model has been de-bugged, ie the code is a correct implementation of the model. The magnitude of this task depends greatly on the package being used. With simulation languages this can be time-consuming. With generic models it is a function of how well the package and the format of its "data deck" has been understood. The quality of the documentation is sometimes less than might be hoped for.

13.1.3 Collect data

Simulation needs data to work on, and that is usually in short supply until well on into an FMS project. The quality of available data may be a key factor in determining the level of detail and accuracy of the model. Rarely will the operation times on conventional methods be a useful guide to times in an FMS. Operation times will probably be revised several times during the commissioning and operation of the system. One of the important skills of a simulation expert is in knowing how to summarise the data to simplify the modelling process and to minimise the sensitivity of the results to errors in data estimates.

13.1.4 Validate the model

In addition to debugging the program, the model must be validated to ensure that the results are reliable. We have referred to such factors as the length of the run and the run-in period, and the random number stream seeds. But there is also the matter of the acceptability of any simplifying assumptions. Usually validation is done by comparing the results from the model with those of the real system. In FMS simulation the system has normally not yet been built, so this is not possible. If the system is already in existence, the need for simulation may have passed by. With generic model packages, the problem of validation does not usually arise, because the user has little or no means of doing anything about it. All he or she can do is to ensure the validity of the data supplied to the package. He or she is, however, at the mercy of the package writer and any assumptions which have been written into the package. There is also the regrettable fact that the package may have bugs in it, irrespective of its type. Thorough testing is essential.

13.1.5 Run the model

Once the model has been validated it can be run to assess the performance of the system. We will probably perform many runs with different data to assess the sensitivity of the system's performance to various factors and to identify the limiting ones.

13.1.6 Revise the model and repeat experiments

At the initial stage answers are required to questions concerning the system's overall performance and capacity, while later on information will be sought about the transporter priority rules, about tool management and other questions of day-to-day operations. This will necessitate revisions to the model. Perhaps these can be accommodated easily, but perhaps not. Perhaps a substantially new model will be required. The package on which the early models were written may be unsuitable for the later, probably more detailed requirements. Also as time progresses, planning will be done in more detail and more accurate data on which the system may be evaluated will emerge. This will enable earlier experiments to be repeated more accurately.

As a result of these considerations, simulating a system is not likely to be a once-off exercise. It will go on throughout the planning, commissioning and operation of the system. It is easy to underestimate the time and man-hours involved.

13.2 EXAMPLE OF A SIMULATION PROJECT

To illustrate some of aspects of FMS simulation, a fairly typical simulation project will be described. This relates to modelling the Anderson Strathclyde FMS. The use of the word "typical" is always questionable, but although the nature of that system and the questions which arose during the project were of course specific to that system, the simulation project as a whole was probably representative of such projects in general. We will consider the nature of the project rather than the system-specific issues, which have been reported in detail elsewhere.[3,4]

As the project progressed there were several stages of modelling, when assumptions and data were gradually refined or different questions were being asked. The stages were:

— concept stage
— preliminary design stage
— firm design stage, developing a valid model
— experimenting with the model
— operational stage, expanded models.

13.2.1   Concept stage

At the concept stage of a project it is necessary to know whether the investment in an FMS will be worthwhile. Management need ball-park figures on which to consider the costs and benefits involved, and the sensitivity of the return on investment to the imponderables of a project. Frequently, there will be a call for a model to help the decision making process. Unfortunately, as we have shown, simulation involves a high degree of detail, which is not available at this stage. In addition, the time scales involved in simulation are usually far too long for management's needs at this stage. Management seldom appreciate these facts. In this project, some preliminary model building was done to get the modellers tuned-in until details of the system design were provided by the supplier.

13.2.2   Preliminary design stage

The supplier's first proposal was for a system of eight machines in which castings would be machined in batches, with one type of boom and one type of gearbox in the system at a time. This would minimise problems of balance of operations between types of part, but at the expense of interruptions to production while the machines were retooled for the next set of parts. At this stage, very little accurate information was available concerning operation times, tooling and how the parts were to be fixtured and palletised. Consequently several simplifying assumptions were made and a simple model was quickly developed using CAPS, the code generator for ECSL. The model was, in fact, somewhat simpler than Model Seven of chapter 10. The results suggested a measure of over capacity, and re-inforced the company's reservations about the design of the system, which was clearly not very flexible. The supplier re-examined the issue and produced an improved design.

This experience is probably quite common, in that first thoughts on a system design will frequently be changed, possibly radically, when subjected to detailed analysis. It should be remembered that the Anderson Strathclyde was one of the earliest systems and that since it was designed much experience has been built up.

13.2.3   Firm design stage—developing a valid model

When the supplier submitted the revised proposal, which was accepted, there was a firm design on which modelling could begin in earnest. The earlier model was revised for the new configuration. No operation times were yet available for the special facing-head machine. This posed a problem for the modellers. Should assumptions be made concerning these times, or should the machine be omitted from the model? The machine was omitted. The supplier was using MAST to model the system in support of his design calculations, and he had omitted this machine. By omitting it in our model also comparability of results would be enhanced.

In running the model it was found that the system could become blocked, just as has been observed in some of the models in chapters 8 to 11. However at this time there was little experience of this problem in real systems, and, naturally, an error in the model logic was suspected. However, on investigating closely, it appeared to be due to the nature of the system itself not the model logic. To resolve this problem, activities were added to the model whereby, when this situation arose, the pallet at the load station would be taken off, and laid aside until the congestion was relieved. By experimentation, it was found that the congestion could be reduced by changing the AGV's priority rules, ie the relative priority of the activities which use the vehicle.

At about this time, information on fixturing methods became available. The model was revised to include this data. The results showed, as have our models in the earlier chapters, that the number of pallets in the system strongly influenced the time in the system, the utilisation of the load–unload station and the number of "take-offs". The results showed good correlation with the supplier's results, with the exceptions of the load–unload area and cart utilisations, where the differences could be attributed to the simplifying assumptions.

The problem of congestion was a matter of concern. The model assumed that pallets could be taken out of the system if necessary at the load–unload area. However, the company did not wish to lift pallets off the system to avoid the risk of damage to locating edges. It was thought that the assumptions in modelling the load–unload area were contributing to the problem, and that in a more detailed model the situation might not arise. At this time more data became available about the procedures at the load–unload area. The model was revised again, and now was somewhat similar to Model Five of chapter 10. The problem of congestion observed in the initial models was found to recur, and was referred to the supplier for comment. It took some time to convince him of the truth of the finding, because he had not encountered it in any of his previous systems. However, when he ran the control system software in simulation mode using the same data, he found the same result, and agreed to amend the control system software to cope with the situation.

The model was revised again to include the movement of empty pallets around the system and to clear blockages. The model now resembled Model Eight of chapter 11, and included features specific to the system which were not available in the supplier's MAST model.

This stage is typical of many projects, that of gradually developing a more elaborate and accurate model. In this case a simulation language was used, so changes took the form of re-writing the code of the model. If a data driven or generic model had been used, the model would have been redefined. This would probably have been a straightforward and quick process, but if the system has unusual features it might be difficult to specify them.

13.2.4 Experimenting with the model

This model was a good working model which could be used for experiments to evaluate various aspects of the system. For example, the effects of failure of a pallet stand were investigated, since it was thought that the micro-switches on them might be easily damaged or clogged by swarf. Another enquiry concerned the inclusion of a vacuum cleaning station where a robot might clean out swarf from the cavities of the castings. Various aspects of system performance were studied:

— the loss of output by having to move empty pallets about the system
— the loss of output due to moving pallets to avoid blockages
— the possibility that additional roughing passes would be needed, with longer roughing but shorter finishing operations, due to the dimensions of the castings being less precise than hoped for
— the possibility that facing-head operation times were underestimated
— the effects of having extra sets of fixtures
— the effects of failure of pallet stands
— the effects of revised operation times as NC tapes were proved-out on the machines. These showed that earlier estimates were optimistic.

These are bread and butter questions of simulation which are easy to assess once a sound model has been developed.

13.2.5 Operational stage—expanded models

Once such a system becomes operational, attention will switch to questions such as scheduling rules. The flexibility provided to the user by the control software is an important parameter. Once operational, additional components will be planned for manufacture on the system. Over the past few years the number of parts being produced on the system has increased from the initial half-dozen to over fifty. This raises new kinds of questions and problems. The factors which now became important may have been overlooked in the earlier, more general models. For example, in the Anderson Strathclyde system one of the operational problems is tool management. The total number of tools need to manufacture the parts far exceeds the magazine capacities. Therefore the system manager has to decide each week or so which of the components on his schedule he will start to produce and which tools must be loaded into the magazines. He must also decide when to revise the tool allocation and which new parts to launch on the system.

The volume of data involved if a simulation model is to include tooling information is very large. Many simulation packages are not designed to cope with this amount of data. In this case, the model was re-written in FORTRAN and mounted on the system's control computer, so that the system manager could enter the schedule, run the simulation, see how it seems to work and revise it if necessary.[6]

13.2.6 Evolution of the model and the data

From this commentary it can be seen that throughout a simulation project, and indeed during any major project, changes take place in system design, operational procedures, and in the issues to be addressed. In this case there was steady evolution of the model from the inital planning stages through to system operation. In parallel with this the data was being redefined. Initially operation times were based upon rough estimates. These were found to have been optimistic when detailed tape preparation was undertaken. Earlier estimates of performance and bottlenecks had to be revised from time to time.

The factors limiting system performance too will change after the system becomes operational. As mentioned above, one of the problems which showed up early on was the problem of congestion and the need to move empty pallets around the system. In practice, as the number of parts planned for production on the system built up, so did the number of fixtures available. By operating with a modest buffer of fresh or partly machined components, it has been found possible to always have a fixtured component ready when a pallet would otherwise have remained empty. Thus one of the early problems has disappeared.

13.3 THE COSTS AND BENEFITS OF SIMULATION

Simulation is a tool which may be used to provide answers to technical problems. However, like any tool there is a cost, and that cost must be justified. The costs of simulation depend on many factors, but the principal two are:

— the cost of the package and the hardware to run it,
— the salary cost of the time of those who do the simulation.

The software costs can be from one or two thousand pounds (or even less with some of the simpler packages) up to twenty thousand or more for the more powerful packages, with a few being much more expensive. Even at the upper end the cost is small compared to the cost of sophisticated simulators such as flight simulators. The hardware cost may be ignored, if suitable hardware exists and can be made available. Most packages are written for IBM-compatible PCs, possibly with enhanced graphics or some other graphics board. Even if hardware has to be purchased the cost is not likely to be an obstacle if valuable results are likely to be obtained.

The salary costs are more difficult to predict. Much depends on the experience of the persons concerned and the ease of use of the package. From the above discussion concerning the project on the Anderson's system, it is clear that some simulation activity was going on over a period of years, rather than a few weeks or months. Of course, it was not a full-time occupation during the whole life of the project. However, it is easy to underestimate the man-hours which can be absorbed by a simulation project. It is tempting to repeat every experiment when some piece of new data emerges. We are more comfortable with results obtained from the "right" data rather than to have results obtained from what we know are obsolete data. Part of the simulation expert's skill is to know whether the change is likely to have any appreciable effect on the results.

Some sources have been quoted as saying that a company should be prepared to spend about 2% of the total project cost on simulation. By way of contrast, there are many instances of FMS installations where simulation was not used. The project manager in such cases may well make a remark such as "I could not see what simulation could tell me which I could not discover from my knowedge of the system". This is undoubtedly a rather shortsighted view of simulation, although it may well be true so far as any one project goes. As the experiments in chapter 8 and thereafter have showed, complex systems have complex interactions which cannot be easily predicted without simulation. There is always the risk when undertaking a new venture that the project will come unstuck because of some unforeseen eventuality. It is wise to spend some time and money in simulation rather than be caught out later.

Many cases have been reported of substantial savings achieved as a result of simulation. For example, in the Hattersley Newman Hender system, initial predictions suggested that twelve vehicles would be required. Simulation showed that seven or eight should be sufficient. Other cases describe how a simulation exercise enabled decision rules to be improved so that more output could be obtained with the same equipment, or the same output with less equipment. A saving of one machining centre is likely to be worth many times the simulation cost.

It is also difficult to say whether a saving arose from the simulation exercise, or from the general atmosphere of enquiry and deeper thought about a system which is often necessary to develop a model. Simulation makes people think about their system and often to challenge the implied assumptions about how it would operate. Thus savings are often attributable, less to the results of the simulation than to the simple fact that a simulation exercise was undertaken.

It is very difficult to state any general rule concerning the costs and benefits of simulation. If the system is simple, or a repeat of a previous installation then there will be few risks. However, if there are novelties or complexities in the system, it would be foolhardy not to undertake a simulation exercise. Even in such simple models as have been presented here, we have shown that simulation can reveal important findings concerning congestion, numbers of pallets and decision rules.

Hollocks[5] summarises this discussion in the following terms:

Evidence shows that companies and organisations which employ simulation in their evaluations of Advanced Manufacturing (and other) Systems find it pays off handsomely and go on to use simulation more widely. Reductions in the capital cost of a development scheme of 5–10% are common, but, more importantly, AMS systems can be installed with confidence that they can meet their design requirements and with a better insight into their operation.

SUMMARY

Simulation projects tend to follow a pattern. In the early stages the overall capacity and bottlenecks need to be identified. later, attention will be focussed on developing operating control strategies. Once the system is in operation, there may be a need for a simulation tool to aid real time operations. The data will also change, in accuracy, reliability and volume, as a project progresses. The costs and benefits of simulation cannot be stated categorically, because of the unique features of every project. Experience suggests that, except perhaps in the simplest of systems, investment in simulation will be worthwhile.

REFERENCES

1. Carrie, A.S., The role of simulation in FMS, in *Flexible Manufacturing Systems: Methods and Studies*, A. Kusiak, editor, pp191–208, North-Holland, Amsterdam, 1986.

2. Carrie A.S., FMS simulation: needs, experience, facilities, *Proc. SIM-1, 1st International Conference on Simulation in Manufacturing*, Stratford-upon-Avon, March 1985, pp205–215, IFS (Publications) Ltd, Bedford, England, 1985.

3. Carrie, A.S., Adhami, E., Stephens, A. and Murdoch, I.C., Introducing a Flexible Manufacturing System, *Int. J. Prod. Res.*, 22, 6, 907–916, 1984.

4. Carrie, A.S. and Adhami, E., Introducing FMS by simulation, *Proc. FMS-2, 2nd Int Conf on Flexible manufacturing Systems*, 1983, and in *Simulation*, edited by R.D. Hurrion, pp173–184, IFS (Publications) Ltd, Bedford, England, 1986.

5. Hollocks, B.W., Simulation of advanced manufacturing systems, *Proc Conf on Planning for Automated Manufacture*, Coventry, September 1986, pp183–185, MEP, London.

6. Perera, D.T.S. and Carrie, A.S., A simulation tool for real time scheduling of FMS, in *Advances in Manufacturing Technology II*, edited by P. F. McGoldrick, pp133–137, *Proc. 3rd National Conference on Production Research*, Nottingham, September 1987, Kogan Page, London, 1987.

CHAPTER 14

Some developments in simulation

In chapter 6, when discussing the various types of software available for simulation, and the trends in their development, some distinct trends were observed:

— the increasing power of personal computers,
— the availability of good quality graphics,
— the development of model builders and other techniques to make simulation easier to use.

These trends will continue and intensify.

Simulation software migrated from mainframe computers, where it had developed and where good computing power was available, to personal computers, on which the user has a self-contained system under his own control, with more user-friendly software available, and with reasonable computing power. Two directions can be predicted. Software will become more readily available on workstations, such as CAD workstations. These systems now provide an order of magnitude more computing power than 16-bit personal computers. They provide excellent graphics resolution. There will be some debate as to whether any further improvements in resolution are really necessary, or are merely cosmetic "sales gimmicks". However, so long as the technology exists and the hardware is available it will be used. Eventually, we can expect simulation software for manufacturing applications which give as good graphics as the most sophisticated computer animation facilities currently available. The other direction is that at the low cost end of the scale, computer power is increasing very rapidly. Computers are now being marketed for the home with many of the capabilities of CAD workstations and the number crunching power of the so-called super minis of a few years ago. For the price of a rather basic business personal computer, one can obtain a machine with a 32-bit processor capable of a speed of over 4 mips (million instructions per second) up to a peak speed of 18 mips, 4 Mbytes of RAM, 20 Mbytes hard disc, a graphics resolution of up to 640 $\times$ 512 and up to 256 colours.[1]

However, perhaps more significant than the trends in the way existing simulation tasks are performed, and the capabilities of the machines on which they run, are some new ways of using simulation. Three will be briefly considered.

14.1 JUST-IN-TIME CONTROL SYSTEMS

Most manufacturing firms are embracing the ideals of Just-in-Time (JIT) manufacture. This requires a new approach to organising manufacturing operations. Flexible manufacturing is a key technique for achieving the reduction in lead times and work in progress levels which JIT methods aim at. One of the effects of this change is that manufacturing systems are now required to operate on a pull method rather than on a push basis.

This change in approach demands a re-think about simulation modelling of manufacturing systems. As with manufacturing systems, many simulation models are based on the push principle. Jobs to be processed arrive in the system—they are pushed into it. Then the logic of the model tests the queues and discovers whether operations can take place, so that the job is pushed one further stage along its way to completion. Finally the job reaches the completion of its last operation and leaves the system. To implement the JIT approach we need a method of pulling work through the system. To do this in a simulation model is no easier or no more complicated than it is in real life—the decision rules must be framed to suit. New packages will be developed which approach the model building process from this standpoint.

14.1.1 Developing a Kanban control rule for an FMS

To illustrate how new approaches can be developed a brief description will be given of a simulation project[7] aimed at developing control rules for an FMS, which lead to a Kanban control rule. For an explanation of the Kanban system, see for example Schonberger.[18]

A flexible manufacturing system was to be installed in one of the Caterpillar company's plants for the production of a small group of components required in relatively large quantities, which had previously been made on a group of conventional machines. There were two basic types of component, each of which was made in left-hand and right-hand versions. The quantities of each type were similar. In the FMS the main machining operations were done on specially designed machining cells with several spindles for boring and drilling operations. In addition there were broach, gauge and wash operations. There was a machine cell and a broaching machine for each type of component, but only one gauge and wash, which were shared by both types of component. Components were manually loaded into trays at the load–unload area, which had two stations, one for each type of component. The trays were to be moved to the machining cell for the type of component, then to gauge, broach, wash and back to unload, by a vehicle moving along a linear track. There were buffer positions at each machine, from which the components were lifted out of the tray and placed into the machine's fixture by a robot, which would replace the component into the tray after the operation.

Since the capacity of the system was likely to be slightly below the total requirement for the parts, the system would run for three shifts each day, and a major objective would be to keep the machining cells supplied with work. The machining cells were the bottleneck operation, although the cycle time on broaching was almost as long. The wash, gauge and load–unload operations were much quicker. It was intended to operate the system with two operators during the first shift of each day, one on the second shift, but none on the third shift. Fresh components could be loaded in the trays by the

operators during the manned shifts and stored in a miniature AS/RS positioned at the
end of the vehicle's track. Trays of components machined during the unmanned shift
could be held in the store until the start of a manned shift. Figure 14.1 gives the layout
of the system.

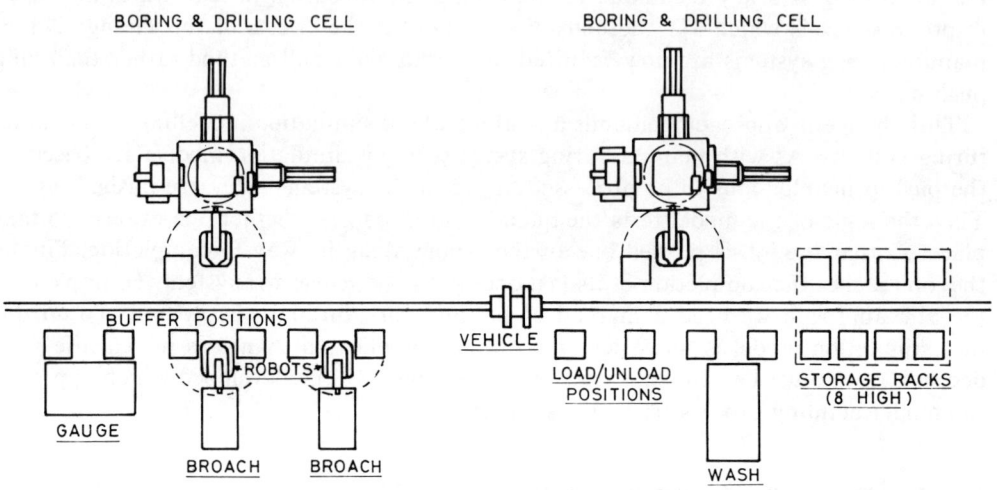

Figure 14.1   Layout of system for which a Kanban control rule was developed.

The simulation project began in the normal way, by defining the activity cycle diagram
for the system, its entities, activities, attributes and priority rules, and so on. When the
model was run, it soon became clear that there would be a problem because it seemed
that during the first manned shift, the vehicle could not cope with moving work to the
machines as well as fetching trays from the store for unloading. Similarly, in the second
manned shift, it did not have enough capacity to keep work moving round the system
as well as putting trays of fresh components into the store for processing during the
unmanned shift. Clearly, in this project the durations of the vehicle movements had to
be computed with great accuracy, this being a key feature of the system. A six-second
time unit was used.

One of the complications of the model was that there could be three types of movement:

— from one machine to the next in the processing sequence,
— from one machine to store, if the buffer at the next machine was not vacant,
— from store to machine, if the tray had been placed in the store instead of being
     delivered directly from the preceding machine.

Since there were five operations (ie load–unload, machine, broach, gauge and wash)
there were 15 different types of move the vehicle could have to make. Setting the
relative priorities was a complex task, especially since, once the chain of direct moves
from machine to machine was broken, two indirect moves would be required instead of
one direct move, which increased the load on the vehicle and the situation deteriorated
quickly. In the initial experiments virtually every tray was placed in store between

operations. The problem was to devise a control rule which avoided these indirect moves as much as possible. A further complication was that the mix of activities changed over the working day. The model had to allow for lunch breaks of the operators, and other disturbances. In the end a rule was developed which achieved full utilisation of the boring and drilling cells, did not overload the vehicle and still ensured that all the processed components could be unloaded during the manned shifts, and also fresh components could be loaded and placed in store to keep the system going during the unmanned shift. This rule was to form the basis of the rule in the system's controller.

However, during the course of the study it was realised that if indirect movements were to be avoided, then a tray should only be moved away from a machine when there was a vacant buffer position at the following machine. Similarly, whenever a buffer at a machine became vacant a tray should be brought to it as soon as possible. In other words, the existence of a vacant buffer could be taken as a signal that a tray should be brought to the machine. There was thus a clear analogy between a vacant buffer at a machine and a Kanban card being passed back from one workcentre to its supplying workcentre. It was decided to see whether a "Kanban" rule could be formulated for controlling the vehicle's movements.

There was a major problem, however. If there were vacant buffers at two or more machines simultaneously, then there would be conflict over which the vehicle should service next. This led to consideration of the number of trays which should be permitted at the machines. In this system, as in so many, there was a temptation to have a large number of pallets or trays in the system so that, in theory, there should never be a shortage of work at any machine. However, we have shown in the models in the earlier chapters that in general the more pallets in a system the longer will be the lead times, with no improvement in output. A rule was inserted into the model to ensure that the number of trays in the system, disregarding the store, would always be one less than the number of locations for them. This ensured that there would be exactly one free buffer at all times. The model was re-written on these principles and was found to be able to function as effectively as the push-type rule developed earlier.

Experience of applying pull-type control rules to simulation of manufacturing systems is still being built up. This project showed that a new approach to controlling systems can be effective, and can be expressed in formal decision rules.

14.2 SIMULATION AND FMS CONTROL SYSTEMS

One of the problems which is forcing developments in simulation is the need to provide system managers with a simulation tool which they can use to plan the operations of their system. "Traditional" simulation is well able to simulate the FMS at the design stage and for planning operating policies. However, it is not ideal for providing managers with a very short-term and rapid look ahead at their operations for the next hour or so. The time horizon obviously depends on the nature of the operations and variety of parts involved. As has been pointed out, as we get closer to real-time planning the volume of data to be considered becomes very large and the model may not be responsive enough. Developments can be expected in this area.[14] System software designers are also trying to bridge the gap, by providing a look-ahead facility in their system control software. So far these facilities are rather rudimentary. Production control packages are also being developed to provide this type of facility. There is still some way before the three disciplines meet at a standard interface. One of the problems is that the control system software writers are understandably reluctant to allow another person's software access to the real-time data files, in case the files become corrupted and the FMSs come to a halt or damage is caused. This is easy to understand, although it does not seem unreasonable that the type of access to a file which is permitted to the different programs could be controlled so that the simulation program would have read access only.

It may seem misdirection of effort to develop simulation models which mimic the operation of a system, when the system itself contains control software which does this job in the real system. In the project described in chapter 13, the supplier adapted the real control software to operate in simulation mode at 60 times real speed. As suggested by Carrie,[6] FMS manufacturers should provide clients with a version of the real software which could be run in simulation mode. This would enable the performance of the system to be studied at the negotiation stage, would ensure that the model was a faithful mirror of the real system, and avoid the need to bring in a third party to write a model which might contain dubious simplifications.

Another avenue of development, and one which is almost the opposite of the trends discussed above, is the use of simulation to develop control system software. Hutchinson and Clementson[11] have shown the very close analogy between the activities in an ECSL simulation program and the functions of the control system. An activity contains the tests which determine whether its actions can be carried out. Similarly the control system software contains many different actions which are dependent on signals sent in to the control computer by sensors and other forms of data collection. They summarise the comparison as shown in Table 14.1.

Table 14.1

Simulation	Control system
Test the state variables	Test the state variables
If satisfactory (otherwise abandon routine):	If satisfactory (otherwise abandon routine):
change the state variables to indicate the involvement of entities in the activity	change the state variables to indicate the involvement of entities in the activity
determine the activity duration	actions to initiate the activity
schedule the "end event" after the duration	
After duration:	Recognition of "activity completed" signal(s)
change state variables to indicate new availability of entities	change state variables to indicate new availability of entities after "end event"

The subject of software design raises another topic which is undergoing rapid development, namely the software development tools. There are now several structured design techniques, and they are being used by many companies to help in the specification of software systems. Some work on their application to simulation model design is underway,[3] and developments can be expected in the near future.

14.3 SIMULATION AND EXPERT SYSTEMS

One of the most significant developments is the application of expert systems to simulation. The research community is very active and the results of this work are being implemented in commercially available simulation systems. The field of expert systems and artificial intelligence is very broad. Many different aspects of expert systems have relevance to simulation. At the risk of some over-simplification, the main points of similarity between expert systems and simulation and potential applications will be briefly discussed.

14.3.1 Similarities between expert systems and simulation

(1) General analogy between the structure of an expert system and a simulation model

Expert systems differ from traditional computer programs in that the knowledge which is built in to the program is separated from the logic which processes it. Whereas in a FORTRAN program the knowledge of the program writer is embedded in the program itself, in an expert system the "knowledge base" is self-contained outwith the program, and the program provides general routines to process that knowledge. The knowledge base consists of two distinct types of information, declarative knowledge, or facts which are known to be true, and procedural knowledge, or rules from which other facts may be deduced.[17] A trivial example would be that two facts might be that John and James are brothers, and that Joseph is the father of John, then a rule could be defined which would enable the system to deduce that Joseph is the father of James as well. The analogy is that in a simulation program, especially one structured on activities, the rules are within each activity, the facts define the status of the model, and, if the rule "fires", then new facts will result from the actions which can be carried out.

(2) Similarity between simulation rules and PROLOG rules

There are a variety of langauges for artificial intelligence applications, of which the best-known are probably LISP and PROLOG. There is a fairly close similarity between the structure of PROLOG rules and the rules in an ECSL activity. This similarity has lead various authors to develop simulation models in AI languages, eg Adelsberger[2] and Ben-Arieh,[4] or with higher level languages based on LISP or PROLOG, such as OPS-5[15] or KEE, the so-called Knowledge Engineering Environment.[12]

(3) A taxonomy of relationships between simulation and expert systems

How expert systems and simulation should be inter-connected is a complex matter of software systems design. O'Keefe[13] gives a taxonomy for combining expert systems and simulation. Among the variants he suggests are systems in which an expert system is embedded within a simulation program, or vice versa, and where an expert system is used as an intelligent front end for simulation.

14.3.2   Applications of expert systems in simulation

(1)   Intelligent front ends

Although written before expert systems came to be widely known, CAPS,[8] which we have demonstrated, is a good example of an intelligent front end. It does not rely on an AI language, and is written in FORTRAN. It uses terms which are familiar to anyone familiar with simulation. These terms are not familiar however to manufacturing engineers. It is to be expected that front end programs tailored to FMS design will become available in the near future.[9,10]

(2)   Selection of simulation program modules for FMS simulation

The form of such a program can be deduced from the content of the preceding chapters. In these, various features of FMS and how they can be modelled were discussed. If we restrict our discussion to ECSL, then it should have been observed that at each stage the program modules for the various activities were revised to take the new feature into account. A program could be developed to ask questions about the FMS to be modelled and, from the information received, construct a simulation program assembled from the various segments of program which have been presented in this book.[19] Whether this would strictly be an expert system might be the subject of debate between a simulation engineer and an knowledge engineer.

(3)   Specifying the data for the initial conditions of a simulation run

One of the complications of building a simulation model, such as those in chapters 10 and 11, is that, when specifying the initial conditions for a model, it is difficult to ensure that the values of the attributes and the membership of the sets of all the entities are consistent. CAPS does this for us, as was demonstrated in the earlier chapters, but for the more complex models this had to be done manually. It was a time-consuming and error-prone process! An expert system can readily be conceived to handle this work. The rules are simple to state. For example, we knew that each pallet could have only one component, and that only one pallet could occupy any location, and so on. Thus it would be relatively simple to develop a system which would check the validity of any proposed set of initial conditions and report invalid attribute values.

(4)   Design of simulation experiments

One of the features of simulation likely to benefit from application of expert systems, is that of design of experiments. As has been indicated several times in this text, simulation results have to be assessed for validity rather than accepted at face value. Frequently, this means performing several runs to ensure that the observed effect is not due to some quirk of the sampling process. Simulation experiments have to be designed to ensure that sufficient, and preferably the minimum number of, runs are carried out. The conditions under which they are run have to be specified. An expert system could be valuable in determining these parameters.

(5) Probabilistic reasoning

Reference was made above to the ability of some expert systems to deduce new information from existing knowledge. There are many different approaches to designing software for this task. Among them is a form of probabilistic reasoning, which results in a conclusion being deduced with a stated probability of its truth. Some of the famous early expert systems were of this type. Software of this type can conceivably be of great potential value in designing FMS control software, specifically with regard to the production control rules. Although the discussion on the topic in chapter 7 was brief, we know that certain job sequencing rules work reasonably well under certain circumstances. In FMS we have also tooling, fixtures and pallet availability to consider, and the problem is much more complex. It is conceivable that the probabilistic reasoning capabilities of expert systems could be used to deduce from the FMS status and the objectives of the system manager at any time the rules most likely to achieve the objectives and the probability that any rule is most suitable. It is expected that there will be a growing interest in applying expert systems to the management and control of FMSs.

(6) Capturing knowledge

One of the difficulties in developing expert systems is capturing the knowledge of the human experts. The rules by which human experts operate are seldom articulated concisely. The development of methods for capturing this knowledge is one of the main thrusts of expert system research. In a very similar way, the knowledge about how FMSs should be operated which has been gained by system managers, needs to be captured for use in simulation models and in control software. It is likely that the techniques developed for expert systems will be applicable.

Although significant results from applying AI concepts can be expected, it is possible that some the most important achievements will not be widely reported, due to commercial sensitivity. The companies exploiting these ideas most effectively should gain advantages over their competitors.

SUMMARY

Although the trends in improved computing power, speed and graphics will continue, there are several new fields for development of simulation. Three have been discussed in this chapter. The modelling of new styles of decision rules is one area of application. Another is the use of simulation, and simulation thinking, to help develop control software. It was suggested that there is a clear correlation between simulation rules and real control rules. The third topic is the application of expert systems to simulation. Several potential forms of application were suggested.

REFERENCES

1. Archimedes Series Computers, Acorn Computers Ltd, Cambridge, England, 1987.

2. Adelsberger, H.H., Prolog as a simulation language, Winter Simulation Conf., 1984.

3. Banerjee, S.K., Al-Maliki, I., Carrie, A.S., Chan, T.S. and Christie, N.R., A modular simulation model for FMS using structured techniques in *Advances in Manufacturing Technology II*, edited by P. F. McGoldrick, pp117–122, 3rd National Conference on Production Research, Nottingham, September 1987.

4. Ben-Arieh, D., A knowledge-based system for simulation and control of FMS, *2nd International Conference on Simulation in Manufacturing*, Chicago, June 1986, and in *Simulation*, R.D. Hurrion, editor, pp287–295, IFS (Publications) Ltd, Bedford, England. 1986.

5. Ben-Arieh, D., A knowledge based control system for automated production and assembly, in *Modelling and Design of Flexible Manufacturing Systems*, A. Kusiak, editor, pp347–368, Elsevier Science Publishers, Amsterdam, 1986.

6. Carrie, A.S., The role of simulation in FMS, in *Flexible Manufacturing Systems: Models and Studies*, A. Kusiak, editor, pp191–208, North-Holland, Amsterdam, 1986.

7. Carrie, A.S. and Purves, J., Development of control rules for a high-volume low-variety flexible manufacturing system, 2nd National Conference on Production Research, Edinburgh, September 1986.

8. Clementson, A.T., *ECSL User's Manual*, Cle-Com Ltd, Birmingham, 1985.

9. Gaines, B.R., Expert systems and simulation in the design of an FMS advisory system, *2nd International Conference on Simulation in Manufacturing*, Chicago, June 1986, and in *Simulation*, R.D. Hurrion, editor, pp311–324, IFS (Publications) Ltd, Bedford, England, 1986.

10. Haddock, J., An expert system framework based on a simulation generator, *Simulation*, 48, 2, 45-53, 1987.

11. Hutchinson, G.K. and Clementson, A.T., Manufacturing Control Systems: an approach to reducing software costs, *Robotics and Computer-integrated Manufacturing*, 1, 3/4, 271–281, 1984.

12. KEE Software Development System, Intellicorp, Menlo Park California.

13. O'Keefe, R., Simulation and expert systems—a taxonomy and some examples, *Simulation*, 46, 1, 10–16, 1986.

14. Perera, D.T.S. and Carrie, A.S., A Simulation Tool for Real Time Scheduling of FMS in *Advances in Manufacturing Technology II*, edited by P. F. McGoldrick, pp133–137, 3rd National Conference on Production Research, Nottingham, September 1987.

15. Reddy, Y., Fox, M., Husain, N. and McRoberts, M., KBS: A knowledge based simulation system, Research Report, Intelligent Systems Laboratory, Carnegie-Mellon University, 1985.

16. Shannon, R.E. Mayer, R. and Adelsberger, H.H., Expert systems and simulation, *Simulation*, 44, 6, 275–284, 1985.

17. Shivnan, J. and Browne, J., AI-based simulation of advanced manufacturing systems, *2nd International Conference on Simulation in Manufacturing*, Chicago, June 1986, and in Simulation, R.D. Hurrion, editor, pp297–309, IFS (Publications) Ltd, Bedford, England, 1986.

18. Schonberger, R.J., *Japanese Management Techniques*, Free Press, New York, 1982.

19. Tang, K.W., An intelligent front end for FMS simulation, MSc dissertation, University of Strathclyde, Glasgow, 1987.

Index